Vol. 32. **Determination of Organic Compounds: Methods and Procedures.** By Frederick T. Weiss

Vol. 33. **Masking and Demasking of Chemical Reactions.** By D. D. Perrin

Vol. 34. **Neutron Activation Analysis.** By D. De Soete, R. Gijbels, and J. Hoste

Vol. 35. **Laser Raman Spectroscopy.** By Marvin C. Tobin

Vol. 36. **Emission Spectrochemical Analysis.** By Morris Slavin

Vol. 37. **Analytical Chemistry of Phosphorus Compounds.** Edited by M. Halmann

Vol. 38. **Luminescence Spectrometry in Analytical Chemistry.** By J. D. Winefordner, S. G. Schulman, and T. C. O'Haver

Vol. 39. **Activation Analysis with Neutron Generators.** By Sam S. Nargolwalla and Edwin P. Przybylowicz

Vol. 40. **Determination of Gaseous Elements in Metals.** Edited by Lynn L. Lewis, Laben M. Melnick, and Ben D. Holt

Vol. 41. **Analysis of Silicones.** Edited by A. Lee Smith

Vol. 42. **Foundations of Ultracentrifugal Analysis.** By H. Fujita

Vol. 43. **Chemical Infrared Fourier Transform Spectroscopy.** By Peter R. Griffiths

Vol. 44. **Microscale Manipulations in Chemistry.** By T. S. Ma and V. Horak

Vol. 45. **Thermometric Titrations.** By J. Barthel

Vol. 46. **Trace Analysis: Spectroscopic Methods for Elements.** Edited by J. D. Winefordner

Vol. 47. **Contamination Control in Trace Element Analysis.** By Morris Zief and James W. Mitchell

Vol. 48. **Analytical Applications of NMR.** By D. E. Leyden and R. H. Cox

Vol. 49. **Measurement of Dissolved Oxygen.** By Michael L. Hitchman

Vol. 50. **Analytical Laser Spectroscopy.** Edited by Nicolo Omenetto

Vol. 51. **Trace Element Analysis of Geological Materials.** By Roger D. Reeves and Robert R. Brooks

Vol. 52. **Chemical Analysis by Microwave Rotational Spectroscopy.** By Ravi Varma and Lawrence W. Hrubesh

Vol. 53. **Information Theory As Applied to Chemical Analysis.** By Karel Eckschlager and Vladimir Stepánek

Vol. 54. **Applied Infrared Spectroscopy: Fundamentals, Techniques, and Analytical Problem-solving.** By A. Lee Smith

Vol. 55. **Archaeological Chemistry.** By Zvi Goffer

Vol. 56. **Immobilized Enzymes in Analytical and Clinical Chemistry.** By P. W. Carr and L. D. Bowers

Vol. 57. **Photoacoustics and Photoacoustic Spectroscopy.** By Allan Rosencwaig

Vol. 58. **Analysis of Pesticide Residues.** Edited by H. Anson Moye

Vol. 59. **Affinity Chromatography.** By William H. Scouten

Vol. 60. **Quality Control in Analytical Chemistry.** By G. Kateman and F. W. Pijpers

Vol. 61. **Direct Characterization of Fineparticles.** By Brian H. Kaye

Vol. 62. **Flow Injection Analysis.** By J. Ruzicka and E. H. Hansen

Vol. 63. **Applied Electron Spectroscopy for Chemical Analysis.** Edited by Hassan Windawi and Floyd Ho

Vol. 64. **Analytical Aspects of Environmen** ch and Philip K. Hopke

Vol. 65. **The Interpretation of Analytical C** **s.** By D. Luc Massart and Leonard Kau

Vol. 66. **Solid Phase Biochemistry: Analyti** m H. Scouten

Analytical Solution Calorimetry

CHEMICAL ANALYSIS

A SERIES OF MONOGRAPHS ON
ANALYTICAL CHEMISTRY AND ITS APPLICATIONS

VOLUME 79

A WILEY-INTERSCIENCE PUBLICATION

JOHN WILEY & SONS

New York / Chichester / Brisbane / Toronto / Singapore

Analytical Solution Calorimetry

Edited by

J. KEITH GRIME
The Procter & Gamble Company
Ivorydale Technical Center
Cincinnati, Ohio

A WILEY-INTERSCIENCE PUBLICATION

JOHN WILEY & SONS

New York / Chichester / Brisbane / Toronto / Singapore

Library of Congress Cataloging in Publication Data:

Main entry under title:

Analytical solution calorimetry.

(Chemical analysis, ISSN 0069–2883 ; v. 79)
"A Wiley-Interscience publication."
Includes index.
1. Thermometric titration 2. Enthalpimetric titration. I. Grime, J. Keith, 1947– . II. Series.
QD111.A53 1985 543′.086 84–28424
ISBN 0-471-86942-2

Printed in the United States of America

10 9 8 7 6 5 4 3 2 1

CONTRIBUTORS

DELBERT J. EATOUGH
Thermochemical Institute and the Department of Chemistry
Brigham Young University
Provo, Utah

J. KEITH GRIME
The Procter & Gamble Company
Ivorydale Technical Center
Cincinnati, Ohio

LEE D. HANSEN
Thermochemical Institute and the Department of Chemistry
Brigham Young University
Provo, Utah

JOSEPH JORDAN
Department of Chemistry
The Pennsylvania State University
University Park, Pennsylvania

EDWIN A. LEWIS
Thermochemical Institute and the Department of Chemistry
Brigham Young University
Provo, Utah

R. S. SCHIFREEN
E. I. DuPont de Nemours & Co.
Clinical Systems Division
Wilmington, Delaware

JOHN W. STAHL
Department of Chemistry
The Pennsylvania State University
University Park, Pennsylvania

To Anne for her unfailing encouragement
and to Rebecca and Sarah
for tolerating their father's frequent preoccupation
during preparation of this text

PREFACE

The title of this book is intended to convey, as succinctly as possible, the scope of the text. However, in view of the somewhat confused state of calorimetric nomenclature, a brief explanation is warranted. If used generically, the term analytical calorimetry encompasses a wide variety of instrumental techniques used to determine the extent or rate of a physical, chemical, biochemical, or biological process by measuring the temperature change or rate of heat change associated with the process. In this text, the term analytical calorimetry is used more specifically to describe a quite distinct group of thermochemical techniques in which temperature (or rate of heat change) is the *dependent* variable of the experiment. This definition allows a clear distinction between the techniques discussed in this text and the other major groups of techniques based on a thermal measurement and usually given the designation thermal analysis. This delineation is in keeping with a recent report from the International Union of Pure and Applied Chemistry (1981), which defines thermal analysis as "the group of techniques in which a physical property of the substance is measured as a function of temperature whilst the substance is subjected to a controlled temperature program." In other words, temperature is the *independent* variable of a thermal analysis experiment. This distinction between analytical calorimetry and thermal analysis is reinforced by the inclusion of the modifier "solution," since thermal analysis techniques are usually (although not exclusively) applied directly to solid samples. The term enthalpimetric analysis, if defined rigorously, describes any technique in which an enthalpy change is used directly or indirectly to obtain an analytical result. In the analytical literature, however, the term has become associated specifically with isoperibol calorimetry, a connotation used extensively by some authors in this text, but which is too restrictive for it to be considered as a title option.

The objective of this volume is to provide a comprehensive guide for the application of calorimetry to analytical chemistry and the determination of thermodynamic parameters. My intention in organizing the contents was to include the information necessary for the reader to understand the principles of the technique, to decide on the instrumental variation best-suited to a particular application, and, finally, to interpret and treat the data. Accordingly, the essential features of three major instrumental approaches, namely, isoperibol, heat conduction, and isothermal (heat-compensation) calorimetry, are discussed (Chapter

3). Inevitably, the overall emphasis is on isoperibol calorimetry, reflecting the popularity of this technique in the analytical sector.

In response to the most often quoted limitation of analytical calorimeters, namely, poor sample throughput, I have included a section on flow enthalpimetry (Chapter 4). Design and construction specifications have been included for those who may wish to construct this type of instrumentation.

For the specialist, Chapter 6 contains a comprehensive review of applications in many fields of chemistry. A large section is devoted to advances in catalytic methods, primarily those based on polymerization reactions. Chapter 7 focuses on the largest single growth area for calorimetry in the last decade—biochemical and clinical analyses. The appropriate information has been included in the latter so that the non-biochemist can understand the kinetic principles underlying the design of any quantitative experiment incorporating an enzyme-catalyzed reaction.

I would like to express my sincere thanks and appreciation to my co-authors and colleagues Delbert Eatough, Lee Hansen, Joseph Jordan, Ed Lewis, Richard Schifreen, and John Stahl for their cooperation and patience with my editorial attempts to make this multiauthor volume as cohesive as possible. Thanks are also due to Ted Greenhow, Gerry Kasting, and Gordon Kresheck for reading various chapters and making valuable suggestions.

Finally I would like to thank my Management at the Procter & Gamble Company for providing secretarial, graphics, and, not least, moral support without which this book could not have been produced.

J. KEITH GRIME

Cincinnati, Ohio
January 1985

CONTENTS

LIST OF SYMBOLS

$\mathring{a}$	Empirically determined parameter in the Debye–Hückel expression describing the mean distance of approach of ions in solution
a	Activity
α_{κ}	Change in κ with volume in the reaction vessel
A	Temperature-dependent constant in the Debye–Hückel expression
A_{TOT}	Area under the curve produced in a batch heat conduction of isothermal calorimetric experiment
b_B	Calibration constant relating bridge voltage E_B and temperature T
β_c	Dimensionless parameter to describe equilibrium curvature in isoperibol titration plots
β_k	Dimensionless parameter to describe kinetic curvature in isoperibol titration plots
β_{ω}	Change in ω with volume in the reaction vessel
$\mathcal{B}$	Constant dependent on solution dielectric constant in the Debye–Hückel expression
$\mathcal{C}$	Empirical constant in the Debye–Hückel expression
[ ], c	Concentration
CMC	Critical micelle concentration
C_P^*	Heat capacity at constant pressure per unit volume
C_P	Heat capacity at constant pressure
$C_{P,mp}^*$	Heat capacity per unit volume of mobile-phase solution in a flow isoperibol calorimeter
$\bar{C}_P$	Mean heat capacity at constant pressure
C_V	Heat capacity at constant volume
D	Determinant
$d\epsilon_E/dt$	Calibration constant relating q_E to the output signal of heat-conduction or isoperibol flow calorimeters
D_{S}	Diffusion coefficient of substrate
$\mathbf{E}$	Voltage
$\mathbf{E}_B$	Voltage across a Wheatstone bridge

$\mathbf{E}_C$	Voltage output from a heat-conduction calorimeter
ER	Excess reagent region
$\mathbf{E}_{TH}$	Voltage across a thermistor
ϵ	Energy equivalent of an isoperibol reaction vessel and contents
ϵ_C	Energy equivalent of a heat-conduction calorimeter
ϵ_η	Energy equivalent of the control heater in an isothermal calorimeter
ϵ_{RV}	Energy equivalent of an empty isoperibol calorimeter reaction vessel
ϵ_p	Energy equivalent of an isoperibol titration calorimeter vessel plus contents at some point p in a titration
[E]	Concentration of enzyme
ΔE	Internal energy change
EA	Enzyme activity
F	Flow rate, final region in calorimetric data plots
$\mathcal{F}$	Stoichiometric factor
f	Fraction titrated or reacted
γ	Activity coefficient
ΔG_{MC}	Gibbs free energy of micellization
ΔG_R	Gibbs free energy of reaction
ΔH	Enthalpy change
$\Delta H^\ominus$	Standard enthalpy change
ΔH_D	Enthalpy of dilution
ΔH_d	Enthalpy of dissociation
ΔH_f	Enthalpy of formation
ΔH_{MC}	Enthalpy of micellization
ΔH_N	Enthalpy of neutralization ($= \Delta H_d + \Delta H_W$)
ΔH_R	Enthalpy of reaction
ΔH_W	Enthalpy of reaction for $H^+ + OH^- = H_2O$
i	Degree of inhibition
$[I]$	Inhibitor concentration, initial region in calorimetric data plots
IU	International unit of enzyme activity
J	Joules
J_q	Heat flux (thermal enzyme probe)
J_{S}	Substrate flux (thermal enzyme probe)
κ	Heat leak modulus
k	Rate constant
K_{eq}	Equilibrium constant

K_I	Inhibition constant
K_m	Michaelis constant
k_N	Newtonian heat-leak constant
K_{sp}	Solubility product
λ	Thermal conductivity
L	Length (thermal reactor)
μ	Ionic strength
n	Number of moles of chemical species
n'	Adsorption capacity of a molecular sieve
$\mathcal{N}$	Number of thermal plates in a flow reactor
η	Pulse rate of the control heater in an isothermal calorimeter
η_H	Pulse rate of the control heater in an isothermal calorimeter during calibration sequence
P	Pressure
Φ_L	Partial molal enthalpy
q	Heat
$q_{C,p}$	Corrected total heat measured at point p and corrected for all non-chemical effects
q_D	Heat change due to reagent dilution
q_E	Heat added from an electrical calibration heater
q_{HL}	Heat change due to non-chemical effects
q_{IN}	Heat input into isothermal reaction cell by control heater
q_n	Heat per mole
q_{OUT}	Heat removed from isothermal reaction cell by cooling
q_p	Total (non-corrected) heat change to point p
$q_{R,p}$	Total measured heat at point p due to the reactions for which K_{eq} and H_R values are to be calculated
q_{TC}	Heat change due to temperature differences between reactant solutions
R	Universal gas constant
R	Resistance
R_{TH}	Resistance of a thermistor
ρ	Density
rsd	Relative standard deviation
S_I, S_R, S_{ER}, S_F	Initial, reaction, excess reagent, and final regions of calorimetric data plots
ΔS	Entropy change
[S]	Substrate, sample concentration

s	Specific heat
σ	Standard deviation
σ^2	Variance
ΔS_{MC}	Entropy of micellization
$[S_0]$	Initial substrate concentration
$[S_t]$	Substrate concentration at time t
θ	Input plug width in time units for an isoperibol flow injection calorimetry experiment
T	Temperature
t	Time
τ_C	Calorimeter time constant
τ_R	Isoperibol flow reactor time constant
T_{ex}	External temperature
T_T	Titrant temperature
t_H	Activation time for calibration heater
ΔT_{max}	Height of the temperature pulse in an isoperibol flow injection calorimetry experiment
t_R	Residence time of a sample plug in a flow reactor
T_t	Instantaneous temperature deviation from the baseline in an isoperibol flow injection calorimetry experiment
$t_{1/2}$	Peak width at half-maximum peak height (time units) for an isoperibol flow injection calorimetry experiment
U	Error square sum relating q_R to K_{eq} and ΔH_R
v	Rate of reaction
V	Volume
V_{el}	Volume of mobile phase required to "elute" a heat peak through a flow reactor
V_{max}	Maximum rate (substrate saturated) of an enzyme-catalyzed reaction
V_R	Volume of an isoperibol flow reactor cell
V_S	Volume of sample in an isoperibol flow injection reactor
w	Work done (mechanical, electrical, chemical)
$W_{1/2,cl}$	Peak width at half-peak height due to column dispersion in an isoperibol flow injection calorimetry experiment
ω_{EV}	Power term for evaporative cooling
ω_S	Power term for stirrer heating
ω_{TM}	Power term for thermistor heating
Z	Ionic charge

Analytical Solution Calorimetry

CHAPTER

1

THERMODYNAMICS, THERMOCHEMISTRY, AND CALORIMETRY

J. KEITH GRIME

The Procter & Gamble Company
Ivorydale Technical Center
Cincinnati, Ohio

1.1. THERMODYNAMICS

In order to understand the principles of calorimetric measurement it is appropriate to review the significance of the fundamental thermodynamic parameters involved.

1.1.1. Internal Energy, Work, and Heat

Thermodynamics deal with changes of state, that is, with processes in which a system in some initial state undergoes a change to achieve a final state. A convenient way to describe such a change in state is to compare the values of the *state functions* in the initial and final states. In general, four state functions are used to describe a change in state—internal energy (E), enthalpy (H), entropy

(S), and free energy (G). If the system is not at equilibrium, then it becomes necessary to use two or three state functions to describe the system.

Joule's experiments showed that under *adiabatic* conditions, that is, where the heat transfer into and out of the system is zero, a given amount of work would evolve (or absorb) a certain amount of heat. The amount of heat evolved is dependent only on the total amount of work done (w) and not on the origin of the work (electrical, mechanical, or chemical). Accordingly, a change in state of a system, under adiabatic conditions, can be expressed in terms of the work done without reference to the nature of the work involved. The state function used to describe a system in this way is the internal energy (E). In simple terms, E describes the total of all possible kinds of energy in the system. As with all state functions, E is not reported for a system in a given state, but rather as the difference in E between the initial and final states, ΔE. Therefore, for an adiabatic process,

$$\Delta E = w \tag{1.1}$$

As the internal energy of a system is increased by doing work on it, the final-state value is subtracted from the initial-state value or

$$\Delta E = (E_F - E_I) \tag{1.2}$$

where the subscripts F and I are the internal energy terms for the final and initial states, respectively. Therefore, by convention, if a system does work on its surroundings, w and ΔE are negative and conversely if the system is adiabatic.

A more often encountered situation in the laboratory is a change in state of a system resulting from a transfer of heat, q, into or out of the system. This is termed an *isothermal* process when the temperature of the system remains constant as it exchanges heat with its surroundings. Conventionally, if heat flows from the surroundings to the system thereby increasing its internal energy, q is given a positive sign. Such a process is said to be *endothermic*. Similarly, if heat flows out of the system into its surroundings, then q is negative and the process is designated as *exothermic*.

1.1.2. Enthalpy

The relationship between the transfer of heat and work done is expressed by the first law of thermodynamics,

$$\Delta E = q + w \tag{1.3}$$

Equation (1.3) represents the energy balance for all processes. Adherence to the first law demands that the total change in internal energy for all systems taking part in a given change is equal to zero. There are three limiting versions of Eq. (1.3): (1) when no work is done ($w = 0$); (2) when work is performed under adiabatic conditions ($q = 0$); (3) when both q and w are finite. For the purposes of this text, work done will be confined to volume-expansion work, since most (although not all) analytical solution calorimetry experiments are performed in a reaction vessel open to the atmosphere. Since work done by a system expanding against a constant pressure (P) is given by $P\,dV$, then

$$w = P\,\Delta V \tag{1.4}$$

where P is the atmospheric pressure and ΔV is the volume change. From Eq. (1.3) and (1.4) we get

$$E = q_P - P\,\Delta V \tag{1.5}$$

where q_P is the heat transfer under conditions of constant pressure. Rearranging and expanding the equation to include the initial and final states of the system gives

$$q_P = (E_F + PV_F) - (E_I + PV_I) \tag{1.6.}$$

Equation (1.6) leads to another state function, *enthalpy* (H), which can be represented

$$H = E + PV \tag{1.7.}$$

As with all state functions, it is more convenient to describe a change in enthalpy between the initial and final states, ΔH, where

$$\Delta H = (H_F - H_I) \tag{1.8}$$

Therefore,

$$q_P = \Delta H = \Delta E + P\,\Delta V \quad \text{(at constant pressure)} \tag{1.9}$$

An arithmetic function, ΔH, has the same sign convention as q and is defined as the heat transferred in a process at constant pressure if the work done is volume expansion.

As mentioned earlier, heat can be transferred under conditions where no volume expansion occurs, in which $P\,\Delta V$ and therefore w are equal to zero.

Such is the case when a reaction is carried out in thick-walled, sealed containers (bomb calorimeters). However, all forms of calorimetry discussed in this volume are based on measurements made under conditions of constant (usually atmospheric) pressure. Accordingly, the term heat transfer will be consistent with its meaning as defined by Eq. (1.9).

1.2. THERMOCHEMISTRY

Almost every transformation, whether chemical, physical, or biochemical, occurs with an attendant evolution or absorption of heat and a corresponding temperature change. The study of these thermal effects is known as *thermochemistry*. The direct measurement of heat changes associated with thermochemical processes is termed *calorimetry*.

Enthalpy is an extensive property; it is dependent on the size of the system. Quantitatively, its value depends on the extent of the process in going from one state to another. Consequently, a determination of the heat change associated with a chemical reaction can lead to analytical information regarding the number of moles of reactant (analyte) taking part in the reaction. This is the fundamental principle of *analytical calorimetry*. Calorimetry is not normally considered as a qualitative technique; however, the sign and magnitude of an enthalpy change are characteristic of a particular reaction under defined experimental conditions. This information can be used qualitatively to elucidate reaction mechanisms. The heat change occurring in a calorimeter is typically calibrated by comparison of the signal obtained as a result of the chemical process with that obtained by the passage of an electrical current. It is appropriate, therefore, that the principal unit of thermochemistry is the *joule*, which is defined as the amount of energy dissipated when 1 A flows through a resistance of 1 Ω for 1 s. Thermochemical data are usually expressed in kilojoules per mole.

1.2.1. Enthalpy of Formation, ΔH_f

Perhaps the most fundamental parameter of thermochemistry, the enthalpy of formation of a compound, is usually defined as the increase in enthalpy when one mole of a substance is formed from its constituent elements. By convention, the elements should be in the standard state, that is, in their stable form at 298 K and at a pressure of 1 atm or 101.325 kPa. Thermodynamic quantities relating to processes occurring under standard-state conditions are identified by the superscript $\ominus$, for example $\Delta H_f^{\ominus}$. By definition, the enthalpies of all elements in their standard states are zero at all temperatures. The enthalpy of a compound is then equal to the enthalpy of formation, ΔH_f.

1.2.2. Enthalpy of Reaction, ΔH_R

The enthalpy of reaction is defined as the difference in the enthalpies of the reaction products and of the reactants at constant pressure and at a definite temperature with every substance in a specified physical state. ΔH_R can be calculated from the tabulated ΔH_f values of the compounds involved in the reaction. For example, from the information in Table 1.1, the enthalpy of precipitation of silver chloride and the enthalpy of neutralization of a strong acid by a strong base can be calculated as follows:

Reaction $$Ag^+_{(aq)} + Cl^-_{(aq)} = AgCl_{(s)} \tag{1.10}$$

$$\Delta H_R = \Delta H_f \text{ (products)} - \Delta H_f \text{ (reactants)} \tag{1.11}$$

$$\Delta H_R = (-127.07) - (-167.15 + 105.58)$$
$$\Delta H_R = -65.50 \text{ kJ mol}^{-1} \tag{1.12}$$

Reaction $$H^+_{(aq)} + OH^-_{(aq)} = H_2O \tag{1.13}$$

$$\Delta H_R = (-285.83) - (0.00 - 229.99)$$
$$\Delta H_R = -55.84 \text{ kJ mol}^{-1} \tag{1.14}$$

If tabulated thermochemical information is not available, the reaction enthalpy can be determined calorimetrically by measuring the heat change associated with the reaction of known amounts of reactants.

TABLE 1.1. Enthalpy of Formation Data

	ΔH_f (kJ mol^{-1})[a]
$Ag^+_{(aq)}$	+105.58
$Cl^-_{(aq)}$	−167.15
$AgCl_{(s)}$	−127.07
$OH^-_{(aq)}$	−229.99
$H_20_{(l)}$	−285.83
$H^+_{(aq)}$	0.00

Source. Reference 1.
[a]Standard state 1.0 mol L^{-1}.

1.2.2a. Relationship between $\Delta H_R^\ominus$ and ΔH_R

For the most part, reaction enthalpies discussed in this text are those associated with reactions in aqueous solution. Consequently, it is important to consider the relationship between the experimental enthalpy of reaction, ΔH_R, corrected for all attendant chemical and non-chemical thermal events and the standard enthalpy of reaction, $\Delta H_R^\ominus$. The standard state for a solute is unit molality (molarity is also used). However, at such a concentration, a solute inevitably behaves in a non-ideal fashion resulting in activity coefficients considerably less than unity. Therefore, in order to standardize thermochemical data, standard enthalpy changes are calculated by determination of an enthalpy of reaction at several concentrations of reactants and the data are extrapolated to infinite dilution. The logic of this approach can be seen by considering the absolute enthalpy H_A of a component A in a chemical reaction:

$$H_{\mathrm{A}} = H_{\mathrm{A}}^\ominus + \mathbf{R}T^2 \frac{\partial}{\partial \mathrm{T}} \ln a_{\mathrm{A}} \tag{1.15}$$

where a_{A} is the activity of component A. Equation (1.15) can be adapted to include the activity coefficient, γ_{A}:

$$H_{\mathrm{A}} = H_{\mathrm{A}}^\ominus + \mathbf{R}T^2 \frac{\partial}{\partial T} \ln c_{\mathrm{A}} + \mathbf{R}T^2 \frac{\partial}{\partial \mathrm{T}} \ln \gamma_{\mathrm{A}} \tag{1.16}$$

where c_{A} is the concentration of the solute in mol L^{-1}. Similar equations can be developed for all the other components of the reaction in which A is involved, both products and reactants, so that ΔH_R can be calculated. Obviously $H_{\mathrm{A}}^\ominus$ approaches H_{A} as $c_{\mathrm{A}} \rightarrow 0$ and $\gamma_{\mathrm{A}} \rightarrow 1$. Therefore, the dilute conditions preferred with most calorimeters validate the approximation that $H_{\mathrm{A}}^\ominus \approx H_{\mathrm{A}}$ (and therefore $\Delta H_R^\ominus \approx \Delta H_R$). It also becomes apparent from Eq. (1.16) that if reaction heats are used directly to obtain analytical data (i.e., non-end-point methods), the maintenance of a constant, high-ionic-strength environment can minimize variations in data caused by fluctuations in activity coefficients from sample to sample.

1.2.2b. Variation of ΔH_R with Temperature

The enthalpy change associated with a chemical or physical transformation is temperature dependent as shown by Eq. (1.17):

$$\left(\frac{\partial H}{\partial T}\right)_P = C_P \tag{1.17}$$

where C_P is the heat capacity (see Section 1.2.3) of the system at constant pressure.

For the process A → B, the enthalpy change is given by

$$\Delta H = H_B - H_A \tag{1.18}$$

The variation of ΔH with temperature is obtained by differentiating Eq. (1.18) with respect to temperature at constant pressure,

$$\left(\frac{\partial(\Delta H)}{\partial T}\right)_P = \left(\frac{\partial H_B}{\partial T}\right)_P - \left(\frac{\partial H_A}{\partial T}\right)_P \tag{1.19}$$

Substituting Eq. (1.17) gives

$$\left(\frac{\partial(\Delta H)}{\partial T}\right)_P = C_{P(B)} - C_{P(A)} = \Delta C_P \tag{1.20}$$

Therefore, the change in enthalpy of reaction at constant pressure per degree rise in temperature is equal to the change in heat capacity of the system as a result of the reaction. Since heat capacity itself is a temperature-dependent parameter, the variation of C_P over the experimental temperature range must be incorporated into Eq. (1.20). This is usually achieved by use of an equation of the form,

$$C_P = a + bT + cT^2 \tag{1.21}$$

where a, b, and c are tabulated constants characteristic of a component. The variation of ΔH_R with temperature can usually be ignored in a typical calorimetric experiment in which the temperature change does not exceed 0.1°C. Any changes in ΔH_R with this magnitude of temperature change will be less than the overall precision of the measurement, which is typically $\leq$ 1%.

1.2.3. Heat Capacity

By definition, the molar heat capacity of a system (C) is the heat absorbed per mole (q_n) per degree rise in temperature, that is,

$$C = \lim_{\Delta T \to 0} \left(\frac{q_n}{\Delta T} \right) \tag{1.22}$$

As q_n can be defined under conditions of constant volume or pressure, C must be defined accordingly. Therefore, at constant pressure, where $q_n = \Delta H$,

$$C_P = \lim_{\Delta T \to 0} \left(\frac{\Delta H}{\Delta T} \right)_P \tag{1.23}$$

or at constant volume, where $q_n = \Delta E$,

$$C_V = \lim_{\Delta T \to 0} \left(\frac{\Delta E}{\Delta T} \right)_V \tag{1.24}$$

C_P will be used exclusively in the text in accordance with the experimental conditions prevailing in most solution calorimeters. The units of heat capacity can vary. In SI terminology, joules per degree per mole (J K^{-1} mol^{-1}) are preferred; however, it is often convenient to refer to the heat capacity per unit volume when discussing aspects of solution calorimetry. In general, any units consistent with

$$\text{heat capacity} = (\text{energy}) \cdot (\text{deg})^{-1} \cdot (\text{amount of material})^{-1} \tag{1.25}$$

are acceptable. The term *specific heat* found in older texts, refers to the heat capacity of 1 g of a substance and is generally reserved for systems of unknown molar mass.

The heat capacity of a chemical system has fundamental significance since the energy absorbed by a gas, liquid, or solid as it is heated is utilized for the most part to initiate translational, rotational, and vibrational motion within the material. In other words, C_P is a thermodynamic property that can be related to atomic or molecular structure. In practical terms, C_P can be regarded as a proportionality constant relating the enthalpy change occurring in a calorimeter, under a particular set of experimental conditions, to the temperature change generated, that is, in this usage, the *operational heat capacity* includes terms for the heat capacity of the calorimeter itself, for example, cell, stirrer, temperature-measurement device, as well as C_P for the reaction solution. The operational heat capacity of the calorimeter is sometimes referred to as the *energy equivalent*, designated ε, in order to delineate this term from the rigorously defined thermodynamic parameter C_P:

$$\varepsilon = \frac{\text{heat change in the calorimeter}}{\Delta T} \tag{1.26}$$

ε can be used in a very broad sense to relate the energy generated in the calorimeter to the signal ultimately recorded whatever its form (voltage, chart units, number of heater pulses, etc.). The operational heat capacity or energy equivalent of a calorimeter is typically determined by calibration with the passage of a known amount of electrical current (see Chapter 3).

As evidenced by Eq. (1.21), the heat capacity of a system will depend on the temperature at which it is determined. Strictly speaking, therefore, the heat capacity term shown in Eq. (1.25) represents an average of the heat capacities over the temperature range ΔT. In practice, there are several reasons why the temperature change, ΔT, should be minimized in a calorimetric experiment. Accordingly, it is usually a good approximation to consider that ε remains invariant over a typical temperature change in a calorimetric experiment, all other parameters (particularly volume) remaining constant. The effect of volume changes from added reagents on the operational heat capacity is discussed in detail in Chapter 2.

The heat capacity of the solvent system in a calorimetric cell can significantly affect the sensitivity of the experiment. For example, C_P for many non-aqueous solvents is only about one-half that of water as shown in Table 1.2. All else being equal, one would predict sensitivity increases of up to twofold (versus water) for a temperature change generated by a reaction in non-aqueous media. However, the experimental problems associated with the use of non-aqueous solvents, for example, increased thermal noise from evaporation, can offset the advantage of reduced heat capacity. Moreover, it does not necessarily follow

TABLE 1.2. Heat Capacities of Some Common Solvents

Solvent	Heat Capacity (J °C^{-1} mL^{-1})
Acetic acid	2.18
Acetone	1.67
Acetonitrile	2.26
Benzene	1.55
Carbon tetrachloride	1.38
Chloroform	1.46
Ethanol	1.92
Ether	1.54
Pyridine	1.76
Water	4.18

that ΔH_R will be the same in two different solvent systems. Solvent effects on reaction enthalpies are discussed in more detail in the context of acid–base chemistry in Chapter 6.

1.3. NOMENCLATURE

To quote from a recent article by Tanaka (2), "there is no authorized rational system of classification of calorimeters." Unfortunately, this is an accurate statement: calorimetric nomenclature in popular usage is based on several criteria including tradition, instrument sophistication, heat-transfer properties, the method of measuring the heat effect, and the environment in which the calorimeter is immersed. This state of affairs extends beyond the primary classification of calorimeter type to the methodological variants which differ only in the mode of addition of one reactant to another. A simplified nomenclature system is proposed in Fig. 1.1. The three principal types of calorimeter, namely isoperibol, heat conduction, and isothermal, can be subclassified as titration, batch, or flow instruments depending on the method of reactant mixing. For reference, historical terms and others in common use have also been included.

1.3.1. Adiabatic/Isoperibol Calorimetry and Enthalpimetric Analysis

Traditionally, calorimeters have been classified according to the degree of heat transfer p occurring between the reaction vessel and its surroundings and, related to this, the magnitude of the temperature change generated in the reacting solution. The least controversial denotation is the adiabatic calorimeter in which

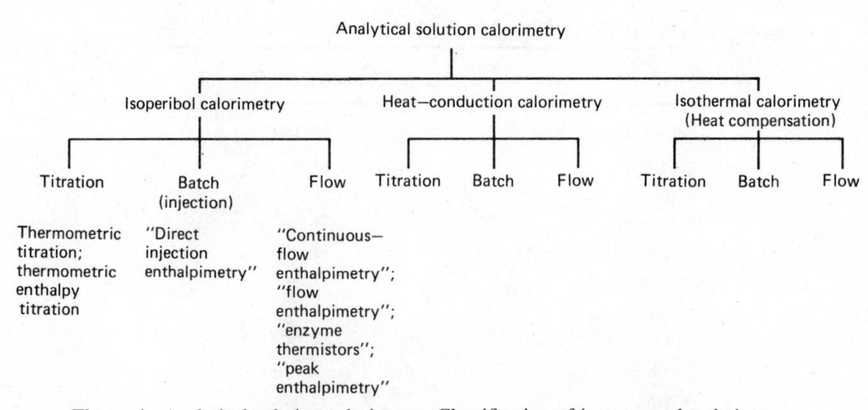

Figure 1. Analytical solution calorimetry. Classification of instrumental techniques.

$p = 0$ or is very small. The heat generated by the transformation is contained within the calorimeter and little or no heat transfer occurs between the cell and its surroundings. For an exothermic reaction, the ideal situation ($p = 0$) is approximated by activation of an electrically heated adiabatic jacket that nullifies the temperature differential between the cell and its surroundings. This type of instrumentation has found no application in analytical chemistry and will not be discussed further.

Isoperibol (3) calorimetry is an important variant of the adiabatic principle. Here, finite heat leaks, minimized by efficient cell construction and experiment design, are compensated for by calculation and/or extrapolation. Isoperibol and adiabatic instruments are quite different and can be delineated clearly by the temperature of the immediate environment enveloping the cell, which remains constant with the former and is the same as the cell contents in the latter. In this text, the term isoperibol has been used accordingly; however, it must be accepted that it is not definitive. Some ambiguity is introduced because the word isoperibol derives from "constant (temperature) environment" which is also a feature of some non-adiabatic (heat-conduction) instruments incorporating a high-thermal-mass heat sink. Hansen and Eatough (4) have circumvented this problem by classifying the calorimeters primarily according to the way the heat measurement is made. Consequently, the adiabatic/isoperibol instruments can be defined as devices that measure the temperature change in the reacting solution, as distinct from heat-conduction calorimeters which measure the voltage output at a thermoelectric transducer between the calorimeter and a heat sink (Chapter 3). In the absence of an acceptable alternative this operational delineation has been adopted, accepting the obvious flaw that the word isoperibol does not denote the element of temperature measurement. As with most nomenclature systems total definition gives way to expediency in some instances.

The terms *enthalpimetry* and *enthalpimetric analysis* are used extensively in the literature to describe isoperibol measurements of varying degrees of sophistication.

1.3.2. Isothermal (Heat-Compensation) and Heat-Conduction Calorimetry

Isothermal calorimeters operate on the principle that the temperature of the reacting solution remains essentially constant throughout the experiment. This term is usually reserved for calorimeters that operate on the heat-compensation principle. Isothermal conditions are achieved by either Joule heating or Peltier cooling compensation of the reaction enthalpy (see Chapter 3).

Quasiisothermal conditions can also be obtained by rapid heat exchange between the calorimeter vessel and an essentially isothermal heat sink surrounding it. Instruments that function in this manner are termed *heat-conduction* calorimeters.

1.3.3. Thermometric Titrations, (Direct) Injection, and Flow Enthalpimetry

Three types of reactant mixing can be identified for each form of calorimetry. These are titration, batch addition, and flow mixing as shown in Fig. 1.1. Several terms are in common use within the isoperibol classification that warrant discussion.

Since the isoperibol titration utilizes temperature change as an endpoint indicator and does not, of necessity, require the quantitation of heat, the designation thermometric is appropriate. In 1957, following an American Chemical Society meeting, it was recommended that the term *thermometric titration* be used to designate an adiabatic titration yielding plots of temperature versus volume of titrant (5). This term is entirely appropriate for those titration experiments in which no attempt is made to determine the heat change associated with the analytical reaction. *Thermometric enthalpy titration* is also used in the same context. In isoperibol calorimeters, batch addition (the mixing of the total volume of both reactants in one operation) is usually achieved by syringe injection. The term *(direct) injection enthalpimetry* is widely used to describe this technique as in Chapter 2.

Most commercial flow calorimeters are designed so that two flowing reactant streams mix in a reaction zone to form a product stream. The term *flow enthalpimetry* is used to describe an isoperibol flow-injection technique in which a discrete sample of reactant is introduced into a flowing stream of reagent (see Chapter 4).

1.4. HISTORICAL DEVELOPMENT OF ANALYTICAL SOLUTION CALORIMETRY

Calorimetry is one of the oldest reported scientific measurement techniques, having its roots in the pioneering animal respiration studies of Crawford, Lavoisier, and Laplace (6) at the end of the 18th century. Although long established as a technique for the determination of thermodynamic data, the concept of using solution calorimetry as a routine analytical tool has only gained significant momentum in the last 30 years. The reason for this slow development is the long equilibration time associated with many calorimeters designed primarily for precise thermodynamic measurements. Even modern heat-conduction calorimeters have a thermal equilibration time, after loading the calorimetric vessels, of 1–2 h. When the appropriate calibration experiments are included, only two to three experiments are possible in a working day, a prohibitive limitation for many analytical laboratories. As a result, the development of analytical applications of calorimetry is dominated by the advances in isoperibol instrumentation, which has the largest sample throughput capacity. This topic has been covered in several texts (7–10) and, therefore, discussion will be limited to key developments. The paper of Bell and Cowell in 1913 (11), introducing discontinuous thermometric titrations, is credited as the first quantitative analytical application of a solution calorimetric technique. However, this rudimentary apparatus, consisting primarily of a Beckmann thermometer, a Dewar flask, and a volumetric buret, did not address the issue of analysis time. The slow response of the thermometer meant that each titration took almost 1 h. Accordingly, analytical solution calorimetry went almost unnoticed until the landmark report of Linde, Rogers, and Hume in 1953 (12), which heralded the appearance of the thermistor, the constant delivery buret, and continuous recording of data in isoperibol calorimetry. In the next 20 years, analytical applications of this type of equipment proliferated, the primary emphasis being on instrumental simplicity and sample throughput. Direct-injection enthalpimetry, first reported in 1964 (13), epitomized this trend; reagent introduction was reduced to the injection of a small volume of solution resulting in analysis times of only a few minutes. It can be argued that solution calorimetry made the transition into the analytical laboratory during this period. In recent years, the emphasis on instrumental simplicity has diminished as calorimetric technology has reacted to the demand for increased sensitivity and small sample volume capacity as well as larger sample throughput. All these criteria have been satisfied with the development of isoperibol calorimeters incorporating the flow-injection principle popularized by Rusicka and Hansen (14). Key advances in the development of these instruments included the "peak enthalpimeter" reported in 1974 (15), a high-sensitivity flow calorimeter described by Schifreen et al. in 1979 (16), and the "enzyme thermistor" concept pioneered by Mosbach and co-workers at the University of Lund in 1974 (17). The isoperibol flow calorimeter is now established as the most sensitive instrumental variant (see Chapter 3). Moreover, up to 60 100μL samples can be determined per hour by this method. A batch isoperibol calorimeter, which will operate with 2 mL of solution in the reaction cell, has been developed extending the scope of batch or titration instruments for sample-limited analyses (18).

The fundamental design of batch and flow heat-conduction calorimeters has not changed radically since 1968 (19, 20). The isothermal (heat-compensation) calorimeter designed by Christensen et al. in 1968 (21) was adapted for small (2-mL) volumes in 1973 (22). More recently, an elevated-temperature, high-pressure version was introduced (23), which allows the measurement of endo- or exothermic reaction enthalpies from 273 to 423 K and from 0.1 to 40.5 MPa. Recent advances in all forms of calorimetric instrumentation have been reviewed extensively elsewhere (24–27).

Chemically catalyzed end-point reactions of the type first introduced by Vaughan and Swithenbank in 1965 (28) remain a realistic option to increase the sensitivity of thermometric titrations. The scope of this technique has been considerably extended by Greenhow who has utilized acid- and base-catalyzed vinyl polymerizations as indicator reactions for the non-aqueous titration of a wide range of strong, weak, and very weak acids and bases (29). The power of this technique is illustrated by the fact that micromole amounts of analyte can be determined with only rudimentary instrumentation.

In summary, a process of diversification has been taking place in the development of solution calorimetry as an analytical technique. The simple, but restrictive, instrumentation of the 1960s has given way to a comprehensive range of calorimeters that can address most analytical problems. The universal detection system, no longer hindered by a prohibitive lack of sensitivity or sample throughput capacity, can now be used with effect on a much broader scale.

REFERENCES

1. J. A. Dean, Ed., *Langes Handbook of Chemistry,* McGraw-Hill, New York, 1978.
2. S. Tanaka, *Thermochim. Acta* **61**, 147 (1983).
3. O. Kubaschewski and R. Hultgren, in *Experimental Thermochemistry,* H. A. Skinner, Ed., Wiley-Interscience, New York, 1962.
4. L. D. Hansen and D. J. Eatough, *Thermochim. Acta* **70**, 257 (1983).
5. D. N. Hume and J. Jordan, *Anal. Chem.* **30**, 2064 (1958).
6. M. Kleiber, *The Fire of Life,* Wiley, New York, 1961.
7. L. S. Bark and S. M. Bark, *Thermometric Titrimetry,* Pergamon Press, Oxford, 1969.
8. H. J. V. Tyrrell and A. E. Beezer, *Thermometric Titrimetry,* Chapman and Hall, London, 1968.
9. G. A. Vaughan, *Thermometric and Enthalpimetric Titrimetry,* Van Nostrand Reinhold, London, 1973.
10. J. Barthel, *Thermometric Titrations,* Wiley, New York, 1975.
11. J. M. Bell and C. F. Cowell, *J.A.C.S.* **35**, 49 (1913).
12. H. W. Linde, L. B. Rogers, and D. N. Hume, *Anal. Chem.* **25**, 404 (1953).
13. J. C. Wasilewski, P. T. S. Pei, and J. Jordan, *Anal. Chem.* **36**, 2131 (1964).
14. J. Rusicka and E. H. Hansen, *Flow-Injection Analysis,* Wiley, New York, 1981.
15. A. C. Censullo, J. A. Lynch, D. H. Waugh, and J. Jordan, in *Analytical Calorimetry,* J. F. Johnson and R. S. Porter, Eds., Plenum Press, New York, 1974, Vol. 3.
16. R. S. Schifreen, C. S. Miller, and P. W. Carr, *Anal. Chem.* **51**, 278 (1979).
17. K. Mosbach and B. Danielsson, *Biochim. Biophys. Acta* **364**, 140 (1974).
18. L. D. Hansen, R. M. Izatt, D. J. Eatough, T. E. Jensen, and J. J. Christensen, in *Analytical Calorimetry,* R. S. Porter and J. F. Johnson, Eds., Plenum Press, 1974, Vol. 3.
19. P. Monk and I. Wadso, *Acta Chem. Scand.* **22**, 927 (1968).
20. P. Monk and I. Wadso, *Acta Chem. Scand.* **22**, 1842 (1968).
21. J. J. Christensen, H. D. Johnston, and R. M. Izatt, *Rev. Sci. Instrum.* **39**, 1356 (1968).

22. J. J. Christensen, J. W. Gardner, D. J. Eatough, R. M. Izatt, P. J. Watts, and R. M. Hart, *Rev. Sci. Instrum.* **44**, 481 (1973).
23. J. J. Christensen, L. D. Hansen, R. M. Izatt, and D. J. Eatough, *Rev. Sci. Instrum.* **52**, 1226 (1981).
24. J. Brandstetr, *J. Therm. Anal.* **14**, 157 (1978).
25. D. J. Eatough, *J. Therm. Anal.* **14**, 45 (1978).
26. I. Wadso, *Quart. Rev. Biophys.* **3**, 383 (1970).
27. J. Brandstetr, *J. Therm. Anal.* **21**, 357 (1981).
28. G. A. Vaughan and J. J. Swithenbank, *Analyst (London)* **90**, 594 (1965).
29. E. J. Greenhow, *Chem. Rev. (London)* **77**, 835 (1977).

CHAPTER

2

FUNDAMENTALS OF ANALYTICAL SOLUTION CALORIMETRY

JOSEPH JORDAN and JOHN W. STAHL

Department of Chemistry
The Pennsylvania State University
University Park, Pennsylvania

2.1. THEORETICAL ASPECTS

2.1.1. Conceptual Basis

This chapter is rather narrowly restricted to isoperibol (1) methodology, specifically to thermometric enthalpy titration (TET) and direct-injection enthalpimetry (DIE). This is appropriate since TET, DIE, and the flow methods described in Chapter 4 (which together are also known under the broader generic designation of "enthalpimetric analysis") represent the overwhelming majority of calorimetric applications in the analytical literature. Their common feature is use of the heat of reaction ΔH_R as a means of quantitation in solutions (or gases). The conceptual basis of both methods is that the heat effect q_R represents a quantitative measure of the amount of product formed, n_P, namely,

$$q_R = n_P \Delta H_R \tag{2.1}$$

where *both* q_R and ΔH_R are assigned positive values for endothermic reactions and negative values for exothermic reactions. A consequence of this sign convention is that $q > 0$ corresponds to a *decrease in temperature* under adiabatic conditions, and $q < 0$ corresponds to a temperature increase. For a reaction that proceeds to virtual completion, n_P is directly related by a stoichiometric factor to the amount of sample present.

The measurement of heat effect q (which is in practice ultimately accomplished by a temperature measurement) is relatively unaffected by non-reacting impurities and heterogeneous matrices. This is an oft-cited advantage of calorimetric methods. However, the heat of reaction ΔH_R may be dependent on various solvent effects such as ionic strength.

Calorimetric procedures are very widely applicable, because nearly every reaction has a non-zero heat ($\Delta H_R \neq 0$). In Section 2.2.2, techniques for using reactions with $\Delta H_R \simeq 0$ are discussed. This universal nature of heats of reaction is advantageous on considerations of simplicity and versatility but can be a disadvantage because of poor selectivity. *Any* side reactions occurring will yield an interfering heat effect. The selectivity of an enthalpimetric analysis is thus dependent on *chemical* rather than instrumental factors. Selective reagents and/or catalysts are required to obtain meaningful analytical results.

Both DIE and TET use similar instrumentation (described in Chapter 3) for measuring q, the primary difference being the mode of reagent introduction. In DIE, excess reagent and sample are rapidly combined by an instantaneous *injection* of one or the other. The experimental readout consists of a plot of q (or temperature, T) versus time, as depicted in Fig. 2.1. The analytical information is contained *primarily* in the magnitude of the total heat or temperature "step" (ΔT in Fig. 2.1b).

In contradistinction, TET is a calorimetric titration. The reagent is added gradually to obtain a volumetric endpoint that corresponds to a stoichiometric equivalence point for the titration reaction. As depicted in Fig. 2.1a, the endpoint is indicated by a change in slope of the temperature (or heat) versus volume of titrant curve. In nearly all modern instruments, the titrant is added continuously at a constant rate. Thus, the abscissa in Fig. 2.1a may equally well be "read" as moles of titrant, volume of titrant, or time. The analytical information in a

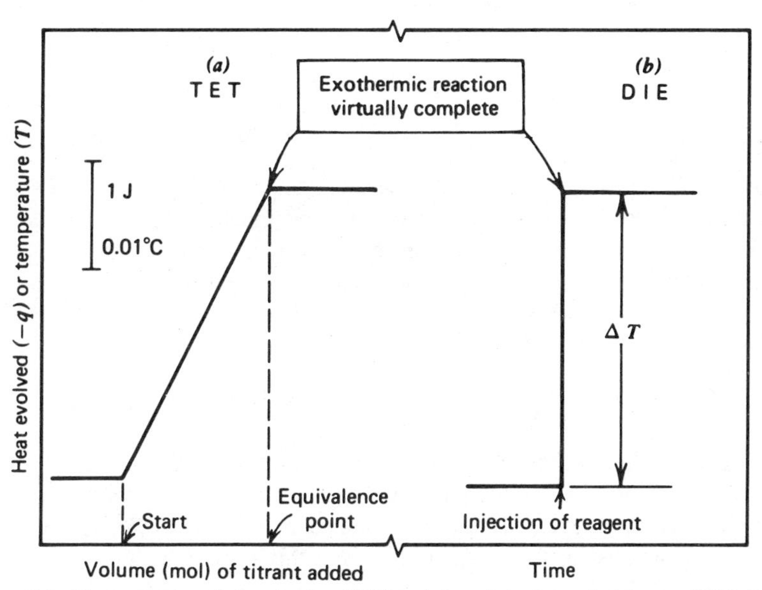

Figure 2.1. Thermometric enthalpy titration (TET) and direct-injection enthalpimetry (DIE). Idealized readouts. From Ref. 39, with permission of publisher.

TET curve is contained in both the volumetric endpoint and the overall magnitude of the heat effect.

Thermometric enthalpy titration is a linear titration technique wherein the signal is proportional to amounts reacted. Linear titration techniques, in general, are superior to logarithmic titration methods (such as potentiometric titrations) for titration reactions having small equilibrium constants (see Section 2.2.3a). The successful titration of boric acid, a very weak acid, with sodium hydroxide documents this ability of TET (2).

2.1.2. Theory of Chemical Effects

To facilitate the discussion at this point, consider an ideal adiabatic titration calorimeter. All physical sources of heat loss or gain that are extraneous to the chemical reaction of interest are taken as null. This situation is closely approximated by an isothermal heat-compensation-type calorimeter (see Chapter 3). In the ideal case, the instrumental output is controlled entirely by the thermodynamics and kinetics of the chemical reactions occurring.

2.1.2a. Thermodynamic Considerations in Direct-Injection Enthalpimetry

Consider the general chemical reaction:

$$A + B \underset{k_b}{\overset{k_f}{\rightleftharpoons}} P \tag{2.2}$$

where A is the sample specie, B is the reagent, and P is the product formed. All three species are in solution. In a DIE experiment, an excess of reagent is rapidly combined with the sample in the calorimetric cell. A and B react until chemical equilibrium is attained in accordance with Eq. (2.3):

$$\frac{k_f}{k_b} = K_{eq} = \frac{\gamma_P[P]}{\gamma_A \gamma_B [A][B]} \tag{2.3}$$

In Eq. (2.3), the brackets denote molar concentrations and γ is the appropriate activity coefficient. Generally, in the relatively dilute solutions employed in analytical calorimetry, activity coefficients are near unity. They will subsequently be ignored in this chapter, although this approximation is not always valid.

Assuming that no P was initially present, the total heat evolved will be

$$q_R = n_P \Delta H_R = V[\mathrm{P}]\Delta H_R \tag{2.4}$$

where V is the volume of the solution in the calorimeter reaction cell. The following substitutions can be made into Eq. (2.3):

$$[\mathrm{P}] = \frac{q_R}{V\Delta H_R} \tag{2.5}$$

$$[\mathrm{A}] = [\mathrm{A}]_0 - \left(\frac{q_R}{V\Delta H_R}\right) \tag{2.6}$$

$$[\mathrm{B}] = [\mathrm{B}]_0 - \left(\frac{q_R}{V\Delta H_R}\right) \tag{2.7}$$

and solved for q to yield

$$q_R = \Delta H_R V \left\{ \frac{([\mathrm{A}]_0 + [\mathrm{B}]_0 + K_{eq}^{-1} \pm \{([\mathrm{A}]_0 + [\mathrm{B}]_0 + K_{eq}^{-1})^2 - 4[\mathrm{A}]_0[\mathrm{B}]_0\}^{1/2}}{2} \right\} \tag{2.8}$$

In Eqs. (2.6)–(2.8), $[\mathrm{A}]_0$ and $[\mathrm{B}]_0$ represent the concentrations of sample and reagent immediately after injection, but before reaction has taken place.

In the limiting case of $K_{eq} = \infty$, Eq. (2.8) simplifies to

$$q_R = \Delta H_R V\, [\mathrm{A}]_0 \tag{2.9}$$

or

$$q_R = \Delta H_R V[\mathrm{B}]_0 \tag{2.10}$$

depending on whether B or A is present in stoichiometric excess. In the "lower limiting case" of $K_{eq} = 0$, Eq. (2.8) becomes $q_R = 0$. In order for the reaction to be analytically useful, the equilibrium constant must be large enough for the reaction to attain virtual completion, that is, the condition of Eq. (2.9) must prevail. A straightforward calculation shows that for $[\mathrm{A}]_0 = [\mathrm{B}]_0 = 10^{-3}$ mol L^{-1}, K_{eq} must be greater than 10^8 in order for reaction (2.2) to attain 99.9% completion. A DIE experiment will not yield any information about the completeness of reaction, so that K_{eq} must be independently known. Reactions with stoichiometries other than that of reaction (2.2) yield equations similar to Eq. (2.8), although more complicated. The general conclusions remain the same. An advantage of DIE is that a large excess of reagent can be used to drive reactions toward completion.

2.1.2b. Thermodynamic Considerations in Thermometric Enthalpy Titrations

Conceptually, a TET curve is equivalent to a series of incremental DIE experiments using increasing reagent/sample ratios. Therefore, provided the titration reaction attains equilibrium instantaneously, each point on a TET curve is described by Eq. (2.8) where $[B]_0$ denotes the moles of titrant added divided by the total volume of solution in the calorimetric cell. The resulting curve shape is illustrated in Fig. 2.2, where theoretical titration curves are shown for exothermic reactions of various equilibrium constants. Decreasing the magnitude of the equilibrium constant leads to an increasing degree of curvature at the endpoint. Obviously, this degrades the precision of endpoint location slightly, but the endpoint is still *accurately* located by linear extrapolation of the regions before and after the

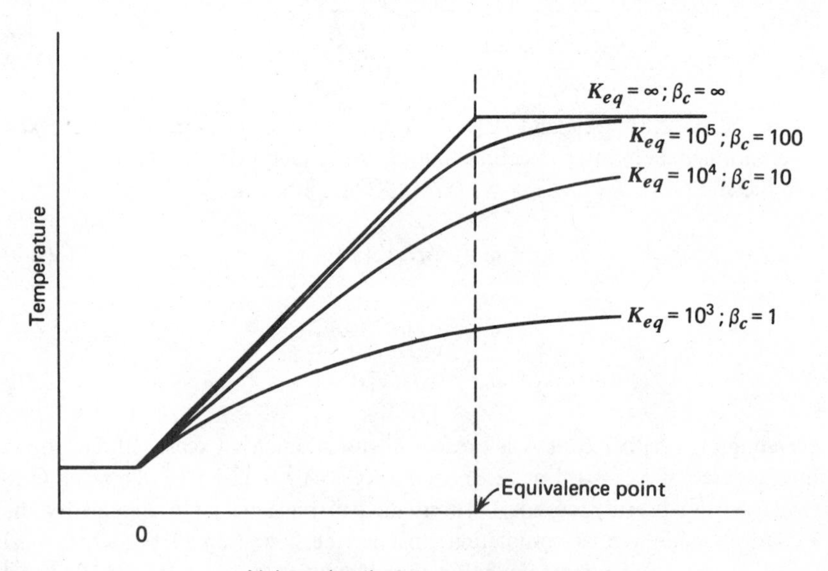

Figure 2.2. Theoretical thermometric titration curves for a reaction of the type $A + B = P$. Curves are calculated according to Eq. (2.8) for $[A]_0 = 10^{-3}$ mol L^{-1} and specified values of the equilibrium constant. The dimensionless parameter β_c is defined as $K_{eq} \cdot [A]_0$.

endpoint. This was shown to be true in general for linear titration methods by Rosenthal et al. (3).

The quantitative treatment of equilibrium incompleteness in linear titration methods is somewhat simplified by the use of reduced or normalized variables (3–5) and a summary of the approach is given here. The concentrations of each reactant and product specie are normalized by dividing by the initial concentration of the sample specie. Thus, for reaction (2.2), the reduced parameters $\overline{A}$, $\overline{B}$, and $\overline{P}$ are given by

$$\overline{A} = [A]_t/[A]_0 \qquad (2.11)$$

$$\overline{B} = [B]_t/[A]_0 \qquad (2.12)$$

$$\overline{P} = [P]_t/[A]_0 \qquad (2.13)$$

where $[A]_t$, $[B]_t$, and $[P]_t$ represent the concentrations of each specie at any time t during the titration. In addition, the amount of titrant added is normalized by defining the fraction of A titrated, f_A, as

$$f_A = \frac{\text{moles B (titrant) added}}{\text{moles B required at equivalence point}} \qquad (2.14)$$

so that f_A progresses from 0 to 1 from the start of the titration to the equivalence point. In terms of reduced variables, the equilibrium expression for reaction (2.2) is

$$K_{eq}\,[A]_0 \equiv \beta_c = \frac{\overline{A}}{\overline{B}\,\overline{P}} \qquad (2.15)$$

where β_c is introduced as a variable that is a function of both K_{eq} and initial sample concentration. When combined with the appropriate mass balance equations, the result is

$$\beta_c\overline{A}^2 + [\beta_c(f_A - 1) + 1]\overline{A} - 1 = 0 \qquad (2.16)$$

Equation (2.16) is in substance no different from Eq. (2.8). Carr (5) has pointed out that the use of reduced concentrations and the parameter β_c more generally defines the *shape* (= analytic geometry) of a titration curve and facilitates intercomparison of data and the consideration of accuracy and precision. Accordingly, in Fig. 2.2, the curve labeled β_c = 100 equally well describes the *shape* of the titration curve when $K_{eq} = 10^5$ and $[A]_0 = 10^{-3}$ and when $K_{eq} = 10^4$ and $[A]_0 = 10^{-2}$.

Stoichiometries more complicated than that of reaction (2.2) (two reactants going to one product) are also readily handled using reduced parameters. Table 2.1, taken from Carr (5), gives the appropriate equations to describe the analytic geometry of the titration curve in those cases. In each instance, a unique titration curve is defined for a given value of β_c. These equations are primarily useful in predicting the shape of an experimental TET curve and evaluating its analytical utility based on known thermodynamic parameters. Calorimetric titrations are also a valuable tool for performing the complementary task, that is, determining thermodynamic parameters from an experimental curve. This topic is treated in detail in Chapter 5.

2.1.2c. Kinetic Considerations in Direct-Injection Enthalpimetry

The hypothetical reaction (2.2) has, to this point, been assumed to achieve equilibrium instantaneously ($k_f = \infty$). An idealized DIE curve corresponding to this situation has been shown in Fig. 2.1*b*. "Real" reactions will, of course, proceed at a finite rate. Because the heat evolved (or absorbed) is directly related to the amounts reacted, a DIE experiment yields a direct plot of amounts reacted versus time. As an example, several DIE plots for the deprotonation of various nitroalkanes according to Eq. (2.17) are shown in Fig. 2.3 (6):

TABLE 2.1. Dimensionless Equations Describing Linear Titration Curves Controlled by Equilibrium[a]

Reaction Type[b]	Equation	Definition of β_c
A + B = P	$\beta_c\bar{A}^2 + [\beta_c(f_A - 1) + 1]\bar{A} - 1 = 0$	$K_{eq}[A]_0$
A + B = P + R	$(1 - \beta_c)\bar{A}^2 + [\beta_c(1 - f_A) - 2]\bar{A} + 1 = 0$	K_{eq}
A + B = P↓ (ppt)	$\bar{A}^2 + (f_A - 1)\bar{A} - \beta_c = 0$	$K_{sp}[A]_0^{-2}$
A + 2B = P	$4\beta_c\bar{A}^3 + 8\beta_c(f_A - 1)\bar{A}^2 + 1 = 0$	$K_{eq}[A]_0^3$
A + 2B = P↓ (ppt)	$4\bar{A}^3 + 8(f_A - 1)\bar{A}^2 + 4(f_A - 1)\bar{A} - \beta_c = 0$	$K_{sp}[A]_0^{-3}$
2A + B = P	$\beta_c\bar{A}^3 + \beta_c(f_A - 1)\bar{A}^2 + \bar{A} - 1 = 0$	$K_{eq}[A]_0^3$
2A + B = P↓ (ppt)	$\bar{A}^3 + (f_A - 1)\bar{A}^2 - 2\beta_c = 0$	$K_{sp}[A]_0^{-3}$

[a]Taken from Ref. 5.
[b]A, titrate; *B*, tritrant; P,R, products; $\bar{A}$, $[A]_t/[A]_0$; f_A, fraction of A titrated.

$$\mathrm{RCH_2NO_2} + \mathrm{OH^-} \underset{k_b}{\overset{k_f}{\rightleftharpoons}} \mathrm{RCHNO_2^-} + \mathrm{H_2O} \tag{2.17}$$

The rate equation for this reaction is first order in each reactant. If $k_f >> k_b$, the rate law may be expressed as

$$\frac{-d[\mathrm{RCH_2NO_2}]}{dt} = k_f[\mathrm{RCH_2NO_2}]\,[\mathrm{OH^-}] \tag{2.18}$$

This type of rate equation is fairly common in solution chemistry and, in general, may be expressed in terms of heat evolved:

$$\frac{dq_R}{dt}\frac{1}{\Delta \mathrm{H}_R \mathrm{V}} = k_f[\mathrm{A}][\mathrm{B}] \tag{2.19}$$

The corresponding integrated expression, with initial conditions $[\mathrm{A}] = [\mathrm{A}]_0$ and $[\mathrm{B}] = [\mathrm{B}]_0$, is

$$\frac{1}{[\mathrm{A}]_0 - [\mathrm{B}]_0} \ln \frac{[\mathrm{B}]_0([\mathrm{A}]_0 - q_R \Delta H_R^{-1} V^{-1})}{[\mathrm{A}]_0([\mathrm{B}]_0 - q_R \Delta H_R^{-1} V^{-1})} = k_f \Delta t \tag{2.20}$$

where Δt is the time elapsed after reagent injection. In the special case when $[\mathrm{A}]_0 = [\mathrm{B}]_0$,

$$\frac{1}{([\mathrm{A}]_0 - q_R \Delta H_R^{-1} V^{-1})} = k_f \Delta t + \frac{1}{[\mathrm{A}]_0} \tag{2.21}$$

In the presence of a large excess of reagent B, Eq. (2.19) becomes pseudo-first-order and upon integration yields

$$q_R \Delta H_R^{-1} V^{-1} = [\mathrm{A}]_0 \{1 - \exp(-k_f[\mathrm{B}]_0 \Delta t)\} \tag{2.22}$$

The curves displayed in Fig. 2.3 are all pseudo-first order due to the excess of hydroxide present.

The experimental DIE curve accurately represents the actual reaction course only within a certain "time window." If the rate of the reaction is very fast, the slope of the DIE readout (Fig. 2.1*b*) is determined entirely by the response time of the calorimeter. For a well-designed isoperibol calorimeter, this is on the order of a few seconds. For an isothermal calorimeter, the response time is

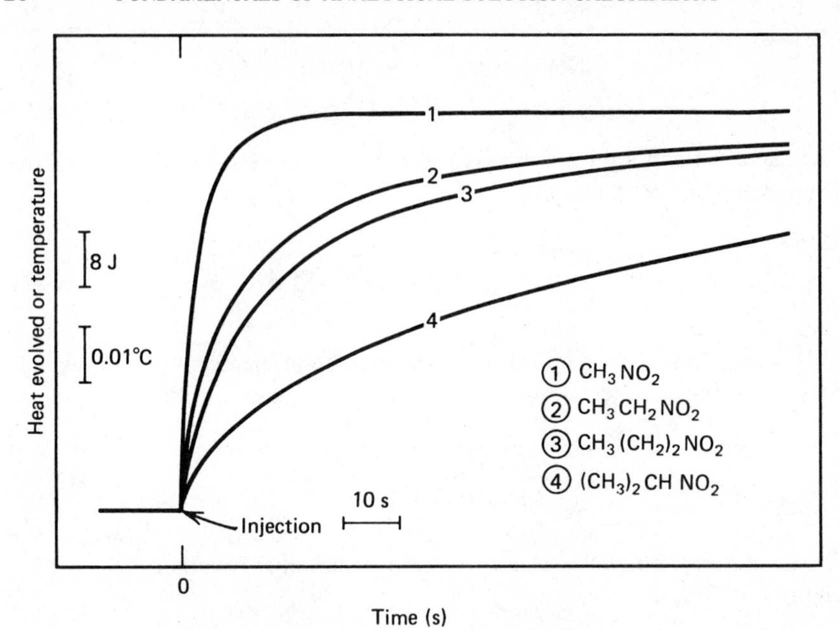

Figure 2.3. Idealized plots of heat evolved versus time for the reaction

$$RCH_2NO_2 + OH^- \text{ (excess)} \rightleftharpoons RCHNO_2^- + H_2O$$

RCH_2NO_2 at approximately millimolar concentrations. Reprinted from Ref. 6.

several minutes. On the other hand, for very slow reactions, non-chemical thermal effects and heat losses introduce large errors when an isoperibol system is used (in lieu of a genuine adiabatic calorimeter). Chapter 3 contains a discussion of these considerations. Within these constraints, DIE is useful as a tool for determining kinetic parameters (6, 7), enzyme activities (8), and sample concentrations (9, 10).

2.1.2d. Kinetic Considerations in Thermometric Enthalpy Titrations

A reaction having sluggish kinetics can cause misleading results in any titration technique. In conventional titrations, where the titrant is added discontinuously, this problem can, in part, be overcome by allowing sufficient time between titrant addition intervals for the reaction mixture to reach chemical equilibrium. Thermometric titrations, however, generally use the continuous addition of titrant at a constant rate. If the rate of titrant addition is of comparable magnitude to the rate of chemical reaction, the experimental endpoint will display curvature and will lag behind the theoretical equivalence point. This situation is depicted in Fig. 2.4*b*, where *Z* denotes the spurious nonstoichiometric endpoint obtained.

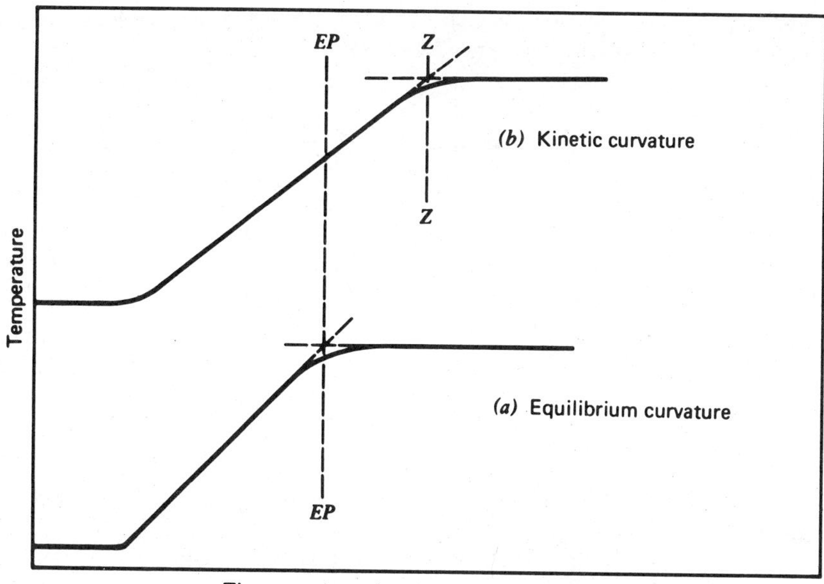

Figure 2.4. Endpoint rounding of thermometric titration curves caused by chemical factors. (Curves idealized with respect to nonchemical heat effects.) Titrant delivered at a constant rate: *EP*, equivalence point; *Z*, spurious endpoint. From Ref. 39, with permission of publishers.

A definitive treatment of kinetic effects in linear titration methods with continuous titrant addition has been presented by Carr and Jordan (11). Consider reaction (2.2), whose rate law is unimolecular in each of the two reactants, A and B [Eq. (2.19)]. The reaction rate during a titration can be expressed in the form:

$$\frac{-d[\mathrm{A}]}{dt} = k_f[\mathrm{A}]\left(\frac{F[\mathrm{B}]t}{V} - [\mathrm{A}]_0 + [\mathrm{A}]\right) \tag{2.23}$$

where F is the titrant (species "B") addition rate in L s^{-1}, [B] in the titrant concentration, and t is time in seconds. Equation (2.23) is put into a more convenient form by the introduction of reduced concentrations [as in Eqs. (2.11)–(2.13)] and a dimensionless parameter which we will call β_k, defined below:

$$\beta_k = k_f[\mathrm{A}]_0 t_{EP} = \frac{k_f[\mathrm{A}]_0^2 V}{F[\mathrm{B}]} \tag{2.24}$$

The time required to titrate to the stoichiometric endpoint is given by t_{EP}. β_k

uniquely defines dimensionless characteristics of the titration curve, including the relative titration error (see Fig. 2.4):

$$\% \text{ titration error} = \frac{Z - EP}{EP} \times 100 \tag{2.25}$$

By solving Eq. (2.23), the authors were able to calculate theoretical titration curves and from these derive values of percentage titration errors versus β_k. These values are used to plot Fig. 2.5. The use of that plot is illustrated in the following example. For the reaction

$$CH_3NO_2 + OH^- \xrightarrow{k_f} CH_2NO_2^- + H_2O \tag{2.26}$$

the forward rate constant is $k_f = 18 \text{ L mol}^{-1} \text{ s}^{-1}$. When the experimental conditions are such that $[A]_0 = 0.01 \text{ mol L}^{-1}$ nitromethane and $t_{EP} = 100$ s, then $\beta_k = 18$. In Fig. 2.4, this corresponds to a positive error of 20% for the titration endpoint Z. By slowing the titrant addition rate by an order of magnitude, so

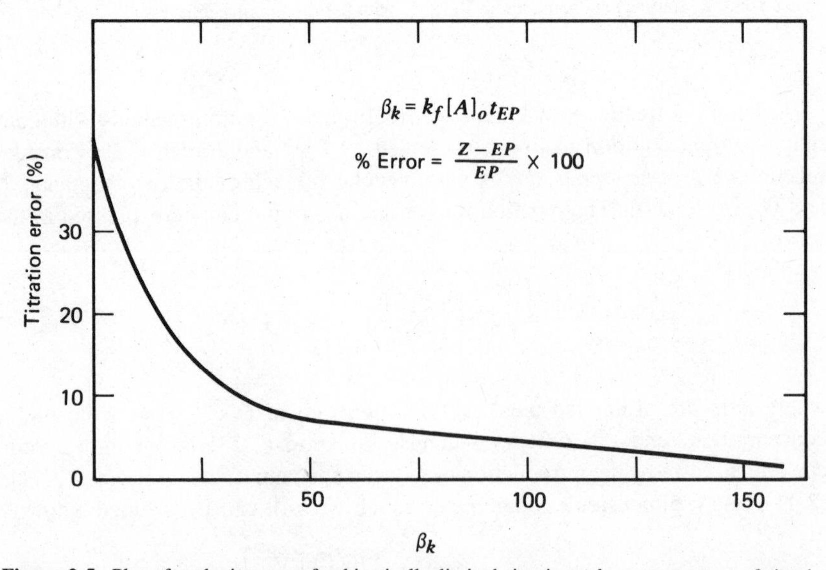

Figure 2.5. Plot of endpoint error for kinetically limited titrations. k_f, rate constant of titration reaction (2.2); $[A]_0$, sample (titrate) concentration; t_{EP}, time required to reach equivalence point as titrant is added at a constant rate; Z, Apparent endpoint illustrated in Fig. 2.4. (Taken from Reference 11. Copyright, 1973, American Chemical Society.)

that $t_{EP} = 1000$ s, β_k becomes 180 and the titration error drops to $\simeq 1\%$ (however, the analysis time now becomes disadvantageously long).

For reactions of known rate, Fig. 2.5 can thus be used to assess the analytical utility of a reaction. It can also be used to correct spurious endpoints (Z in Fig. 2.4) to obtain meaningful endpoints. For reactions of unknown rate, the presence of kinetic effects can be tested by varying the titrant addition rate. A shift in the position of the apparent endpoint indicates sluggish kinetics. This information can also be used to evaluate unknown rate constants.

2.1.3. Theory of Non-Chemical Thermal Effects

The shape of a DIE or TET curve is *primarily* controlled by the thermodynamic and kinetic parameters of the relevant chemical reaction, as just outlined. However, other sources of heat loss or gain are inherent in both the equipment and the experimental process that can modify the resulting data. For the isothermal calorimeter (Chapter 3), these effects are inconsequential with the exception of heat of dilution effects. Therefore, the ensuing discussion applies mainly to the isoperibol-type calorimeter (constant-temperature environment), which has generally been more popular for analytical uses.

The thermal effects to be considered are summarized in Table 2.2, which is largely derived from Ref. 5. The discussion of these thermal effects is facilitated by breaking the TET curve into four segments, which are illustrated in Fig. 2.6. In DIE, the "titration" or addition of reagent (the reaction and excess reagent segments in Fig. 2.6) is compressed into a single point in time, which is the injection. However, the same considerations apply to the initial and final segments.

2.1.3a. Stirring, Thermistor Heating, and Evaporation

The frictional drag and turbulence created by the stirring of the solution generate heat, as discovered by James Joule long ago. Thoughtful stirrer design is required to minimize this heat source while maintaining rapid and thorough mixing. Provided the stirrer is driven at a constant speed and volume changes during the experiment are minimal, this heating effect can be considered a constant, ω_S, through all segments of a TET or DIE curve, namely,

$$\frac{dq}{dt} = -\omega_S \tag{2.27}$$

Thermistors are almost universally used as sensors in analytical calorimetry. The current passing through the thermistor itself generates power as heat. As the temperature changes during the course of an experiment, the thermistor's

TABLE 2.2. Background Effects Contributing to Nonideal Thermometric Titration Curves[a]

Process	Mathematical Representation for Heat Flux of Process[b]	Affected Regions (Fig. 2.6)	Comments
1. Stirring	$\frac{dq}{dt} = -\omega_S$	All	Exothermic
2. Thermistor heating	$\frac{dq}{dt} = -\omega_{TH}$	All	Exothermic
3. Evaporation	$\frac{dq}{dt} = \omega_{EV}$	All	Endothermic, depends on atmosphere above solution
4. Heat transfer (imperfect adiabacity)	$\frac{dq_{HL}}{dt} = k_N(T_t - T_{ex})$	All	k_N may be time and volume dependent
5. Temperature mismatch of titrant	$\frac{dq_{TC}}{dt} = C^*_{P,T}F(T_t - T_T)$	*R*, *ER*	Assumes additivity of heat capacities
6. Heat of titrant dilution	$\frac{dq_D}{dt} = H_D F[B]$	*R*, *ER*	May be exo- or endothermic
7. Increase in heat capacity	$C_p = C_{P,0} + \frac{dC_P}{dV} F(t - t_0)$	*R*, *ER*	Not a heat loss per se, but causes nonlinearity in observed variable ΔT

[a]Derived from Ref. 5.

[b][B] = titrant concentration; C^*_P = heat capacity of titrant per unit volume; T_T = temperature of titrant and also temperature of calorimeter environment; k_N = Newtonian cooling constant. *R* indicates the reaction or titration segment of Fig. 2.6; *ER* indicates the excess reagent segment of the same figure.

resistance R_{TH} changes and, thus, the power dissipated in the thermistor changes. However, for a typical R_{TH} change of 4%/°C and a total temperature change of perhaps 0.1°C, the power of thermistor heating only changes by 0.4%. Generally, then, this is also considered a constant, ω_{TH}, throughout the entire calorimetric experiment:

$$\frac{dq}{dt} = -\omega_{TH} = -\mathbf{E}^2_{TH}R^{-1}_{TH} \tag{2.28}$$

In Eq. (2.28), $\mathbf{E}_{TH}$ is the voltage across the thermistor. To minimize problems of a steeply sloping baseline and, more importantly, thermal noise caused by

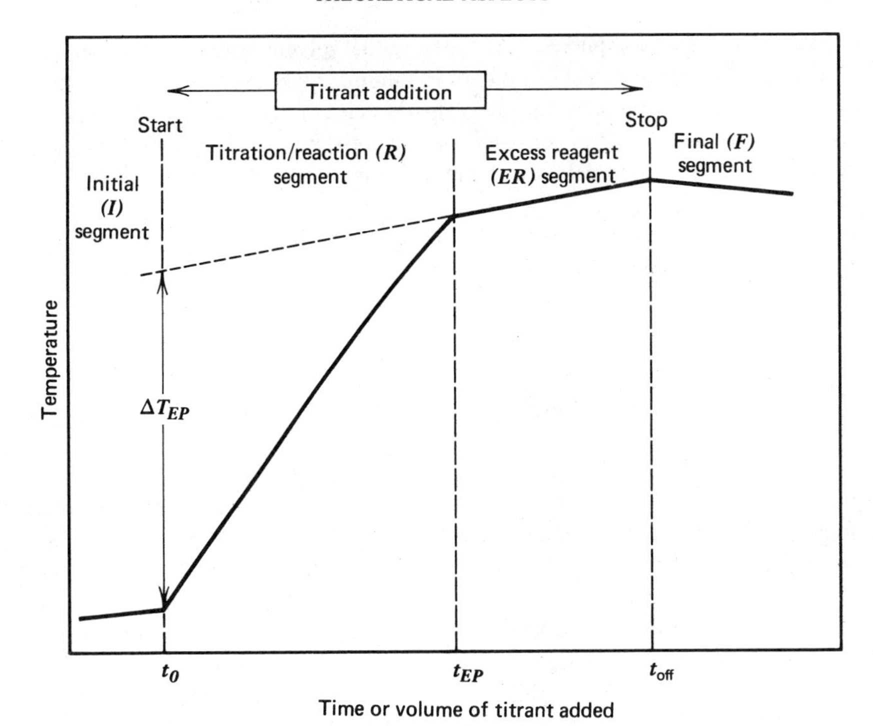

Figure 2.6. Thermometric enthalpy titration (TET) curve, illustrating various regions of the curve (see Table 2.2). Time is proportional to volume of titrant added between t_0 and t_{off} because titrant is added at a constant rate. From Ref. 39, with permission of publisher.

irregular heat dissipation from thermistor to solution, ω_{TH} is best kept to a relatively low value ($\simeq$50 μW).

The evaporation of the solvent in the calorimetric vessel is a third factor that is generally considered as constant throughout all segments of a calorimetric experiment, namely,

$$\frac{dq}{dt} = \omega_{EV} \tag{2.29}$$

Because evaporation is an endothermic process, $\omega_{EV} > 0$, in accord with the sign convention used throughout this chapter. The magnitude of ω_{EV} depends on solvent, on temperature, and also on the design of the calorimetric vessel. If the air space inside the vessel were completely sealed off from the external atmosphere, this inside air space would rapidly become saturated with solvent vapor and no further evaporation would occur ($\omega_{EV} = 0$). Generally, however, some exchange of atmospheres does occur, so that solvent evaporates continually. Goals in calorimetric vessel design are to minimize this exchange and to ensure that it does occur at a constant rate. This is particularly important for work at elevated temperatures or in non-aqueous solvents. Within these constraints, and provided the overall temperature change during an experiment is small ($<1°C$), Eq. (2.29) is valid.

All three of the constant thermal effects defined in Eqs. (2.27), (2.28), and (2.29) can be combined into one constant, namely,

$$\omega = \omega_S + \omega_{TH} - \omega_{EV} \tag{2.30}$$

2.1.3b. Heat Transfer with the Environment

Any real calorimetric vessel cannot be perfectly adiabatic whenever a temperature difference exists between the interior and exterior of the vessel. Some finite heat transfer is bound to occur. This creates a variable thermal effect in all segments of TET or DIE curves. Heat transfer by radiation and convection may be considered zero if a Dewar-flask-type vessel is used, owing to the low emissivity of the silvered wall and the absence of convection in the evacuated space. Heat transfer by conduction through the walls and supporting material of the vessel is the dominant mode, and this may be accounted for at various levels of sophistication.

Fourier's Law. The problem of modeling heat transfer at a finite rate through the conducting material(s) that supports the vessel could in theory be addressed by the use of Fourier's law:

$$\nabla^2 T(x,t) = \rho s \lambda^{-1} \frac{\delta T(x,t)}{\delta t} \tag{2.31}$$

In Eq. (2.31), ρ, s, and λ are the density, specific heat, and thermal conductivity of the appropriate materials, and x is distance. Explicit solution of Eq. (2.31) is practically impossible for a real calorimeter, since the boundary conditions are unknown and several different materials may be used in the construction. Wachter, contributing in reference (12), has provided a solution for the simplified case of conduction in a glass tube with insulated lateral surfaces, Fig. 2.7. The glass tube in that figure is initially at a uniform temperature T_0. At time greater than zero, the temperature along the surface $x=0$ is suddenly stepped to $T_1 = T_0 + \Delta T$. Solving Eq. (2.31) for this situation and substituting the values (for Pyrex glass) $\lambda = 0.0113$ J cm^{-1} s^{-1} K^{-1}, $s = 0.75$ J g^{-1} K^{-1}, $\rho = 2.23$ g mL^{-1}, and a value of $\Delta T = 0.1°C$, the temperature "profile" of the glass bar at

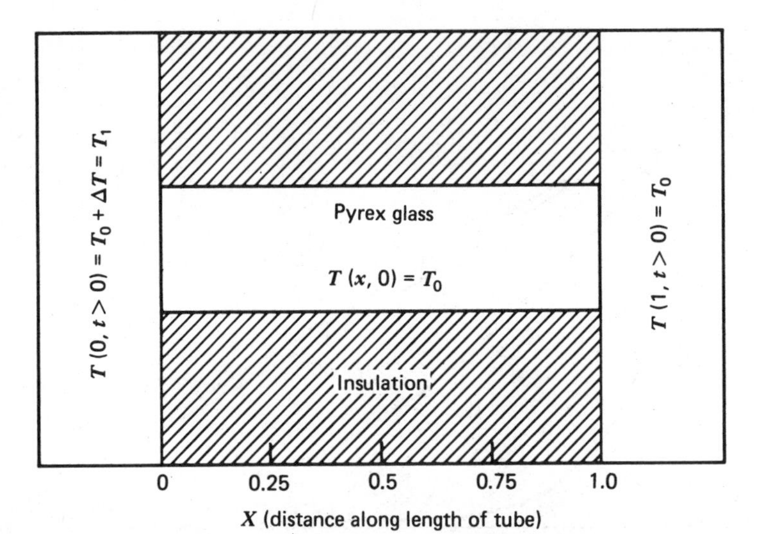

Figure 2.7. Heat transfer in a tube with completely insulated lateral surfaces. Temperature at $x = 0$ is stepped to T_1 at $t > 0$.

various times may be obtained. These temperature "profiles" at times of 10, 100, 500, and 1000 s are plotted in Fig. 2.8.

At 1000 s, the temperature gradient within the conducting tube has reached a constant value, a condition we will call "thermal steady state." At times closer to $t = 0$, the thermal gradient is changing in time, and thus the rate of heat transfer is changing. The practical effect of this with regard to a real calorimetric vessel is that the apparent heat capacity of the vessel and its contents will be somewhat variable, depending on the time scale of the experiment.

Newton's Law. The example portrayed in Figs. 2.7 and 2.8 is a somewhat extreme case, since the attainment of thermal steady state is very slow. With good experimental design, the heat-transfer process is generally modeled fairly well by Newton's law:

$$\frac{dq_{HL}}{dt} = k_N(T_t - T_{ex}) \tag{2.32}$$

where k_N is the Newtonian heat-leak constant, T_{ex} is the external temperature, and T_t is the internal temperature of the vessel at any time t in a calorimetric experiment. This equation actually represents the rate of thermal conduction when the system is in thermal steady state (1000 s in Fig. 2.8). Some workers

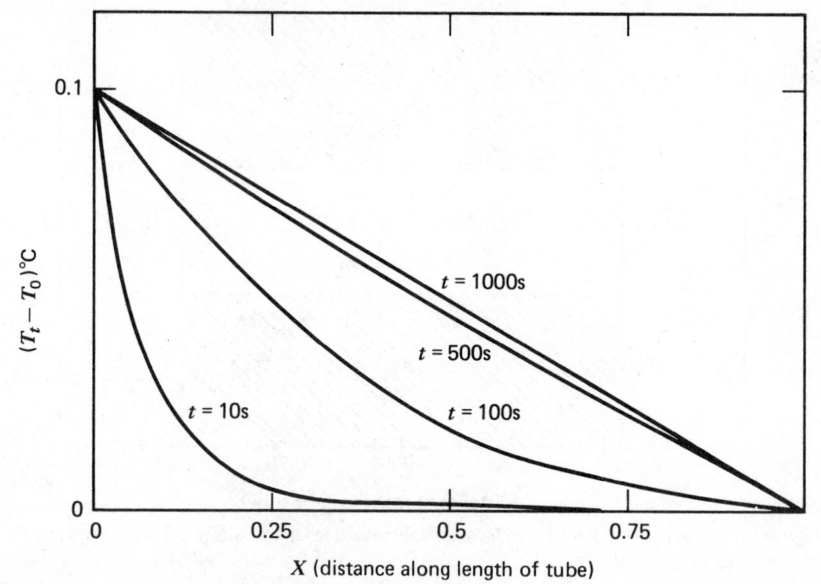

Figure 2.8. Profile of temperature gradient in glass tube (Fig. 2.7) at various times (in seconds) after temperature at $x = 0$ is stepped to $T_0 + 0.1$°C.

have preferred to represent Newton's law in terms of the rate of temperature change within the vessel, namely,

$$\frac{dT}{dt} = \kappa(T_{ex} - T_t) \tag{2.33}$$

where κ is the "heat transfer modulus." κ is equal to k_N/ε, where ε is the operational heat capacity or energy equivalent of the calorimetric vessel and its contents.

Modifications to Newton's Law. Equation (2.33) is only valid to the extent that κ (or k_N) is truly a constant. In the case of small vessels (<5 mL), κ has been found to vary considerably with the volume of solution inside the vessel (13). This is reasonable, considering that the heat-leak path decreases as the level of solution in the vessel rises (for a vessel which is supported from the top). A linear dependence of κ on volume was found to be obeyed over a working range of volumes, namely,

$$\kappa = \kappa_0 + \alpha_\kappa \Delta V \tag{2.34}$$

where $\alpha_\kappa = d\kappa/dV$. The initial Newtonian heat-loss constant κ_0 and its volume dependence α_κ were determined by independent experiments for a given calorimetric vessel. An analogous equation can be derived to represent the linear dependence of ω with titrant volume (13). The correction of titration data for variation of κ and ω over a limited titrant volume range is described in Chapter 3.

The use of Eq. (2.32) or (2.33) assumes that the experiment is done on a time scale such that the vessel remains in thermal steady state. When this is not the case, the correction for heat loss (or gain if the chemical reaction is endothermic) becomes less accurate. Martin et al. (14) have been able to reintroduce the time dependence of the heat-transfer process in a parametric way without explicit solution of Eq. (2.31). A "two-hole" heat-transfer model was used to describe successfully the characteristics of small vessels when carrying out very rapid reactions. The parameters of the model were obtained by independent calibration experiments.

Implications for Equipment and Experimental Design. In the design of the calorimetric vessel, heat-transfer characteristics mandate two major considerations. First, the vessel should be as nearly adiabatic as possible [i.e., k_N in Eq. (2.32) or κ in Eq. (2.33) should be small], so that the heat exchanged with the environment is small compared to the heat of the chemical reaction in the cell. Thus, even if the correction for the heat exchange is inaccurate, the relative error introduced will be minimized. Second, it is desirable for the vessel to have a rapid thermal response time (rapid attainment of thermal steady state), so that the Newtonian heat-loss model is generally followed. The construction and operating characteristics of isoperibol calorimetric cells based on these principles is described in detail in Chapter 3.

2.1.3c. Titrant Temperature Mismatch

In a DIE experiment, the temperatures of the solution to be injected and the solution inside the vessel can and should be closely matched. Assuming additive heat capacities, the temperature change, $T_F - T_1$, upon mixing *non-isothermal* solutions is

$$T_F - T_1 = (T_2 - T_1)\frac{V_2}{(V_1 + V_2)} \tag{2.35}$$

where the subscripts 1, 2, and F denote the solution initially in the vessel, the injected solution, and the mixture of the two, respectively. If the chemical reaction of interest is essentially instantaneous, this mismatch temperature change, $T_F - T_1$, is indistinguishable from the temperature change due to the reaction heat. The "mismatch" is minimized by making $T_2 - T_1$ small ($\simeq 0$) and also by injecting a relatively small volume ($V_2 << V_1$).

During a TET experiment, the titrant and titrate solutions may be isothermal initially (at t_0 in Fig. 2.6) but inevitably will be mismatched as the reaction proceeds. Thus, some of the chemical heat will be expended in warming up (or cooling down in the case of an endothermic reaction) the titrant as it is added and the overall temperature change will be diminished. The heat flux, dq_{TC}/dt, which corresponds to this process at a given time t, is

$$\left(\frac{dq_{TC}}{dt}\right)_t = C^*_{P,T}F(T_t - T_T) \tag{2.36}$$

where $C^*_{P,T}$ is the unit volume heat capacity of the titrant in J °C^{-1} mL^{-1}, F is the rate of titrant addition, T_T is the (constant) titrant temperature, and T_t is the temperature of the reaction mixture at a given time. Clearly the overall magnitude of the mismatch depends on the total volume of titrant added.

2.1.3d. Heat of Dilution

Consider an aqueous solution of an electrolyte of concentration c. The ionic strength μ is given by Eq. (2.37),

$$\mu = \frac{1}{2}\sum_i c_i Z_i^2 \tag{2.37}$$

where Z_i is the charge on the ith ion and the summation is over all ions in solution. Owing to differing degrees of solvation and ion–ion interactions, the heat of formation of a solute depends on its concentration. Therefore, the dilution of a solute has a related heat effect. Tabulated values of heats of dilution to infinity, $-\Phi_L$ (where Φ_L denotes the partial molal enthalpy), are available for many common electrolytes (15, 16). If the solution of concentration c_1 is diluted to concentration c_2, the heat of dilution ΔH_D will be equal to the integral heat of dilution of c_1 from μ_1 to infinite dilution $(-\Phi_{L,1})$ *minus* the integral heat of dilution of c_2 from μ_2 to infinite dilution $(-\Phi_{L,2})$:

$$\Delta H_D = \Phi_{L,2} - \Phi_{L,1} \tag{2.38}$$

In practice the titrant is generally many times ($\simeq$50) more concentrated than the titrate solution to minimize volume changes during the titration and titrant-temperature-mismatch effects. Therefore, a considerable dilution of the titrant specie B often occurs, accompanied by a commensurate heat of dilution effect. Philosophically, this thermal effect is "chemical" in nature. However, it is secondary to the chemical reaction of interest, which is occurring between titrant and titrate. The titrate specie A is also diluted to some extent by the addition of the titrant volume, but this is negligible in comparison to *titrant* dilution.

If the ionic strength of the titrate solution μ_A changes appreciably during the titration, then ΔH_D will also change as the titration progresses. However, often μ_A remains essentially constant during the titration. For example, some titration reactions, such as the neutralization reaction (2.39), do not change the number of ions present in the titrate solution:

$$\underbrace{Na^+_{aq} + OH^-_{aq}}_{\text{titrant}} + \underbrace{H^+_{aq} + Cl^-_{aq}}_{\text{titrate}} = \underbrace{H_2O + Na^+_{aq} + Cl^-_{aq}}_{\text{postreaction mixture}} \quad (2.39)$$

Or, if the titrate contains an excess of an inert electrolyte, the change in μ_A wrought by the addition of titrant will be negligible. In these cases, ΔH_D will be nearly constant through the reaction and excess reagent segments of the titration (Fig. 2.6), and the heat flux due to titrant dilution is given by

$$\frac{dq_D}{dt} = \Delta H_D F[B] \quad (2.40)$$

After correction for ω and heat losses, the excess reagent region of the titration curve can be used to estimate and correct for the heat of dilution when this assumption is valid. Otherwise, point by point calculation of ΔH_D from tabulated data or theoretical models (5,17) is necessary. Alternatively, a blank experiment can be used to evaluate these effects.

Non-aqueous solvents can, of course, also be used. Heats of dilution or heats of mixing are often large in this case, and careful experimental design is required.

In a DIE experiment, exactly the same considerations apply. However, no excess reagent region is available to estimate ΔH_D. The heat of dilution will be indistinguishable from the heat of reaction.

2.1.3e. Volume Change During Titration

As a thermometric titration progresses, the volume of the titrate solution, and thus the heat capacity of the system, increases. Over a working range of volumes, a linear dependence of C_P on volume is obeyed

$$C_{P,t} = C_{P,0} + \frac{dC_P}{dV} V_t \quad (2.41)$$

where $C_{P,0}$ is the initial heat capacity of the titrate solution and vessel, dC_P/dV is the dependence of heat capacity on volume, and V_t is the volume of titrant added at time t. dC_P/dV and $C_{P,0}$ must be determined for a given calorimetric cell, usually by electrical-heating calibration experiments. When aqueous titrate and titrants are used, dC_P/dV will be somewhat larger than the heat capacity of water per unit volume, 4.184 J °C^{-1} mL^{-1}, due to the heat capacity of calorimeter materials and inserts. A TET curve (plot of T versus volume) in which a considerable volume of titrant is required to attain the endpoint (>5% of the titrate volume) will display a noticeable curvature in the reaction region (Fig. 2.6) owing to the increase in heat capacity in accordance with Eq. (2.41). Under these circumstances determination of the endpoint becomes inaccurate and imprecise. The effect is minimal if concentrated titrants are used. Replotting of the data as q versus volume, applying the correction of Eq. (2.41), will eliminate this effect.

2.1.4 Working Equations

2.1.4a. Thermometric Titration Curves

In the preceding sections, the various "non-chemical" thermal effects have all been expressed as heat-flux processes. These can be collected into unified equations that describe the TET curve in each of the four segments of Fig. 2.6. Note that the sign of q has been included to be consistent with the actual slopes in the figures.

Initial Region (Slope S_I)

$$S_I = \left(\frac{dT}{dt}\right)_t = -\left(\frac{dq}{dt}\right)_t C_{P,0}^{-1} = \frac{\omega + k_N(T_{ex} - T_t)}{C_{P,0}} \tag{2.42}$$

In Eq. (2.42), T_{ex} is the temperature of the outside environment of the calorimeter, which by design will be very near T_t in this region. The pretitration baseline will then be essentially a straight line of slope $S_I = \omega/C_{P,0}$, provided the calorimetric vessel has attained a thermal steady state (unchanging k_N).

Reaction and Excess Reagent Regions (Slopes S_R *and* S_{ER})

$$\left(\frac{dT}{dt}\right)_t = -\left(\frac{dq_R}{dt} + \frac{dq_{\text{thermal}}}{dt}\right)_t C_P^{-1}$$

$$= \frac{-(dq_R/dt)_t + \omega + k_N(T_{ex} - T_t) + C^*_{P,\text{B}}\,F\,(T_{ex} - T_t) - \Delta H_D\,[\text{B}]\,F}{C_{P,0} + (dC_P/dV)\,F\,\Delta t} \tag{2.43}$$

In Eq. (2.43), $(dq_R/dt)_t$ is the rate of heat evolution associated with the chemical reaction, as discussed in Section 2.1.2. For a rapid and complete reaction, this is given by ΔH_R [B] F in the reaction segment. T_{ex} is again the temperature of the environment, which is now also assumed to be the temperature of the titrant. The elapsed time from the beginning of the titration is Δt.

To obtain the experimental curve of T versus time, Eq. (2.43) must be integrated, as has been done in rigorous detail by several authors (5,12). However, Eq. (2.43) itself, the differential form, can be used to correct experimental TET curves for non-chemical thermal effects and obtain the desired net chemical heat q_R. The usual treatment is to *sum* Eq. (2.43) over a number of experimental points (t_i, T_i), which yields a very good approximation to the integral. For given values of (t_i, T_i), Eq. (2.43) can be solved analytically (provided the constants have been independently determined) and summed to yield a corrected curve of q_R versus t. Details of this process are given in Chapter 3.

Final Region (Slope S_F)

$$S_F \cong \left(\frac{dT}{dt}\right)_t = -\left(\frac{dq}{dt}\right)_t C_{P,F}^{-1} = \frac{\omega + k_N(T_{ex} - T_t)}{C_{P,0} + (dC_P/dV)Ft_{\text{off}}} \tag{2.44}$$

$C_{P,F}$ represents the final heat capacity. With the cessation of titrant addition, the thermal effects are the same as those in the initial region (provided that chemical equilibrium has been attained). Of course, the heat capacity in the final segment of the titration is larger than in the initial region and so is the heat exchange with the ambient environment [because T_t and, consequently, $(T_{ex} - T_t)$ differ in the initial and final segments]. To a good approximation, the final region is also linear. By combining Eq. (2.42) and (2.44) with the following experimental data,

$S_I = (dT/dt)_I \equiv$ the slope of the initial region

$S_F = (dT/dt)_F \equiv$ the slope of the final region

$\Delta T' = \overline{T}_F - \overline{T}_I \equiv$ the difference in temperature between the midpoints of the final and initial segments

the parameters T_{ex} and ω can be eliminated, and the constant k_N solved for, yielding

$$k_N = \frac{C_{P,0}S_I - C_{P,F}S_F}{\Delta T'} \tag{2.45}$$

Because k_N may be a function of volume [Eq. (2.34)], the value of k_N obtained by Eq. (2.45) from a TET curve will reflect some sort of average. Generally k_N is independently evaluated from electrical-heating calibration experiments, where no volume change occurs.

2.1.4b. Direct-Injection Enthalpograms

The pre- and postreaction regions of a DIE curve are subject to the same considerations as in TET and can be represented mathematically by Eqs. (2.42) and (2.44). The temperature change ΔT associated with the injection is given by

$$\Delta T_t = C_{P,F}^{-1}\left[-\int\left(\frac{dq_R}{dt}\right)dt + \omega\int dt + k_N\int(T_{ex} - T_t)dt - \Delta H_D\,\Delta V[B] + (T_{ex} - T_t)\frac{\Delta V}{(V_0 + \Delta V)}\right] \tag{2.46}$$

where ΔV is the volume injected, and the integrals are evaluated over the short length of time required for injection and mixing. Generally, over such a short time interval, the heat-transfer term in Eq. (2.46) is not accurate (owing to the time dependence of k_N) and is so small as to be insignificant. Note that if the reaction is complete and rapid, then

$$\Delta H_R\, n_A = \int\left(\frac{dq_R}{dt}\right)\mathrm{dt} \tag{2.47}$$

where n_A denotes moles of sample when the reagent is in excess. When the reaction is rapid (complete in 2s or less), the heat effects of the chemical reaction, of dilution, and of temperature mismatch are indistinguishable.

Slower reactions continue after the short time required for injection and mixing and result in a temperature–time plot described by Eq. (2.48):

$$\left(\frac{dT}{dt}\right)_t = C_{P,F}^{-1}\left[-\left(\frac{dq_R}{dt}\right)_t + \omega + k_N(T_{ex} - T_t)\right] \tag{2.48}$$

An approach similar to that used in TET (described in Section 2.1.4a) can be utilized to correct for the extraneous thermal effects in Eq. (2.48) and obtain a plot of q_R versus t.

2.2. ANALYTICAL ASPECTS

2.2.1. Precision, Accuracy, and Dynamic Range

2.2.1a. Direct-Injection Enthalpimetry

In any DIE experiment, the magnitude of the "analytical signal," ΔT, obviously depends on both ΔH_R and sample concentration. Therefore, for an exothermic reaction,

$$\Delta T = \frac{-q_R}{C_P} = \frac{(-)n_P\,\Delta H_R}{C_P} \simeq \frac{(-)n_P\,\Delta H_R}{V\,C_P^*} = \frac{(-)\,\mathcal{F}\,[\mathrm{A}]_0\,\Delta H_R}{C_P^*} \tag{2.49}$$

where $\mathcal{F}$ is a stoichiometric factor and C_P^* is the heat capacity of the solvent per unit volume (4.18 kJ °C^{-1} L^{-1} for water at 25°C). The ultimate precision attainable depends on the noise in the ΔT signal measured. Thermal inhomogeneities in the solution apparently limit this to 5–10 μ°C (18). Routinely, there will be some additional electrical noise pickup in the signal, so that the noise in ΔT is about 50 μ°C. The signal-to-noise ratio will thus be

$$\frac{\Delta T_{\text{signal}}}{\Delta T_{\text{noise}}} = \frac{-\mathcal{F}\,[\mathrm{A}]_0\,\Delta H_R}{2 \times 10^{-4}} \tag{2.50}$$

Thus, for example, a signal-to-noise ratio of 100 is obtained when $\mathcal{F}\,[\mathrm{A}]_0 = 4 \times 10^{-4}$ mol L^{-1} and $\Delta H_R = -50$ kJ mol^{-1} (a typical value).

Whereas the signal-to-noise ratio affects precision and sensitivity in DIE, other factors can be detrimental to accuracy. Implicit in Eq. (2.49) is the assumption that all of the sample A reacts to form product. The accuracy of the measurement will depend on the actual degree of completion, which is readily calculated from the equilibrium constant and relevant concentrations via Eq. (2.8).

The reaction kinetics also indirectly affect the accuracy of the analytical results. Slow reactions require a longer measurement time, during which extraneous thermal effects (numbers 1, 2, 3, and 4 in Table 2.2) become proportionately larger and thereby degrade the accuracy of the ΔT signal. Generally, isoperibol instrumentation becomes unacceptably inaccurate for reactions requiring more than 20–30 min to reach completion.

Even for rapid reactions, two extraneous thermal effects can be significant, namely, temperature mismatch (between sample and reagent solutions) and heats of dilution. These effects are indistinguishable from the analytical signal and must be minimized to obtain the requisite degree of accuracy. Carr (5) has analyzed these limitations in detail. Routinely, precision and accuracy of 5% can be attained at millimolar concentration levels.

2.2.1b. Thermometric Enthalpy Titration

In thermometric titrations, the *primary* analytical signal is the location of the volumetric endpoint(s). An endpoint is signaled by a change in slope on the plot of T versus titrant added. Under ideal conditions, the precision and accuracy is limited by the titrant delivery system to about 0.2%, typical for volumetric analysis. The precision of the endpoint, being located by extrapolation of straight line segments, depends on the magnitude of the slope change $|S_R - S_{ER}|$ (Fig. 2.6), and on its sharpness. The slopes S_R and S_{ER} are roughly proportional to $(\Delta H_R + \Delta H_D)$ and ΔH_D, respectively. Thus the heat of reaction must not be small compared with the heat of dilution.

The sharpness of the endpoint depends largely on the equilibrium constant of the titration reaction. Figure 2.2 illustrates the curvature due to equilibrium limitations. The precision and accuracy of the endpoint is directly related to the dimensionless parameter β_c discussed in Section 2.1.2b. References 3 and 5 provide a quantitative treatment of this relationship. For a precision and accuracy of 0.5% or better, the requirement is that β_c be >500. β_c values smaller than this yield TET curves whose endpoints are difficult to extrapolate experimentally although they may in theory be still fairly accurate.

Kinetic factors can be a serious detriment to accuracy in TET, as discussed in 2.1.2d and Ref. 5 and 11. Again, a dimensionless parameter, β_k, defines the accuracy of the endpoint (see Fig. 2.5). That figure, combined with experience, suggests that for accuracy of 0.5%, β_k must be >500. From Eq. (2.24) and for a sample concentration of 10^{-2} mol L^{-1}, this requires that

$$\frac{k_f V}{F\,[B]} > 5 \times 10^6 \qquad (2.51)$$

The "non-chemical" thermal effects in Table 2.2 also limit the precision and accuracy of a TET curve. Carr (5) has pointed out that processes 1, 2, 3, and 6 do not seriously detract from the analytical utility of the TET curve and thus may be termed "innocuous." In contradistinction, processes 4, 5, and 7 (in Table 2.2) cause non-linear deviations which make endpoint determination less precise and less accurate, and thus are called "parasitic." These processes must be minimized to negligible proportions (by the use of relatively concentrated titrants and short titration times), and/or corrected for by appropriate data reduction. The latter is discussed in detail in Chapter 3.

2.2.1c. Dynamic Range

As mentioned earlier, the "noise" level in common enthalpimetric measurements is about 5×10^{-5}°C. This establishes for any enthalpimetric procedure an ultimate lower limit of detection that is three times the noise level and is given by Eq. (2.52):

$$\text{Eq. (2.49)} \rightarrow 4.2 \times 3 \times 5 \times 10^{-5} = 6 \times 10^{-4} = |\mathcal{F}[\mathrm{A}]_0 \, \Delta H_R| \qquad (2.52)$$

In practice, the lower limit of detection is about an order of magnitude greater than this, owing to the extraneous thermal effects associated with the introduction of the reagent. For a typical ΔH_R (-50 kJ mol^{-1}) this corresponds to a sensitivity of $\simeq 10^{-4}$ mol L^{-1}.

Two factors tend to impose an upper limit on the range of enthalpimetric analysis. If a simple thermistor bridge is used for temperature measurement, the signal becomes increasingly non-linear with temperature for $\Delta T > \simeq 1$°C. From Eq. (2.49), this corresponds to $[\mathrm{A}]_0 \simeq 0.1$ mol L^{-1} if ΔH_R is assumed to be -50 kJ mol^{-1}.

A second factor arises from the requirement for relatively small volume changes in a DIE or TET experiment. The titrant should normally be 50 times as concentrated as the sample solution. Solubility of the reagent thus imposes a limit.

2.2.2. Special Techniques for Improving the Signal-to-Noise Ratio

2.2.2a. Chemical Amplification of Heat Effects by Coupled Reactions

The signal (ΔT) obtainable for a given sample concentration in DIE or TET depends on the heat of reaction. An increased signal results if the overall heat of reaction can be "chemically amplified" by coupled reactions. The most common type of coupled reaction is one involving a proton transfer. For example, consider the enzyme-catalyzed phosphorylation of glucose:

$$\text{glucose} + \mathrm{Mg(ATP)}^{2-} \xrightleftharpoons{\textit{hexokinase}} (\text{Glucose-6-phosphate})^{2-} + \mathrm{Mg(ADP)}^{-} + \mathrm{H}^{+} \qquad (2.53)$$

$$\Delta H_R = -27.6 \text{ kJ mol}^{-1}$$

In using reaction (2.53) as the basis for a DIE determination of glucose, McGlothlin and Jordan (10) coupled reaction (2.53) with the protonation of the buffer, tris-hydroxymethylaminomethane (THAM), namely,

$$H^+ + C(CH_2OH)_3NH_2 \rightleftharpoons C(CH_2OH)_3NH_3^+ \quad (2.54)$$
$$\Delta H_R = -47.3 \text{ kJ mol}^{-1}$$

Reactions (2.53) and (2.54) combined yield a heat nearly triple that of reaction (2.53) alone. Many applications of this technique are outlined in Chapter 7.

The technique of "chemical amplification" is equally applicable to TET for increasing the slope of the titration or reaction segment and thus "sharpening" the endpoint. For example, solutions of *p*-hydroxymercuribenzoate (HMB) (a reagent useful in sulfide and polysulfide analysis) are conveniently standardized against thiosulfate by TET (19) via the Lewis acid–base reaction:

$$^-OOC\phi HgOH + S_2O_3^{2-} \rightleftharpoons {}^-OOC\phi HgS_2O_3^- + OH^- \quad (2.55)$$
$$\text{(HMB)} \qquad \Delta H_R = -13.7 \text{ kJ mol}^{-1}$$

When performed in borax buffer, the hydroxide ion produced in reaction (2.55) reacts further:

$$B(OH)_3 + OH^- \rightleftharpoons B(OH)_4^- \qquad \Delta H_R = -42.2 \text{ kJ mol}^{-1} \quad (2.56)$$

As a result the overall heat evolved is quadrupled.

Other types of coupled reactions [e.g., H_2O_2 decomposition (20)] have been successfully used as chemical amplifiers of the heat of reaction. They generally also serve to drive the primary reaction to completion.

2.2.2b. Thermochemical Indicators

In thermometric titrations, it has been noted that the primary *analytical* information is in the location of the endpoint. An enhanced heat of reaction will facilitate endpoint determination. However, since it is the *change* in slope between the reaction and excess reagant segments that defines the endpoint, increased precision can also be obtained by adjusting the slope of the region after the endpoint. If the experiment is designed so that a very exo- or endothermic *endpoint "indicator reaction"* will occur after the stoichiometric equivalence point, TET can be applied to reactions with small heats. One approach is to take advantage of a direct reaction between titrant and indicator. For example, Hansen et al. (21) have used excess of sulfate as an indicator in the titration of moieties of the type $RCOO^-$ (salts of carboxylic acids) with a strong acid. Relevant data are contained in Table 2.3. The protonation of most carboxylic acids is weakly exothermic, and, as a result, endpoints are difficult to locate. Sulfate is generally a much weaker base than $RCOO^-$ and has a large endothermic heat of protonation. Figure 2.9 illustrates the use of sulfate as an endpoint indicator in the

TABLE 2.3. Thermodynamic Data Explaining the Use of Sulfate as a Direct Thermochemical Indicator in the Titration of Carboxylate Salts by Strong Acid[a]

Reaction	K_{eq}	ΔH_R (kJ mol^{-1})
$H^+ + RCO_2^- \rightleftharpoons RCO_2H$	10^3–10^5	0 to −5
$H^+ + SO_4^{2-} \rightleftharpoons HSO_4^-$	1.2×10^2	+19.7

[a]From Ref. 39.

titration of sodium acetate. The slope change at the acetate endpoint is increased fivefold by this method.

2.2.2c. Catalytic Thermometric Titrations

The thermochemical indicator just described reacts directly and stoichiometrically with the titrant. While this is a useful technique, the lowest concentration accessible is still about 10^{-3} mol L^{-1}. To titrate a sample which is more dilute requires a relatively dilute titrant for volumetric precision. Correspondingly, the slope of the posttitration region will be flatter, making the endpoint less perceptible. A

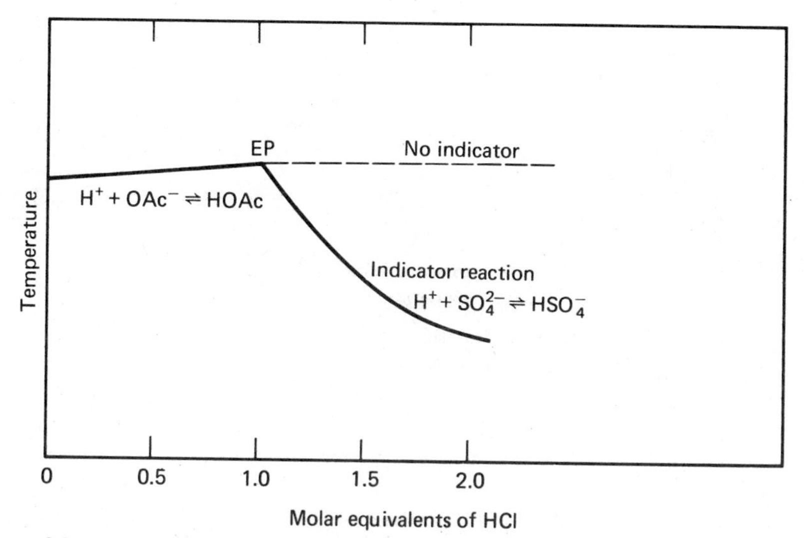

Figure 2.9. Thermometric titration curve for the titration of sodium acetate (NaOAc) with hydrochloric acid using sodium sulfate as a direct thermochemical indicator. EP: endpoint. (Reprinted from Ref. 21, p. 22, by courtesy of Marcel Dekker, Inc.)

novel approach, which overcomes this limitation, is to use an endpoint indicator reaction which is *catalyzed* by the first excess of titrant added upon virtual completion of the primary "determinative" reaction. The general reaction scheme is

Determinative reaction:

$$\underset{\text{Sample}}{A} + \underset{\text{Titrant}}{B} \rightleftharpoons \underset{\text{Products}}{P} \tag{2.57}$$

Indicator reaction:

$$\underset{\text{Indicator Reactants}}{D + E} \xrightarrow{\text{B (catalyst)}} \underset{\text{Products}}{F} \tag{2.58}$$

where, ideally, reaction (2.58) does not commence until reaction (2.57) has reached stoichiometric equivalence. The catalyzed indicator reaction proceeds rapidly and yields a steep slope as illustrated in Fig. 2.10. The slope can, in principle, be endothermic or exothermic, depending on the indicator reaction. However, all catalyzed indicator reactions used to date have been exothermic.

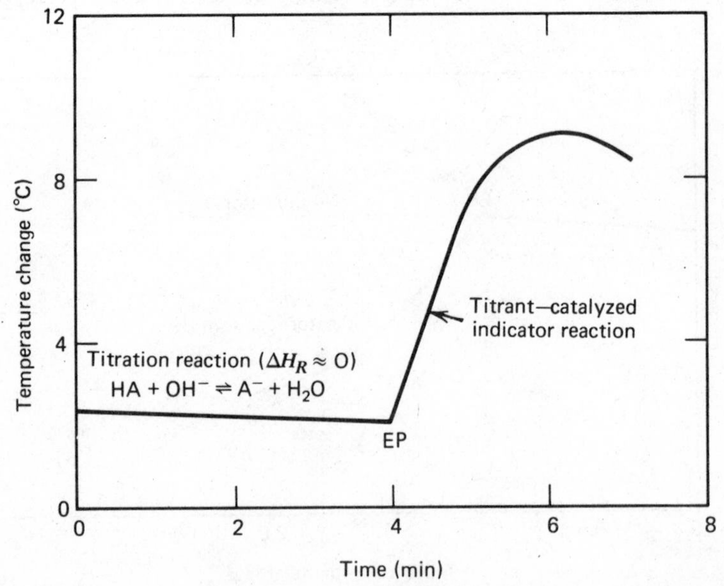

Figure 2.10. Titration of an acidic substance with KOH using acetone as solvent and catalytic thermochemical indicator. Time is proportional to volume because titrant is added at a constant rate. (From Ref. 22.)

The "indicator reactants," D and E, can be present in large amounts, so that the total temperature change may be as much as several degrees. Because the temperature change is so large, endpoints can be determined adequately even with simple manual equipment (23). Catalytic thermometric titrations are amenable to microanalysis, having been used on samples as dilute as 10^{-6} mol L^{-1} (24).

The applicability of catalytic thermometric titration is limited by the availability of suitable indicator reaction/titrant–catalyst combinations. Table 2.4 summarizes representative methods that have been developed to date. These can be divided into two broad categories. One category is the determination of acidic or basic organic compounds in non-aqueous solvents. Endpoint indication is by dimerization, polymerization, or hydrolysis of a component of the solvent system, catalyzed by excess strong acid or base. A second broad category of reactions amenable to catalytic endpoint indication are those involving trace metal ions in aqueous solution. Catalytic endpoint indication methods extend the dynamic range of conventional TET to concentrations as low as 10^{-6} mol L^{-1}. A disadvantage is the inability to obtain multiple endpoints for multicomponent mixtures. Also, the scope of application is limited by the handful of indicator reactions and catalysts presently available. However, the increased sensitivity and the possibility of using simpler equipment may make this a desirable option when applicable. A comprehensive discussion of catalytic thermometric titration methodology and its analytical applications is presented in Chapter 6.

2.2.3. Analysis of Mixtures

2.2.3a. Sequential Endpoints in Thermometric Enthalpy Titrations

In practice a sample may contain a mixture of components that will react with a given titrant. The ability to differentiate between these components by TET depends primarily on the thermodynamics of the relevant reactions and also on the kinetics. For a sample containing the species A_1, A_2, . . . , which all react rapidly with a titrant B according to

$$A_i + B = P_i \qquad \text{where } _i = 1, 2, 3, \text{etc.} \tag{2.59}$$

a thermometric titration will yield discrete sequential endpoints for each species if the following conditions are met:

$$K_{eq,i}/K_{eq(i+1)} \geqslant 50 \tag{2.60}$$

$$\frac{|\Delta H_{R,(i+1)} - \Delta H_{R,i}|}{\Delta H_{R,i}} \geqslant 0.2 \tag{2.61}$$

TABLE 2.4. Catalytic Thermometric Titrations[a]

"Sample" Species	Titrant (Catalyst)	Solvent	Indicator Reaction	Reference
Weak organic acids (phenols)	Alcoholic KOH	Acetone	$2CH_3COCH_3 \xrightarrow{OH^-} CH_3COCH_2C(CH_3)_2OH$	22
Weak organic acids (sulfanilamides)	Alcoholic KOH or tetra-*n*-butyl ammonium hydroxide	Acrylonitrile	$nCH_2{:}CHCN \xrightarrow{OH^-} HO(CH_2CHCN)_nH$	23
Weak bases (amines, alkaloids)	$HClO_4$	Toluene and 2-phenyl-propene	$nCH_2{:}C\varphi\ CH_3 \xrightarrow{H^+} H(CH_2C\varphi\ CH_3)_n^+$	23
Weak bases (pyridine, caffeine)	$HClO_4$ or coulometrically generated H^+	Acetic acid, acetic anhydride, and trace water	$(CH_3CO)_2O + H_2O \xrightarrow{H^+} 2CH_3COOH$	25
Hg^{2+}, Ag^+, Pd^{2+}	KI	Water	$As(III) + 2Ce(IV) \xrightarrow{I^-} As(V) + 2Ce(III)$	24, 26
Hg^{2+}, Ag^+, Pd^{2+}	KI	Water	$As(III) + 2Mn(III) \xrightarrow{I^-} As(V) + 2Mn(II)$	27
EDTA	$Mn(NO_3)_2$	Water	$H_2O_2 \xrightarrow{Mn^{2+}} H_2O + \frac{1}{2}O_2$	28
EDTA	Cu^{2+}	Water	$2H_2O_2 + N_2H_4 \xrightarrow{Cu^{2+}} 4H_2O + N_2$	29

[a]From Ref. 39.

Equation (2.61) requires that heats of reaction differ enough to yield slope changes of 20% on the TET curve. This is a somewhat arbitrary criterion and has been expressed in various forms (5, 30), but obviously the heats of sequential reactions must differ appreciably to resolve an intermediate endpoint. The requirement for equilibrium constants, Eq. (2.60), is less stringent than that for potentiometric titrations of mixtures, namely

$$\frac{K_i}{K_{(i+1)}} \geqslant 10^4 \tag{2.62}$$

This is due to the fundamental differences between linear (TET) and logarithmic (potentiometric) titration curves. A linear titration curve (where the signal, ΔT in this case, is directly proportional to amounts reacted) can be extrapolated from regions where the reaction of one analyte predominates. The corresponding intersection can yield an accurate endpoint, even though reactions of other analytes may be overlapping at the equivalence point. This simple procedure is not possible for logarithmic titration curves (where the signal is logarithmically related to reactant or product concentrations), albeit the so-called "Gran plots" (31, 32) and other procedures (33) provide linearization options for potentiometric titration curves.

An example of a thermometric titration yielding sequential endpoints is shown in Fig. 2.11. That figure represents the titration of calcium and magnesium ions with EDTA (Y^{4-}). The relevant reactions are

$$Ca^{2+} + Y^{4-} = CaY^{2-} \tag{2.63}$$

$$Mg^{2+} + Y^{4-} = MgY^{2-} \tag{2.64}$$

For calcium, reaction (2.63), $K_{eq} = 10^{10.6}$ and $\Delta H_R = -25.2$ kJ mol^{-1}, while for magnesium, reaction (2.64), $K_{eq} = 10^{8.7}$ and $\Delta H_R = +16.8$ kJ mol^{-1}.

In addition to titrations of mixtures, TET curves containing several endpoints may be obtained for titration reactions of a single moiety occurring in multiple steps, such as the alkalimetric determination of polyprotic acids. Reference 21 contains additional examples and discussion of multiple endpoints.

Even if the thermodynamics of reaction of two or more components of a mixture with a given titrant are similar, one may still be determined in the presence of the other(s) if the *kinetics* of reaction differ greatly. An example is the precipitation of calcium and magnesium ions with oxalate (35). Both reactions are thermodynamically favored, but the nucleation of the magnesium oxalate precipitate is very slow. Therefore, a thermometric titration of Ca^{2+} can be carried out even in the presence of excess Mg^{2+}.

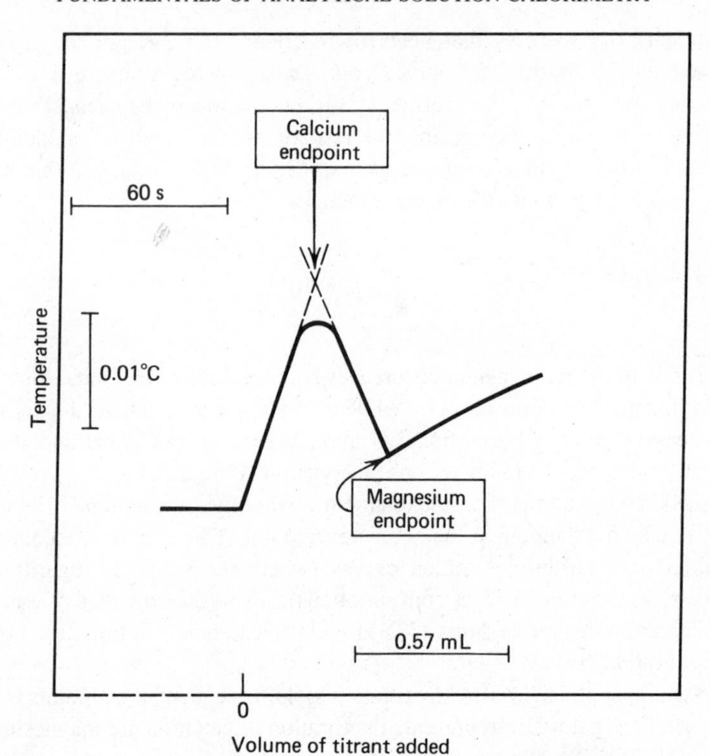

Figure 2.11. Titration curve of a mixture of Ca^{2+} and Mg^{2+} with ethylenediaminetetraacetate. (Reprinted with permission from Ref. 34. Copyright, 1957, American Chemical Society.)

2.2.3b. Combination of Endpoint Determinative and Enthalpimetric Procedures

Even when the equilibrium constants for two components in a mixture are nearly the same, if the heats are different and known, differentiation of the mixture may be possible on the following considerations. For the reactions

$$\mathrm{A} + \mathrm{B} = \mathrm{P} \qquad K_{eq,\mathrm{A}},\ \Delta H_{R,\mathrm{A}} \tag{2.65}$$

$$\mathrm{C} + \mathrm{B} = \mathrm{D} \qquad K_{eq,\mathrm{C}},\ \Delta H_{R,\mathrm{C}} \tag{2.66}$$

where

$$K_{eq,\mathrm{A}} \simeq K_{eq,\mathrm{C}} \qquad \text{and} \qquad \Delta H_{R,\mathrm{A}} \neq \Delta H_{R,\mathrm{C}}$$

only one TET endpoint will be observed. However, two pieces of information are available from the titration curve, namely, the endpoint n_B, corresponding to the sum of A + C, and the total heat evolved by both reactions, q_{tot}. The following equations are applicable:

$$n_B = n_A + n_C \tag{2.67}$$

$$q_{tot} = n_A \Delta H_{R,A} + n_C \Delta H_{R,C} \tag{2.68}$$

If the heats $\Delta H_{R,A}$ and $\Delta H_{R,C}$ are known from prior experiments, Eqs. (2.67) and (2.68) can be simultaneously solved for n_A and n_C. Hansen and Lewis (36) demonstrated that analysis of binary mixtures by this procedure is practicable with a precision of 5%.

2.2.3c. Multiple Reagents in Direct-Injection Enthalpimetry

Conventional DIE yields only one data point (ΔT, the temperature or heat pulse) per sample. Therefore, resolution of multicomponent mixtures by a single experiment is not normally possible. Obviously, if a specific reagent for each component can be found, the mixture's composition can be determined by a series of measurements. Even if the reagents are non-selective, a "j"-component mixture can be resolved by "j" DIE experiments without prior separation provided they react with the various components of the mixture with *different* ΔH_R's. For each reagent i added in excess, the heat evolved will be

$$q_i = n_1 \Delta H_{i,1} + n_2 \Delta H_{i,2} + \cdots + n_j \Delta H_{i,j} \tag{2.69}$$

where n_j denotes the number of moles of the jth component and $\Delta H_{i,j}$ is the heat of reaction between j and i. These heats must be known independently. To solve for the unknown variables n_j, an equal number of data points q_i must be obtained ($i = j$). Furthermore, each expression for q_i, Eq. (2.69), must be linearly independent. The corresponding necessary and sufficient condition is

$$D = \det(M) = \begin{vmatrix} \Delta H_{1,1} & \Delta H_{1,2} & \cdots & \Delta H_{1,j} \\ \Delta H_{2,1} & \Delta H_{2,2} & \cdots & \Delta H_{2,j} \\ & \vdots & & \\ \Delta H_{i,1} & \Delta H_{i,2} & \cdots & \Delta H_{i,j} \end{vmatrix} \neq 0 \tag{2.70}$$

which invokes the coefficient matrix M. Whenever condition (2.70) is met, the system of equations can be solved for the unknowns n_j by any convenient method. For systems where $i = j = 2$ or 3, the method of determinants, using Cramer's rule, is simple to apply. Use of determinants also simplifies an error analysis.

The applicable form of Cramer's rule is

$$n_j = \frac{D_j}{D} \tag{2.71}$$

where D is the determinant defined in Eq. (2.70) and D_j is the determinant of the matrix obtained from M by replacing its jth column by the experimental values of q_i. Thus, for a system where $i = j = 2$, D_1 and D_2 are given by

$$D_1 = \begin{vmatrix} q_1 & \Delta H_{1,2} \\ q_2 & \Delta H_{2,2} \end{vmatrix}, \qquad D_2 = \begin{vmatrix} \Delta H_{1,1} & q_i \\ \Delta H_{2,1} & q_2 \end{vmatrix}. \tag{2.72}$$

Due consideration to sources of error is very important because an indirect method such as this is very prone to propagation of errors. (This is the classical problem of a large relative error in a small difference between two large quantities.) The error analysis given below is based largely on a paper by Marik-Korda et al. (37).

Assume that each of the conditional heats of reaction $\Delta H_{i,j}$ can be accurately measured and that this source of error is small compared to others. The relative variance of each n_j, derived from Eq. (2.71), is given by

$$\left(\frac{\sigma(n_j)}{n_j}\right)^2 = \left(\frac{\sigma(D_j)}{D_j}\right)^2 + \left(\frac{\sigma(D)}{D}\right)^2 \tag{2.73}$$

where $\sigma(n_j)$, $\sigma(D_j)$, and $\sigma(D)$ represent the standard deviations of the bracketed quantities, respectively. Since $\sigma(D)$ is assumed small, the last term in Eq. (2.73) may be neglected, from which results

$$\left\{\frac{\sigma(n_j)}{n_j}\right\}^2 = \left\{\frac{\sigma(D_j)}{D_j}\right\}^2 = \left\{\frac{\sigma^2(D_j)}{D^2 n_j^2}\right\} \tag{2.74}$$

Experimentally, the chief source of error is "noise" $\sigma(q)$ in the signal q, as discussed in Section 2.2.1a. $\sigma(q)$ also contains a contribution from extraneous thermal effects. It can be considered independent of the magnitude of q within a given sensitivity setting of the recording device. Accepting this assumption, $\sigma^2(D_j)$ can be evaluated from the definition of D_j:

$$\sigma^2(D_1) = \sigma^2(q)[\Delta H_{1,2}]^2 + \sigma^2(q)[\Delta H_{2,2}]^2 \tag{2.75}$$

$$\sigma^2(D_2) = \sigma^2(q)[\Delta H_{1,1}]^2 + \sigma^2(q)[\Delta H_{2,1}]^2 \tag{2.76}$$

or

$$\sigma^2(D_{j=1,2}) = \sigma^2(q)\sum_i(\Delta H_{i,k\neq j})^2 \tag{2.77}$$

Equations (2.75)–(2.77) apply to the two-component case identified in Eq. (2.72). In the general case of large dimensionalities,

$$\sigma^2(D_j) = \sigma^2(q)\sum_i(\det{}^2 M_{i,j}) \tag{2.78}$$

where $M_{i,j}$ is the coefficient matrix of dimension $(i-1) \times (j-1)$ obtained from M by deleting its ith row and jth column.

The relative standard deviation of the analytical result is obtained by combining Eqs. (2.74) and (2.78), yielding

$$\frac{\sigma(n_j)}{n_j} = \frac{\sigma(q)}{n_j D}\left(\sum_i \det{}^2\ M_{i,j}\right)^{1/2} \tag{2.79}$$

In the two-dimensional case this simplifies to

$$\frac{\sigma(n_j)}{n_j} = \frac{\sigma(q)}{n_j D}\left(\sum_i \Delta H^2_{i,k\neq j}\right)^{1/2} \tag{2.80}$$

It is apparent from Eq. (2.80) that the uncertainty in the determination of a given component increases as the heats of reaction of the other component increase. Consistent with the initial assumption of $\sigma(q)$ being independent of q, Eq. (2.79) and (2.80) predict that the uncertainty in a determination, $\sigma(n_j)$, is independent of the magnitude of n_j. (Note that n_j can be cancelled from both sides of either equation.) Thus, the relative standard deviation $\sigma(n_j)/n_j$ is inversely proportional to n_j. This relationship is illustrated in Fig. 2.12 from Marik-Korda et al. (37), where $\sigma(n_j)/n_j$ is plotted as a function of $\Sigma_i(\det^2 M_{i,j})^{1/2}/D$ for various values of n_j.

Independent and accurate knowledge of the heats of reaction $\Delta H_{i,j}$ is crucial to the applicability of the method of DIE with multiple reagents. If the presence of one component affects the reaction of another, large errors will will be introduced. Applications of indirect DIE analysis using multiple reagents are given in Chapter 6.

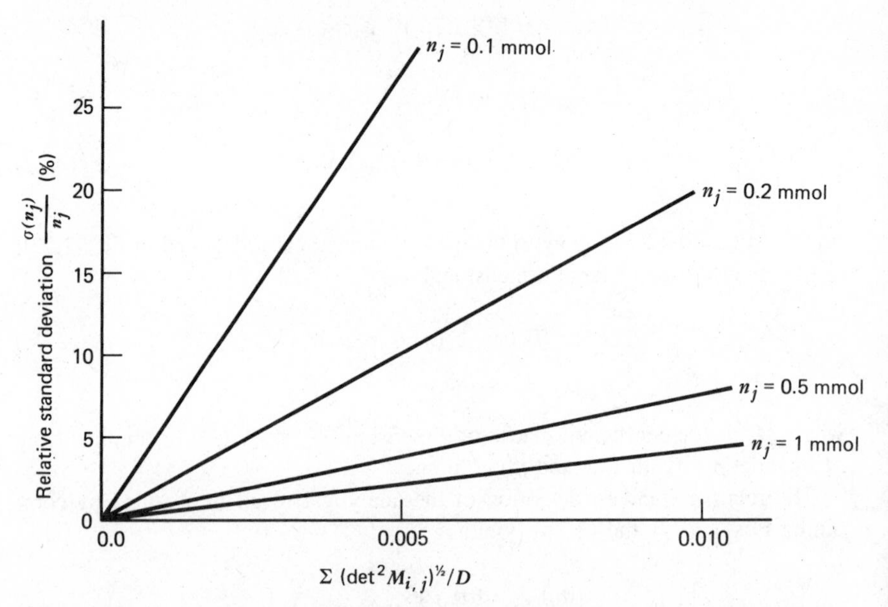

Figure 2.12. Plot of relative standard deviation versus ratio of coefficient determinants [defined in Eqs. (2.70) and (2.78)] for direct-injection enthalpimetry using multiple reagents. Each line represents a given amount of sample. Adapted from Ref. 37, assuming $\sigma(q) = 3$ m°C.

2.2.3d. Differential Reaction Kinetics in Direct-Injection Enthalpimetry

DIE has also been applied to the analysis of mixtures by differential reaction kinetics. Differential kinetic methods of analysis have been widely investigated since first proposed by Lee and Kolthoff (38), and particular application has been found in the functional group analysis of mixtures of similar organic molecules. The use of quasiadiabatic calorimetry to monitor reaction progress in kinetic analysis was first reported by Papoff and Zambonin (7).

Consider a two component mixture, A_1 and A_2, each of which reacts with reagent B, as follows:

$$A_1 + B \xrightarrow{k_1} P_1 \qquad \Delta H_1 \tag{2.81}$$

$$A_2 + B \xrightarrow{k_2} P_2 \qquad \Delta H_2 \tag{2.82}$$

When pseudo-first-order kinetics prevail (large excess of B), Eq. (2.83) represents the total heat evolved (or absorbed).

$$\frac{\Delta T_t}{C_P} = -q_t = -V\Delta H_1[A_1]_0\{1 - \exp(-k_1[B]t)\}$$
$$-V\Delta H_2[A_2]_0\{1 - \exp(-k_2[B]t)\} \qquad (2.83)$$

If the rate constants and heats are known, Eq. (2.83) can be solved for the concentrations of A_1 and A_2 by picking two data points off the plot of q_t versus t. Alternatively, logarithmic extrapolation procedures can be used (6, 7). Binary mixtures can thus be resolved in a single experiment.

ACKNOWLEDGMENTS

Financial support was provided by the U.S. Department of Energy under Grant DE FG22-81PC40783. Some of the material in this chapter has also been incorporated in "Thermometric Titrations and Enthalpimetric Analysis," in I. M. Kolthoff and P. J. Elving (Eds.), *Treatise on Analytical Chemistry,* 2nd ed., Wiley-Interscience, New York, 1985, Part 1, Section H, Chapter 14, which was written by the same authors.

REFERENCES

1. O. Kubaschewski and R. Hultgren, in H. A. Skinner, Ed., *Experimental Thermochemistry,* Vol. 2, Wiley-Interscience, New York, 1962, p. 351.
2. J. Jordan, *Chimia* **17**, 101 (1963).
3. D. Rosenthal, G. L. Jones, and R. Megargle, *Anal. Chim. Acta* **53**, 141 (1971).
4. H. J. V. Tyrrell and A. E. Beezer, *Thermometric Titrimetry,* Chapman and Hall, London, 1968.
5. P. W. Carr, *Crit. Rev. Anal. Chem.* **2**, 491 (1972).
6. R. A. Henry, Ph.D. Dissertation, Pennsylvania State University, University Park, 1967.
7. P. Papoff and P. G. Zambonin, *Talanta* **14**, 581 (1967).
8. J. K. Grime, *Anal. Chim. Acta* **118**, 191 (1980).
9. J. C. Wasilewski, P. T-S. Pei, and J. Jordan, *Anal. Chem.* **36**, 2131 (1964).
10. C. D. McGlothlin and J. Jordan, *Anal. Chem.* **47**, 786 (1975).
11. P. W. Carr and J. Jordan, *Anal. Chem.* **45**, 634 (1973).
12. J. Barthel, *Thermometric Titrations,* Wiley, New York, 1975.
13. L. D. Hansen, T. E. Jensen, S. Mayne, D. J. Eatough, R. M. Izatt, and J. J. Christensen, *J. Chem. Thermodyn.* **7**, 919 (1975).
14. C. J. Martin, B. R. Sreenathan, and M. A. Marini, *Biopolymers* **19**, 2047 (1980).
15. D. D. Wagman et al., *Selected Values of Chemical Thermodynamic Properties,* National Bureau Standards Technical Notes 270-3 through 270-8, U.S. Government Printing Office, Washington, D.C., 1968–1981.

16. V. B. Parker, *Thermal Properties of Aqueous Uni-Valent Electrolytes,* National Bureau of Standards, NSRDS-NBS-2, 1965.
17. H. S. Harned and B. B. Owen, *The Physical Chemistry of Electrolyte Solutions,* 3rd ed., Reinhold, New York, 1958.
18. E. B. Smith, C. S. Barnes, and P. W. Carr, *Anal. Chem.* **44**, 1663 (1972).
19. J. W. Stahl, Ph.D. Dissertation, Pennsylvania State University, University Park, 1983.
20. N. N. Rehak and D. S. Young, *Clin. Chem.* **23**, 1153 (1977).
21. L. D. Hansen, R. M. Izatt, and J. J. Christensen, "Applications of Thermometric Titrimetry to Analytical Chemistry," in J. Jordan, Ed., *New Developments in Titrimetry,* Marcel Dekker, New York, 1974
22. G. A. Vaughan and J. J. Swithenbank, *Analyst* (London) **90**, 594 (1965).
23. E. J. Greenhow and L. E. Spencer, *Analyst* (London) **98**, 90, (1973).
24. K. C. Burton and H. M. N. H. Irving, *Anal. Chim. Acta* **52**, 441 (1970).
25. V. J. Vajgand, F. F. Gaal, and S. S. Brusin, *Talanta* **17**, 415 (1970).
26. H. Weisz, T. Kiss, and D. Klochow, *Fresenius' Z. Anal. Chem.* **247**, 248 (1969).
27. N. Kiba and M. Furosawa, *Anal. Chim. Acta* **98**, 343 (1978).
28. H. Weisz and T. Kiss, *Fresenius' Z. Anal. Chem.* **249**, 302 (1970).
29. H. Weisz and S. Pantel, *Anal. Chem. Acta* **62**, 361 (1972).
30. N. D. Jespersen and J. Jordan, *Anal. Lett.* **3**, 323 (1970).
31. G. Gran, *Acta Chem. Scand.* **4**, 559 (1950).
32. G. Gran, *Analyst* (London)**77**, 661 (1952).
33. L. J. Reed and J. Berkson, *J., Phys. Chem.* **33**, 760 (1929).
34. J. Jordan and T. G. Alleman, *Anal. Chem.* **29**, 9 (1957).
35. J. Jordan and E. J. Billingham, Jr., *Anal. Chem.* **33**, 120 (1961).
36. L. D. Hansen and E. A. Lewis, *Anal. Chem.* **43**, 1393 (1971).
37. P. Marik-Korda, L. Buzasi, and T. Cserfalvi, *Talanta* **20**, 569 (1973).
38. T. S. Lee and I. M. Kolthoff, *Ann. N.Y. Acad. Sci.* **53**, 1093 (1951).
39. J. Jordan and J. W. Stahl, "Thermometric Titrations and Enthalpimetric Analysis," in I. M. Kolthoff and P. J. Elving, Eds., *Treatise on Analytical Chemistry,* 2nd ed., Wiley-Interscience, New York, 1985, Part 1, Section H, Chapter 14.

CHAPTER

3

INSTRUMENTATION AND DATA REDUCTION

LEE D. HANSEN, EDWIN A. LEWIS, and DELBERT J. EATOUGH

Thermochemical Institute and the Department of Chemistry
Brigham Young University
Provo, Utah

3.1. INSTRUMENTATION AND DATA REDUCTION

The purpose of this chapter is to describe the nine fundamentally different types of solution calorimeters along with the data evaluation procedures that have been developed to obtain the heat associated with a given reaction using these instruments. A critical comparison of the capabilities, advantages, and limitations of each type of solution calorimeter is also given. The nine types of solution calorimeters are described by a matrix consisting of three ways of measuring heat and three ways of initiating the reaction. The heat-measurement techniques used in isoperibol, isothermal, and heat-conduction calorimetry are described. Initiating the reaction by batch, titration, or flow techniques are described for each heat measurement method.

3.2. REACTION INITIATION METHODS

The main components of a solution calorimeter are indicated in the block diagram shown in Fig. 3.1. In a solution calorimetric experiment, the two solutions shown in compartments A_1 and A_2 are mixed to give the final solution of composition B in the calorimeter reaction vessel. Either compartment A_1 or A_2 may or may not be external to the calorimeter reaction vessel. The resultant temperature change caused by the reaction is sensed by the temperature sensor T and converted by the output circuit to a signal corresponding to either the temperature change or a heat flow rate. The signal may be amplified and recorded on either a strip-chart recorder or other data-collection system. The temperature of the bath or block containing the reaction vessel and the solution compartments A_1 and A_2 is measured by the sensor S and controlled by the temperature controller G. The temperatures of A_1, A_2, and B must be equal or the differences must be measured.

The various ways of initiating a reaction between A_1 and A_2 in the calorimeter reaction vessel are shown schematically in Fig. 3.2. Details on these various techniques follow.

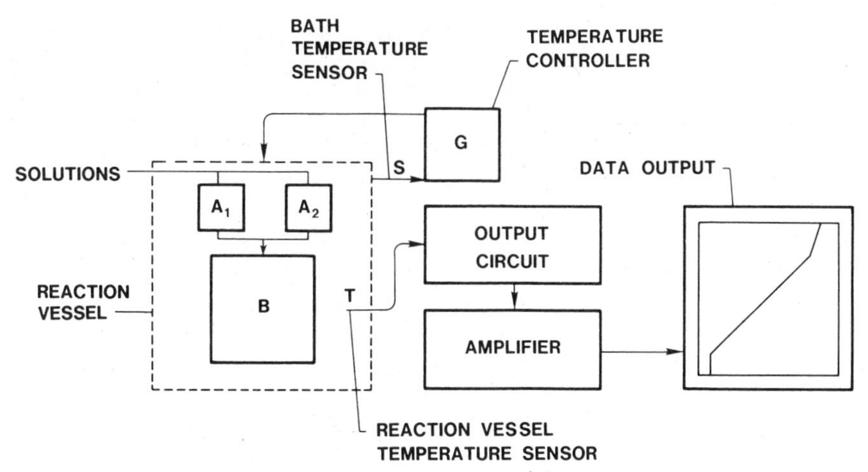

Figure 3.1. Components of a solution calorimeter.

3.2.1. Batch Mixing

In batch solution calorimetry, both reactants are placed in the calorimeter reaction vessel and mixed at the initiation of the reaction period. The calorimetric measurement may consist of either measuring the temperature change or the heat flux associated with mixing the two reactants. The two reactants are in thermal equilibrium at the start of the experiment and changes in the volume and/or heat capacity of the calorimeter reaction vessel during the experiment are usually negligibly small. Mixing of the two reactants is accomplished by breaking a bulb and allowing the reactants to mix, by displacing a seal separating the two reactants in the calorimeter reaction vessel, or by rotating the reaction vessel and allowing the reactants to mix. Batch solution calorimeters that have been described in the literature include isoperibol (1), isothermal (2), and heat-conduction (3) instruments. Details of the instrument designs are given in the referenced papers. Each of these types of batch solution calorimeters have been made available in commercial units.

3.2.2. Incremental Titration or Batch Injection Mixing

Although batch solution calorimetry simplifies data analysis and eliminates the possibility of some experimental errors, it does suffer from design problems in arranging for the mixing of reactant solutions. The difficulty of designing a capsule which is simple to construct, easy to use, has a negligible heat of opening,

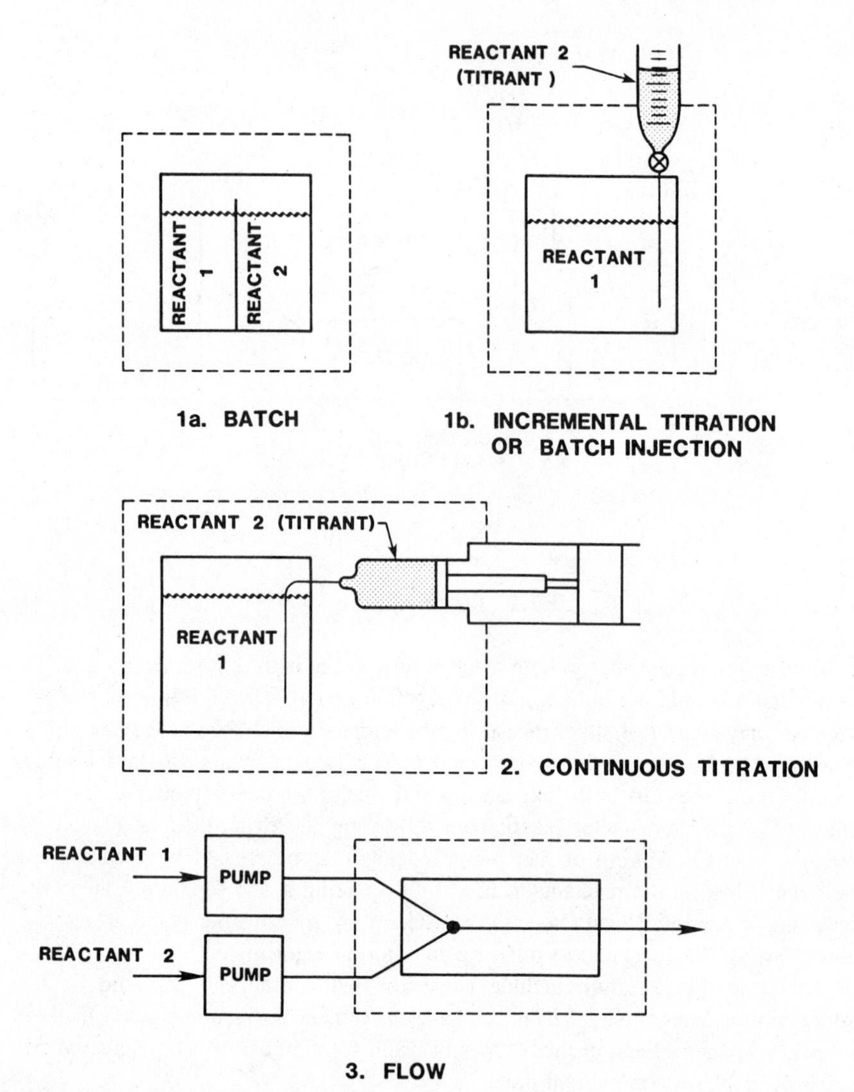

Figure 3.2. Schematic of reaction initiation methods: (1*a*) batch, (1*b*) incremental titration or batch injection, (2) continuous titration, and (3) flow solution calorimetry.

and does not change the stirring power has caused calorimetrists to look for alternative techniques for mixing the reactant solutions. One of these techniques has been to place one of the reactants in a syringe or buret, which is placed in a temperature-controlled environment external to the reaction vessel. If repeated injections are made, the method is called incremental titration, and if only one injection is made, the method is called batch injection calorimetry. This generic designation includes direct injection enthalpimetry (Chapter 2), which is typically associated with the syringe injection of a small volume of reagent, in stoichiometric excess, into a sample solution. The batch injection method of reactant mixing has the advantage of simplicity. However, the temperature of the added reactant may be different from the temperature of the reactant in the calorimeter reaction vessel and the energy equivalent of the calorimeter changes when the reactant is added. These effects must be taken into account in analysis of the calorimetric data. Procedures for these calculations are discussed in detail later. The development of isoperibol (1, 4) and heat-conduction (5) incremental titration or batch injection calorimeters has been described in the indicated references.

3.2.3. Continuous Titration

In continuous titration calorimetry, one reactant is titrated continuously into the other reactant and either the temperature change or heat produced in the system is measured as a function of titrant added. A single titration calorimetric experiment yields thermal data as a function of the ratio of the concentrations of the reactants. Titration data in the form of either temperature change or heat flux versus volume of titrant added can be analyzed to give both analytical (thermometric titrimetry) and thermodynamic (titration calorimetry) information on the reactions taking place during the titration. Isoperibol (6, 7), isothermal (2, 8), and heat-conduction (9) continuous titration calorimeters have been described.

When using a continuous titration calorimeter, the assumption is made that the data obtained are representative of the calorimeter system in both thermal and chemical equilibrium. In practice, exact thermal equilibrium is not obtained. However, isoperibol titration calorimeter reaction vessels can be so designed (6) that the response time for attainment of thermal equilibrium (six time constants, $6\tau_C$) is about 2 s. For either isothermal (8) or heat-conduction titration calorimeters (9) the response time of the instrument is one to two orders of magnitude larger, and some form of data manipulation must be used to simulate thermal equilibrium. Obviously, if the chemical reaction rate is slower than or of the same order of magnitude as the thermal response rate of the calorimeter, an additional complication enters into the data analysis.

There are two advantages of continuous titration calorimetry over incremental titration calorimetry. First, a continuous record of the heat effects produced during the titration is obtained. If the system being studied is complex, the continuous trace of the reaction progress allows one to choose any number of data points for calculation purposes. Second, continuous titration gives much more rapid data acquisition than does incremental titration. A complete continuous titration experiment requires only a little more time than a single injection in an incremental titration. However, apparatus based on the continuous addition of titrant must either respond quickly to temperature or heat flux changes or be accurately modeled to correct for the thermal lag. Also, the addition of a constant-rate buret makes the equipment more complex and requires that more extensive calibration be done than in incremental titration or batch calorimetry.

3.2.4. Flow Mixing

In flow calorimetry, both reactants are simultaneously added to a mixing cell and the heat associated with the mixing process determined by measuring either the temperature change in the product stream or the heat flux associated with the mixing process. Flow calorimetry offers many of the advantages of titration calorimetry in that data can be easily obtained as a function of the ratio of the concentrations of the two reactants. In addition, flow calorimetry is amenable to the study of reactions in the absence of vapor and under high-pressure conditions (10, 11). The construction of isoperibol (12, 13), isothermal (10, 11), and heat-conduction (14) flow calorimeters has been reported in the literature. Isoperibol flow calorimeters containing only one flow line and a fixed solid reactant bed have also been described (15–17). The design, construction, and operating characteristics of analytical flow instruments based on the "flow injection" principle, termed flow enthalpimeters, are discussed in detail in Chapter 4.

The assumption is made in flow calorimetry that the liquid in the flow line exiting the calorimeter reaction vessel is in both chemical and thermal equilibrium. With either heat-conduction or isothermal flow calorimeters, the response time of the instrument is of the order of minutes rather than seconds. This instrument response time may be compensated for by some form of data manipulation (18–21) to simulate thermal equilibrium or a constant flow must be maintained for a length of time equal to at least six time constants so that the data obtained represent a thermal equilibrium condition. The latter approach is usually used.

3.3. ISOPERIBOL SOLUTION CALORIMETRY

3.3.1. Equipment Description

Isoperibol solution calorimetry is a technique in which the temperature of a reaction vessel in a constant-temperature environment is monitored as a function of time. A variety of batch and continuous titration isoperibol solution calorimeters capable of high accuracy and high precision have been described (1, 6, 7). Common features of these calorimeters include a Dewar vessel as a reaction cell, a rapid responding temperature-measuring device (usually a thermistor), and an electrical heater for temperature equilibration and calibration.

The reaction vessel in an isoperibol calorimeter should be designed to minimize the rate of heat exchange between the vessel and its surroundings. A microreaction vessel for precision isoperibol continuous titration calorimetry (6, 22) is shown in Fig. 3.3. The reaction vessel is placed in a constant-temperature

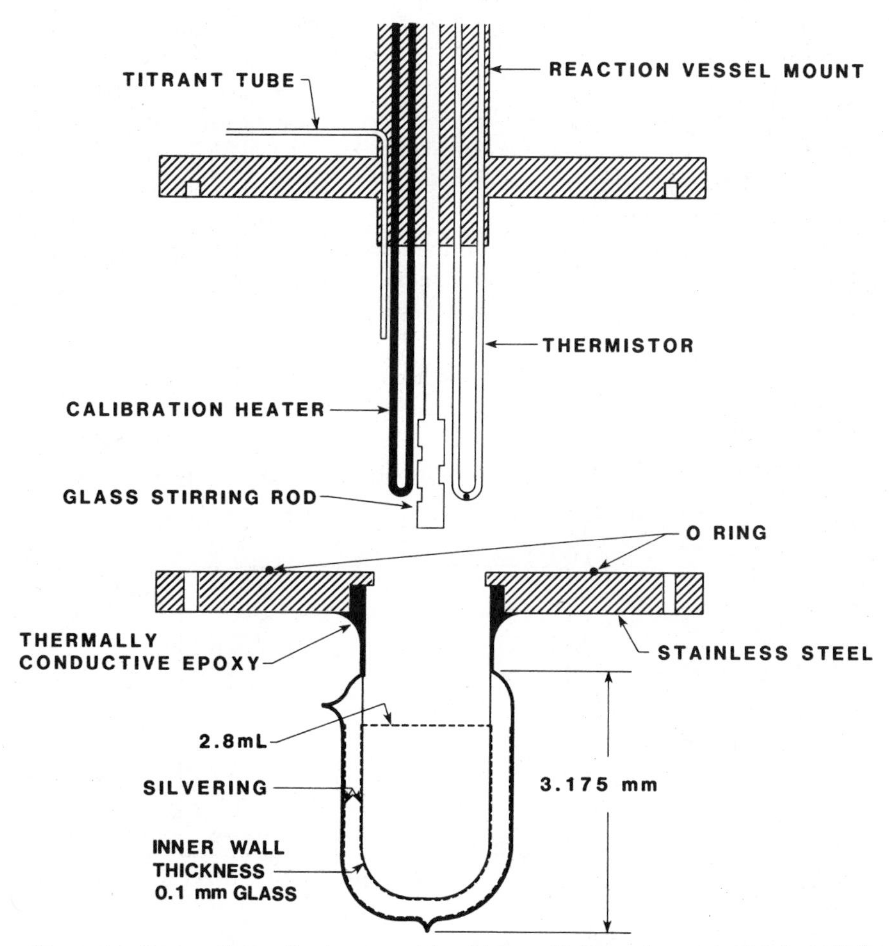

Figure 3.3. Design of a small-volume, precision, high-sensitivity continuous titration isoperibol solution calorimeter reaction vessel.

bath and the run is initiated when the temperature of the reaction vessel and its contents are nearly the same as the bath temperature. Temperature changes are monitored while titrant (at the bath temperature) is continuously added to the reaction vessel. The thin inside wall of the Dewar vessel and the low mass of the various elements inside the vessel (stirrer, thermistor, and heater) are designed to allow rapid thermal equilibration. The total heat capacity of a glass reaction vessel containing 100 mL of water can be as low as 427 J/K with a corresponding heat-leak modulus κ of 2×10^{-3} min^{-1} (6).

Miniaturized glass reaction vessels, similar to the illustration in Fig. 3.3, have been developed with volumes as small as 1.5 mL. A reaction vessel containing 2.0 mL of water can have a heat capacity of only 9.2 J/K and a heat leak modulus of 5×10^{-2} min^{-1} (7, 22). Two problems complicate the design of small-volume isoperibol titration calorimeters. First, κ increases as the volume of the reaction vessel decreases, as illustrated in Fig. 3.4. As a result, the magnitude of the heat-loss correction relative to the heat measured may be large. The uncertainty in the measured chemical heat, $\delta(q_{C,p})$, resulting from the uncertainty in the heat-leak correction is directly dependent on both the measured chemical heat $q_{C,p}$ and time (7) (Fig. 3.5). Second, the fraction of the total heat capacity of the calorimeter due to unstirred material (i.e., Dewar walls, stirrer, thermistor, heater) increases as the volume of the reaction vessel decreases, which may lead to a longer instrument response time if the reaction vessel is improperly designed.

Isoperibol titration calorimeters have been used that do not include an evacuated Dewar and in which the ratio of the heat capacity of the reaction vessel to that of the liquid contents is large (23). As would be expected, both the time response and the thermal heat-leak properties of these vessels are poor. Therefore,

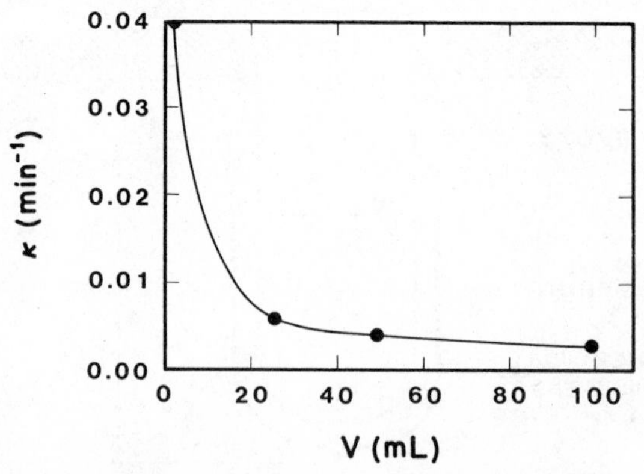

Figure 3.4. Variation of the heat-leak modulus κ of isoperibol solution calorimeters as a function of the total volume in the calorimeter reaction vessel. (Taken from Ref. 7.)

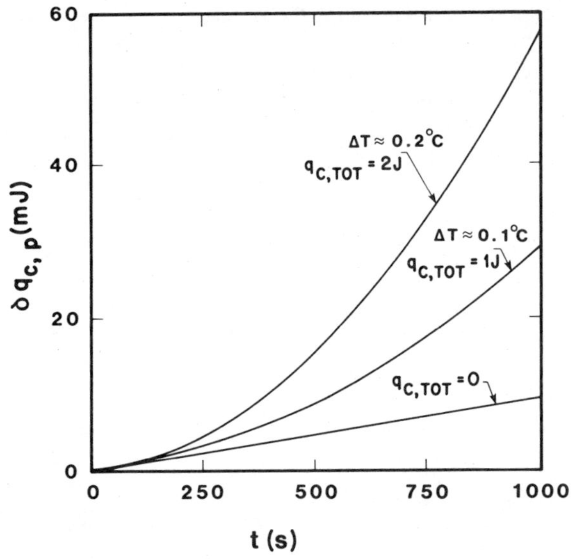

Figure 3.5. Dependence of the uncertainty in the heat-leak correction for an isoperibol solution calorimeter, $\delta q_{C,p}$, on the corrected measured heat, $q_{C,p}$ and the measurement time t. Data are for the reaction vessel shown in Fig. 3.3. (Taken from Ref. 7.)

the data are less suitable for the study of complex systems. Problems associated with this type of calorimetric titration reaction vessel have been described (23, 24).

Data on the heat associated with the neutralization of strong acid by strong base as measured with a 100-mL isoperibol batch calorimeter and with both 100-mL and 3-mL isoperibol continuous titration calorimeters are given in Table 3.1. These data illustrate the precision, accuracy, and sensitivity that can be obtained by isoperibol batch and continuous titration calorimetry.

The data produced by an isoperibol calorimeter are temperature versus time for a batch isoperibol solution calorimeter or temperature versus moles of titrant added during a titration for a continuous titration instrument. A typical plot of temperature versus time for an isoperibol batch calorimeter experiment is shown in Fig. 3.6. Typical data for a continuous titration where a single exothermic reaction is occurring are shown in Fig. 3.7. The data collected from a single continuous titration are equivalent to the data obtained from an incremental titration or from several batch experiments. However, many more experiments must be done in the case of the batch or incremental methods to obtain the same final data set.

The slope of the initial (prereaction or lead) region in Figs. 3.6 and 3.7 indicates the rate of net heat loss or gain of the reaction vessel and contents

TABLE 3.1. Reported ΔH_R Values for the Reaction $H^+(aq) + OH^-(aq) = H_2O(l)$[a]

Type of Calorimeter	Volume (mL)	Total Heat Measured (mJ)	$-\Delta H_R$ (kJ mol^{-1})	Reference
Isoperibol batch	200	55–380	55.79 ± 0.06	33
			55.79 ± 0.07	37
Isoperibol titration	100	280	55.81 ± 0.08	6
	100	100	55.78 ± 0.02	36
	3	2	56.02 ± 0.16	7
		0.4	55.70 ± 0.30	7
Isothermal titration	50	20	55.72 ± 0.02	2
		8	55.73 ± 0.02	2
		2	55.80 ± 0.05	2
		1	55.76 ± 0.11	2
	4	3	55.81 ± 0.22	8
		1	55.77 ± 0.21	8
		0.2	55.80 ± 0.50	8
Isothermal flow	1 mL min^{-1}	0.3 J min^{-1}	55.86 ± 0.20	10

[a]The value assigned by the National Bureau of Standards to the reaction $H^+(aq) + OH^-(aq) = H_2O(l)$ at infinite dilution and 25°C is -55.83 kJ mol^{-1} (32).

before the reaction is begun. The slope in this region (S_I) is a function of heating by stirring, resistance heating in the thermistor, heat effects due to conduction, and, in the case of an unsealed vessel, evaporation or condensation. The final region (postreaction or trail) in Figs. 3.6 and 3.7 is generated after the batch experiment reaction is over or the titrant delivery is stopped, and the slope (S_F) is a function of the same effects as in the initial region. Data from both the initial and final regions are used to calculate the correction for the heat loss from the reaction vessel during the titration or batch reaction. Other corrections must be made to allow for heat effects associated with dilution, breaking a sample bulb, the temperature difference of titrant and titrate solutions, or changes in the energy equivalent due to the addition of titrant. Two additional regions of importance to data analysis for isoperibol titration calorimetry are shown in Fig. 3.7. The increase in the slope in the reaction region (S_R) is caused by the heat of the reaction taking place in the reaction vessel plus the effects of dilution of titrant and titrate and the temperature differential of titrant and titrate. The excess reagent region (slope $= S_{ER}$) is the portion of the curve in which the titration continues but the reaction is complete. This portion of the curve exists only for those

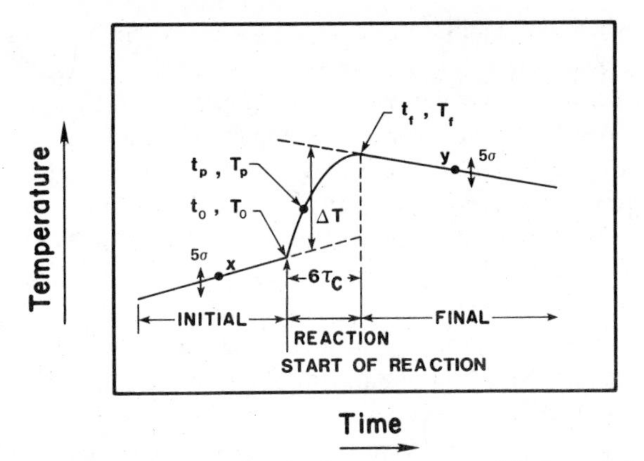

Figure 3.6. Plot of temperature versus time for the measurement of the heat produced by a rapid chemical reaction in an isoperibol batch solution calorimeter with a response time of $6\tau_C$. The double-ended vertical arrows are the peak-to-peak noise in the indicated measurement. The peak-to-peak noise is equal to 5σ.

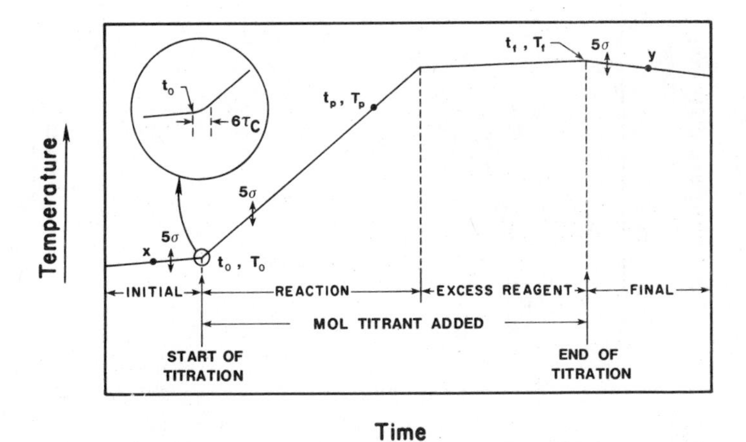

Figure 3.7. Plot of temperature versus time for the measurement of the heat produced by a rapid and quantitative reaction in an isoperibol continuous titration solution calorimeter. The double-ended vertical arrows are the peak-to-peak noise in the indicated measurement. The peak-to-peak noise is 5σ.

systems in which the equilibrium constant for the reaction is large enough that products are quantitatively formed as titrant is added.

In an isoperibol flow calorimeter, the temperature differential is measured between the two input streams and the output stream after mixing. This is the concept used in a flow calorimeter developed by Picker and co-workers (12, 13), by Carr and co-workers (15, 16), and in development of the Microscal instrument (17). Typical data obtained with an isoperibol flow calorimeter are illustrated in Fig. 3.8. The temperature differential measured in the initial region represents the input versus output temperature when only a single stream is flowing through the reaction cell at a constant rate. At point (t_0,T_0), constant flow of the second stream is initiated resulting in the illustrated ΔT shift from the baseline. This ΔT value results from any temperature difference between the two mixed streams, the heat from dilution of the two streams, and heat from chemical reactions.

In order for the data represented by Fig. 3.6 (batch or incremental titration isoperibol calorimeter), Fig. 3.7 (continuous titration isoperibol calorimeter), or Fig. 3.8 (flow isoperibol calorimeter) to be used for the calculation of the thermodynamic quantities associated with the reaction(s) of interest, the equipment must be calibrated and the data analyzed to give the heat produced in the experiment. In the case of a batch experiment, the data illustrated by Fig. 3.6 must be corrected for thermal effects due to non-chemical energy terms. In the case

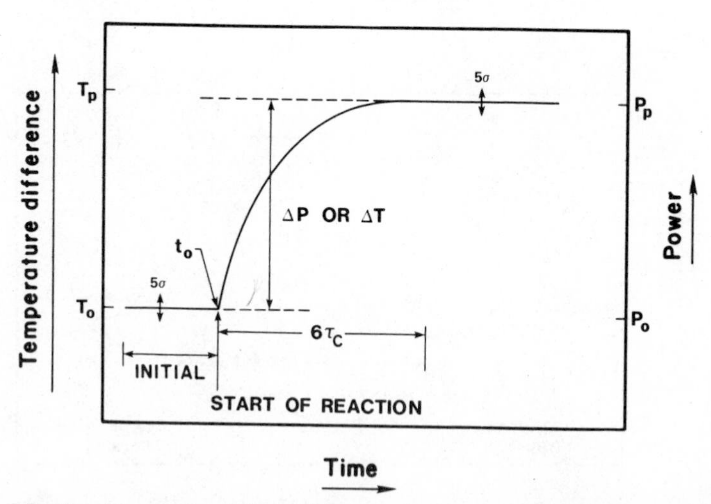

Figure 3.8. Plot of temperature differential between the input and output streams versus time for the measurement of the heat produced at a constant flow rate in an isoperibol flow calorimeter or plot of power versus time for the same measurement in an isothermal or heat conduction flow calorimeter. The double-ended vertical arrows are the peak-to-peak noise in the indicated measurement. The peak-to-peak noise is 5σ.

of titration or flow calorimetry the objective is to produce a data array of heat produced versus the quantity of titrant added or the flow-rate ratio of the two flow streams. Data analysis requires calibration of the titrant or flow delivery system and calibration of the temperature-sensing device as well as the determination of the heat capacity of the reaction vessel. Analysis of continuous titration data, Fig. 3.7, involves taking temperature data at strategic points along the curve, correcting the data for thermal effects due to non-chemical energy terms, and expressing the corrected data in heat units. The greater the complexity of the system, the greater the number of data points required. A large number of points are represented on a strip-chart recording, and the investigator may choose as many points as the complexity of the system dictates. The data from a flow isoperibol calorimeter, Fig. 3.8, may be expressed in terms of the heat flux associated with the reaction that produces the measured temperature change during flow through the reaction cell.

3.3.2. Calibration of Temperature-Sensing Device

The most common device used to measure temperature change in isoperibol reaction vessels is a thermistor incorporated into a Wheatstone bridge circuit (25–27). The change in the output voltage of the bridge may be expressed in terms of energy input to the calorimeter per volt of response or may be related to a standard temperature scale by calibration against a temperature standard (energy input/degree). Calibration against a temperature standard is time consuming and needs to be done only if a wide variety of liquids with different heat capacities are to be used in the reaction vessel without recalibration for each. A calibration procedure that can be used for most batch and titration instruments consists of placing the thermistor and the thermometer in close proximity in a carefully controlled, constant-temperature water bath and measuring changes in the bridge voltage $\mathbf{E}_B$ as a function of bath temperature T (28). Since the response of some thermistors is known to be affected by light intensity, the thermistor should be calibrated and used under identical, preferably dark, conditions to those used during calorimetric measurements. Also, since thermistors are flow sensitive, the flow of liquid over the thermistor should be the same as in the calorimeter reaction vessel. The exact functional dependence of the resistance of a thermistor on temperature is well known (29). Over a small temperature interval (<0.5 K), a linear equation can be used to relate the unbalanced bridge voltage to temperature:

$$\mathbf{E}_B = a_B + b_B T \tag{3.1}$$

The constant b_B is given by

$$b_B = \frac{\left[m\sum_{i=1}^{m} \mathbf{E}_{B_i} T_i - \sum_{i=1}^{m} T_i \sum_{i=1}^{m} \mathbf{E}_{B_i} \right]}{\left[m\sum_{i=1}^{m} T_i^2 - \left(\sum_{i=1}^{m} T_i \right)^2 \right]} \tag{3.2}$$

where m is the number of data points taken (28). If a strip-chart recorder is used to indicate voltage changes, the recorder output can be directly calibrated in terms of chart units per degree of temperature change.

Another way of calibrating the temperature-sensing device that can be used for flow as well as all batch and titration calorimeters is to calibrate the calorimeter system at two different total energy equivalent values, the difference between which is accurately known. This may be accomplished by using (or flowing) two different liquids in the reaction vessel, both of known heat capacities, or by using two different volumes (or flow rates) of water in the reaction vessel. Dividing the difference in the measured values (e.g., J/V) by the known difference in the heat capacity (e.g., J/K) gives the sensitivity of the temperature output circuit (e.g., K/V).

3.3.3. Determination of the Energy Equivalent of the Reaction Vessel and Calculation of Total Uncorrected Heat

Determination of the energy equivalent is done by addition of a known amount of heat either from an electrical heater or from a standard reaction. The energy equivalent of the reaction vessel and its contents are measured electrically by introducing a constant current through a fixed resistance heater over a measured period of time. Details of several heater circuits have been described (30). Because electrical heaters may give erroneous data for several reasons, electrical calibrations should be checked by chemical calibration. For solution calorimeters, the heat of solution of THAM [tris(hydroxymethyl) aminomethane] in 0.1 molal HCl (31), the heat of neutralization of NaOH with HCl(aq) or $HC10_4$ (32–34), or the heat of protonation of THAM (aq) (35–37) or the heat of dilution of NaCl solutions or sucrose solutions (3, 14, 32) are sufficiently well known to be used for this purpose. While the overall features of the determination of the energy equivalent of the reaction vessel are similar for batch, titration, or flow calorimeters, important differences do exist for the three instrumental approaches. Each technique is separately discussed in the following sections.

3.3.3a. Batch Calorimetry

In batch isoperibol calorimetry, both of the reactants are brought to thermal equilibrium in the calorimeter reaction vessel before initiation of the reaction.

As a result, the energy equivalent of the reaction vessel and its contents are approximately the same before and after the reaction. The usual practice is to electrically determine the energy equivalent of the calorimeter reaction vessel and its contents both before and after the reaction. These data may be combined in several ways depending on the desired accuracy and temperature assigned to the final result (38). Typical data that would be obtained for the electrical calibration of a batch isoperibol calorimeter are shown in Fig. 3.9. A reaction vessel which equilibrates rapidly yields the data shown in Fig. 3.9a, while the data obtained using a reaction vessel which equilibrates at a much slower rate are better represented by Fig. 3.9b. The rapid equilibration (Fig. 3.9a) is typical of data obtained using a reaction vessel similar in design to that shown in Fig. 3.3. The energy equivalent of the reaction vessel plus contents for data similar to that illustrated by Fig. 3.9a is approximated by

$$\varepsilon = \frac{q_E}{\Delta T_{\text{tot}} - (S_I + S_F)(t_H/2)} \tag{3.3}$$

where q_E is the electrical energy introduced into the reaction vessel by the heater; ΔT_{tot} is the total uncorrected temperature rise from "heater on" to "heater off" in Fig. 3.9a; S_I and S_F are the initial and final rates, respectively, of temperature rise due to all heat effects other than the heater (stirring, radiation, conduction,

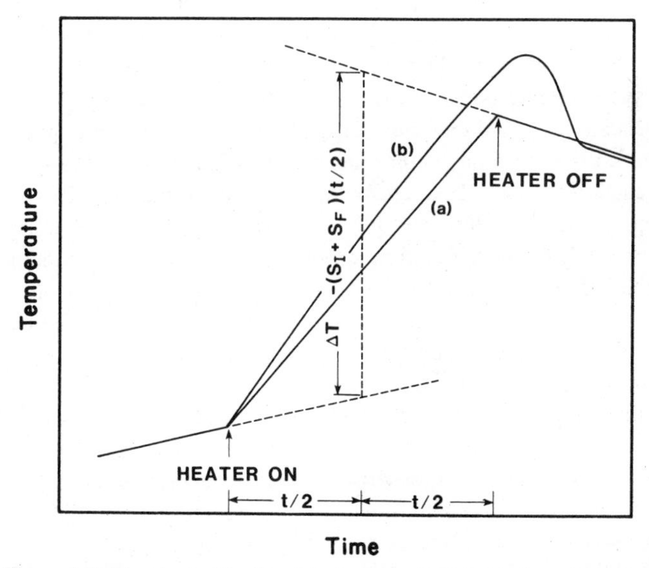

Figure 3.9. Electrical calibration data for an isoperibol solution calorimeter.

etc.); and t_H is the time the calibration heater is on. Equation (3.3) is derived by assuming a constant rate of temperature rise between "heater on" and "heater off." The value in brackets in Eq. (3.3) may be found graphically as illustrated in Fig. 3.9a. For more accurate work, calibration of vessels with large heat leak, or vessels where the electrical calibration data are better represented by Fig. 3.9b, Eq. (3.3) is not valid and an alternate data analysis procedure must be used (7). Accurate methods for data analysis in this case have been discussed in the literature (7, 39) and are detailed in Section 3.3.5a.

The uncorrected total heat q_p produced in the reaction vessel during a batch experiment (Fig. 3.6) is given by Eq. (3.4), where t is the time from the end of the initial to the beginning of the trail region:

$$q_p = \varepsilon\,[\Delta T_{\text{tot}} - (S_I + S_F)(t/2)] \tag{3.4}$$

3.3.3b. Titration and Batch Injection Calorimetry

In continuous titration, incremental titration, and injection batch calorimetry, the energy equivalent of the system is a function of the amount of titrant added at any point during an experiment. The energy equivalent of the reaction vessel and contents can be determined either (1) by measuring the energy equivalent as a function of the amount of titrant added or (2) by calculating the separate energy equivalents of the empty reaction vessel and of the contents and combining them to give the total energy equivalent of the system. Method 1 is the more accurate in that the energy equivalent is measured directly. It is also the more time consuming and laborious procedure if many chemical systems with varying heat capacities are to be studied. Method 2 depends on the energy equivalent of the total system being equal to the sum of the energy equivalents of the parts. It is also usually assumed that the energy equivalent of the empty reaction vessel is independent of the total volume of solution. For dilute solutions and large-volume reaction vessels, these are both good approximations. This method has the advantage that once the energy equivalent of the empty reaction vessel has been determined, no further calibrations are necessary as long as the heat capacities of the titrate and titrant solutions can be evaluated from other sources. Method 2 requires that the temperature-sensing device be calibrated in absolute terms and that the heat capacity of the material added to the reaction vessel be known; method 1 does not.

In method 2 the energy equivalent of the reaction vessel and contents is measured. The heat capacity of the contents is then subtracted to give the energy equivalent of the empty reaction vessel (ε_{RV}). The energy equivalent for a given addition of titrant is then determined by adding the energy equivalent of the empty reaction vessel to the heat capacity of the total contents of the reaction vessel. Method 1 requires the same experiments to be run as method 2. However, the data obtained from method 1 are expressed in units of heat produced/bridge output unit as a function of liquid volume in the reaction vessel. The calibration results obtained by method 1 are valid only for the liquid used during calibration.

The energy equivalent of the reaction vessel plus contents is obtained in a manner identical to that presented in Section 3.3.3a. for batch calorimetry. The quantity ε is valid only for the volume used in the calibration. Therefore, ε must be determined over the volume range from zero to maximum addition of titrant. Usually ε varies linearly with volume so that calibration at two or three volumes is sufficient. The energy equivalent ε_p at any volume of titrant V_p is calculated as

$$\varepsilon_p = \varepsilon_x + V_p\,(C_P + \partial\varepsilon_{RV}/\partial V) \tag{3.5}$$

where ε_x is the energy equivalent at zero addition of titrant (e.g., the initial region in Figs. 3.6 and 3.7). Equation (3.5) is used to describe the data obtained by method 1 assuming $C_P + \partial\varepsilon_{RV}/\partial V$ is constant.

The energy equivalent of the empty reaction vessel ε_{RV} may be calculated from

$$\varepsilon_{RV} = \varepsilon_p - (V_p + V_x)\,C_P \tag{3.6}$$

where V_p, V_x, and C_P are the volume of titrant, initial volume of titrate, and heat capacity of the solution in the reaction vessel, respectively. The quantity ε_{RV} is a constant for a given volume of liquid in the reaction vessel, but varies slightly with liquid volume as titrant is added [i.e., $\partial\varepsilon_{RV}/\partial V$ in Eq. (3.5)]. This change in ε_{RV} is due primarily to more of the reaction vessel wall coming into contact with the liquid as the liquid volume increases. For well-designed calorimeter reaction vessels (6, 7), the change in ε_{RV} for a 10% change in total volume is less than 1% of the value of ε_{RV} (7). The quantity, $\partial\varepsilon_{RV}/\partial V$, can be evaluated by measuring ε_{RV} as a function of the volume of liquid in the reaction vessel. It has been found that for well-constructed reaction vessels, $\partial\varepsilon_{RV}/\partial V$ is a constant (7, 40). As a consequence, a plot of ε_p versus V_p is linear [see Eq. (3.5)]. For a given amount of titrant added, the energy equivalent of the reaction vessel and contents is given by

$$\varepsilon_p = \varepsilon_{RV} + (V_x C_P) + V_p(C_P + \partial\varepsilon_{RV}/\partial V) \tag{3.7}$$

where $V_x C_P$ is the total heat capacity of the initial solution in the reaction vessel, $V_p C_P$ is the total heat capacity of the titrant added, and $V_p(\partial\varepsilon_{RV}/\partial V)$ is the increase in heat capacity due to the increase in the liquid–wall contact area in the reaction vessel due to addition of V_p volume of titrant. Note that the value of C_P is the value for the diluted titrant after it is mixed with the solution in the reaction vessel and not the value for the undiluted titrant. It should be emphasized that the quantities ε_{RV} and $(\partial\varepsilon_{RV}/\partial V)$ are independent of the type of liquid used in the calorimeter.

The uncorrected total heat q_p produced in the reaction vessel during an injection batch or increment of an incremental titration (see Fig. 3.6), or produced from the start of a continuous titration, point (t_0, T_0), to any data point p (see Fig. 3.7), is given by

$$q_p = \varepsilon_p (T_p - T_0) \tag{3.8}$$

where T_p and T_0 are defined in Figs. 3.6 and 3.7. If method 1 was used to calibrate the instrument, ε_p is found from Eq. (3.5). If method 2 was used, then ε_p is found from Eq. (3.7).

3.3.3c. Flow Calorimetry

The data measured in a flow isoperibol calibration experiment for either an electrical or a chemical calibration are illustrated in Fig. 3.8. Data in the literature on chemical calibration experiments are limited because instruments reported in the literature have been used mostly to determine the heat capacities of liquids. These experiments are identical to electrical calibration of a flow isoperibol calorimeter (41).

The temperature rise of the flowing solution due to the addition of heat from the electrical heater is $T_p - T_0$, as shown in Fig. 3.8. The power equivalent of this type of isoperibol reaction vessel is the heater power/output unit or the rate of heat input/kelvin as given by

$$\frac{d\varepsilon_E}{dt} = \frac{dq_E/dt}{T_p - T_0} \tag{3.9}$$

where dq_E/dt is the rate of heat input from the electrical heater. At a given heater power, the temperature rise $T_p - T_0$ is a function of the total flow rate through the calorimeter. If $d\varepsilon_E/dt$ is divided by the flow rate and the density of the liquid, the result is equal to the heat capacity of the liquid. Thus, the calorimeter must be calibrated at each total flow rate unless the flow rate, density, heat capacity of the liquid after mixing, and the calibration constant relating the output of the temperature-sensing device to kelvins are known. The total heat produced during a time Δt is calculated from

$$q_p = \frac{d\varepsilon_E/dt}{(T_p - T_0)(\Delta t)} \tag{3.10}$$

3.3.4. Analysis of Continuous Titration Calorimetry Data

The data from a typical isoperibol titration calorimetric run (Fig. 3.7) was described in Section 3.3.1. Data collected during the initial and final regions are used to correct the titration data in the reaction and excess reagent regions for non-chemical heat effects. In continuous titration calorimetry the titrant is added at a constant rate over the entire period of the run, that is, the reaction and excess reagent regions. The titrant first enters the titrate at point (t_0, T_0). A precision, constant-rate buret is required to obtain accurate continuous calorimetric titration data. The buret is calibrated by weighing the amount of a suitable liquid of known density delivered over measured time intervals. The consistency of the delivery rate over the titration range can be calculated accurately from the mass rate of delivery using the density of the liquid at the calibration temperature. The constancy of the delivery rate over short time intervals should also be checked since the delivery rate may be a waveform of the same periodicity as the buret drive system. This may be done by carefully examining the raw data on temperature or heat flow rate during a titration with a reaction that has a large ΔH_R value. If the buret delivery rate is cycling, the waveform will also appear in the output data. The time at point (t_0, T_0) is not the same as that at which the titrant buret is turned on because a small air space is usually left at the tip of the titrant delivery tube to prevent premixing of the solutions. Point (t_0, T_0) may be found graphically from the data or from independent calibration of the air space.

The following data can be evaluated from the plot of temperature versus time. The slope at point x in the initial ($S_{I,x}$, in K or V/s) and at point y in the final ($S_{F,y}$, in K or V/s) period, and the temperatures of the reaction vessel and its contents (T_0, T_1, T_2, T_3, . . . , T_p, . . . , T_m) at the reaction times (t_0, t_1, t_2, t_3, . . . , t_p, . . . , t_m). A set of q_p values corresponding to the T_p values can be calculated using Eq. (3.8). These q_p values represent the total heat produced in the reaction vessel from point (t_0, T_0) to (t_p, T_p) and must be corrected for all heat effects other than those due to chemical reactions before they can be used to calculate thermodynamic values. These corrections are detailed in the next section.

3.3.5. Calculation of Non-Chemical Energy Correction Terms

Non-chemical contributions to the total heat change measured in any isoperibol solution calorimetric reaction vessel include those heat effects associated with stirring of the solution, heat losses between the reaction vessel and its surroundings, and resistance heating of the thermistor. The slopes in the initial ($S_{I,x}$) and final ($S_{F,y}$) regions can be used to calculate the rates of heat loss at points x and y (see Figs. 3.6 and 3.7) and at any point p if the assumption is made that the rate of heat loss is proportional to the difference between the reaction vessel temperature T_p and the temperature of the surroundings T_{ex}. This assumption is equivalent to saying that the reaction vessel obeys Newton's law of cooling (38). The rate of heat loss from the reaction vessel at points x and y is described by

$$-S_{I,x}(\varepsilon_x) = dq_{HL,x}/dt = -\omega_x - [\kappa_x\varepsilon_x(T_{ex} - T_x)] \tag{3.11}$$

$$-S_{F,y}(\varepsilon_y) = dq_{HL,y}/dt = -\omega_y - [\kappa_y\varepsilon_y(T_{ex} - T_y)] \tag{3.12}$$

where both ω and κ are positive constants and $dq_{HL,x}/dt$ and $dq_{HL,y}/dt$ are rates of heat loss at points x and y (Figs. 3.6 and 3.7), respectively. For many reaction vessels with a large volume (>25 mL), small κ (<0.005 min^{-1}), and small τ_C (<1 s), ω and κ are independent of the volume of solution in the reaction vessel (7). This is assumed to be the case in the development of Eq. (3.13) from Eqs. (3.11) and (3.12). Equations 3.11 and 3.12 may be solved for ω and κ using the experimental values for $S_{I,x}$, $S_{F,y}$, T_x, T_y, ε_x, ε_y, and T_{ex}. These values can then be used to calculate the value of dq_{HL}/dt at any point p,

$$\frac{dq_{HL,p}}{dt} = -\omega - [\kappa\varepsilon_p(T_{ex} - T_p)] \tag{3.13}$$

Equations (3.11), (3.12), and (3.13) can be combined to give an expression for $dq_{HL,p}/dt$ which does not contain the somewhat ambiguous quantity T_{ex} if it is assumed that $\kappa\varepsilon_{ex}T_{ex} = \kappa\varepsilon_pT_{ex} = \kappa\varepsilon_yT_{ex}$:

$$\frac{dq_{HL,p}}{dt} = \frac{dq_{HL,x}}{dt} + \left(\frac{dq_{HL,y}}{dt} - \frac{dq_{HL,x}}{dt}\right)\frac{\varepsilon_pT_p - \varepsilon_xT_x}{\varepsilon_yT_y - \varepsilon_xT_x} \tag{3.14}$$

For batch calorimetry, in those cases where $\varepsilon_x = \varepsilon_p = \varepsilon_y$, Eq. (3.14) reduces to Eq. (3.6) in Ref. 39.

The total contribution of the non-chemical heat effects from the start of the reaction [point (t_0, T_0)] to point (t_p, T_p) (see Figs. 3.6 and 3.7) is given by

$$q_{HL,p} = \int_{t_0}^{t_p} (dq_{HL}/dt) \tag{3.15}$$

Equation (3.15) may be solved by obtaining an analytical expression relating dq_{HL}/dt values calculated from Eq. (3.14) for each data point p and time t_p and integrating the resulting function. A simpler and usually adequate method is to assume for the time interval (Δt) between data points that dq_{HL}/dt is linear in time. This assumption is the basis of the Regnault–Pfaundler derivation described in Ref. 39. Equation (3.15) may then be expressed as a sum over all data points (taken at equal time intervals) to point (t_p, T_p)

$$q_{HL,p} = \sum_{i=1}^{p} \frac{\Delta t_i}{2}\left(\frac{dq_{HL,i}}{dt} + \frac{dq_{HL,i-1}}{dt}\right) \tag{3.16}$$

where $dq_{HL,0}/dt$ is the initial rate of heat loss at point (t_0, T_0). For batch calorimetry, the only $q_{HL,p}$ value of experimental significance is where the point (t_p, T_p) is point (t_f, T_f) in Fig. 3.6. For titration calorimetry several $q_{HL,p}$ values corresponding to each q_p value are calculated. The $q_{HL,p}$ values calculated from Eq. (3.16) are used to correct q_p values obtained from Eqs. (3.4) and (3.8) above for the energy contributed by non-chemical terms.

In those cases in which ω and κ vary significantly with volume, Eqs. (3.13) and (3.14) are not accurate. For example the variation of κ with volume for a small-volume reaction vessel is illustrated in Fig. 3.10. Between certain volumes (dependent on the specific reaction vessel) κ and V will be linearly related. In Fig. 3.10 this is between 2.5 and 3.0 mL. This relationship can be represented by (7)

$$\kappa_p = \kappa_0 + \alpha_\kappa V_p \tag{3.17}$$

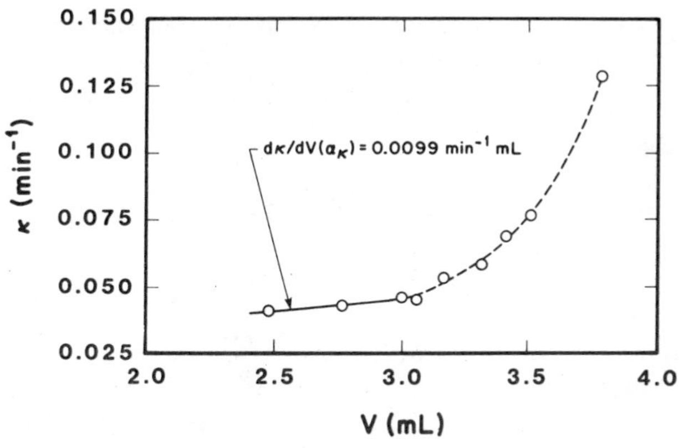

Figure 3.10. Variation of κ with volume for a small volume isoperibol reaction vessel. (Taken from Ref. 7.)

where κ_0 is the heat-leak constant before titrant is added and α_κ is the slope of the linear region shown in Fig. 3.10. It will also be true that the power input in a small-volume reaction vessel will be dependent on the amount of liquid stirred in the vessel. This relationship can be expressed as (7)

$$\omega_p = \omega_0 + \beta_\omega V_p \tag{3.18}$$

where ω_0 is the power input before titrant is added and β_ω is a calibration constant obtained from the slope of the linear portion of a plot of ω versus volume. Combining Eqs. (3.17) and (3.18) with Eq. (3.11) and (3.12) and setting $T_{ex} = T_{\text{bath}}$ gives

$$\frac{dq_{HL,p}}{dt} = \frac{dq_{HL,x}}{dt} - \varepsilon_p \alpha_\kappa V_p \Delta T_p - \beta_\omega V_p - \left(\frac{dq_{HL,x}}{dt} - \frac{dq_{HL,y}}{dt} - \varepsilon_y \alpha_\kappa V_y \Delta T_y - \beta_\omega V_y \right) \frac{\varepsilon_p \Delta T_p - \varepsilon_x \Delta T_x}{\varepsilon_y \Delta T_y - \varepsilon_x \Delta T_x} \tag{3.19}$$

where $dq_{HL,p}/dt$ is the rate of heat loss at point p in the titration; $dq_{HL,x}/dt$ and $dq_{HL,y}/dt$ are the rates of heat loss at times t_x and t_y; ε_p is the energy equivalent of the calorimeter system at point t_p and ε_x and ε_y are the energy equivalents during the initial and final periods, respectively; α_κ and β_ω are calibration constants as shown in Eqs. (3.17) and (3.18); V_p is the volume of titrant added to point p; V_y is the total titrant volume added to point f; and ΔT_p, ΔT_x, and ΔT_y are the temperature differences between the bath and the reaction vessel at times t_p, t_x, and t_y, respectively. Equation (3.19) is used in place of Eq. (3.14) to calculate values for the heat exchange values with the surroundings to be used in Eqs. (3.15) and (3.16). Equation (3.19) or a similar equation involving variations of ω and κ with time should be used to analyze the data from any calorimetric reaction vessel in which ω and κ are not constant during the experiment.

3.3.6. Correction for Temperature Differences Between Titrant and Titrate

Ideally, the temperatures of the titrant T_T and titrate T_0 at point (t_0, T_0) are the same. However, in practice this identity is difficult to achieve. If the quantity $(T_0 - T_T)$ is positive, there will be an endothermic heat effect as the colder titrant is added. Conversely, there will be an exothermic heat effect if the quantity $T_0 - T_T$ is negative. The correction that must be made to the q_p values for this heat effect is

$$q_{TC,p} = (V_pC_P)_t(T_0 - T_T) \tag{3.20}$$

The calculated correction is only that energy resulting from the introduction of the titrant into the reaction vessel at a temperature different from the temperature of the contents of the reaction vessel at the beginning of the titration (T_0). The fact that the temperature changes during the titration does not enter into this calculation (38).

3.4. ISOTHERMAL SOLUTION CALORIMETRY

3.4.1. Equipment Description

Isothermal calorimetry is a technique in which the temperature of the calorimeter reaction vessel and contents are kept constant and heat flux through the system is measured as a function of time or titrant added (8). The temperatures of the calorimeter reaction vessel, its contents, and its environmental surroundings are maintained constant so that radiation heat losses are constant and all changes in heat flux out of the system are due only to chemical or physical changes occurring in the calorimeter reaction vessel. Since heat is measured directly, the amount of required auxiliary data is reduced, and it is not necessary to determine the heat capacity of either the reaction vessel or its contents. Therefore, the isothermal instrument is particularly useful for the study of processes having slow reaction rates, processes in which large amounts of heat are produced, and systems involving large changes in heat capacities during the reaction period. Examples of such systems would include processes associated with microbial growth and metabolism, reactions occurring in concentrated solutions, heats of mixing of organic liquids, and reactions involving two liquid phases.

Whether the experiment is of the batch, titration, or flow type, isothermal instrument calibration and data analysis are simpler than for isoperibol calorimetry because it is not necessary to correct for heat loss or gain to the surroundings. The major drawback to isothermal calorimetry is that the time response of the instrument is on the order of minutes, rather than seconds, as is the case with isoperibol calorimetry. Thus, rapid changes in heat production rate are difficult to follow.

The heat fluxes in an isothermal titration calorimeter are represented schematically in Fig. 3.11. Heat is removed from the calorimeter reaction vessel at a constant rate dq_{out}/dt using a cooling device. This heat effect is balanced against the heat inputs of stirring, thermistor self-heating, and the control heater power, dq_{in}/dt, necessary to maintain a constant temperature in the reaction vessel. The temperatures of the reaction vessel and the surrounding constant-temperature

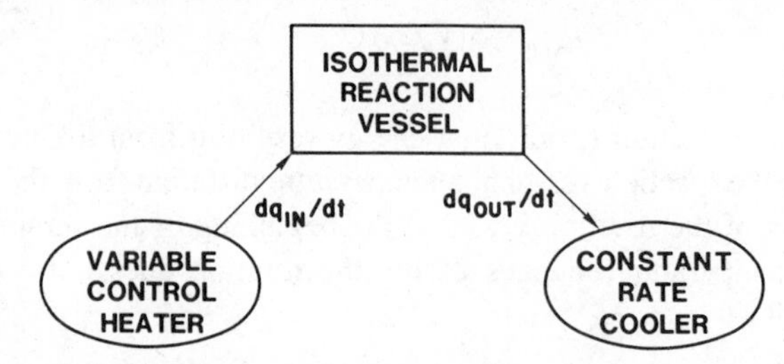

Figure 3.11. Schematic of the heat fluxes in an isothermal calorimeter.

bath are held constant, thereby maintaining constant passive heat transfer between the bath and the reaction vessel. As a reaction proceeds, dq_{in}/dt is adjusted to maintain the isothermal condition. The heat due to the reaction is thus given by the negative of the integral of the change in dq_{in}/dt with time.

An example of an isothermal titration calorimeter reaction vessel is shown in Fig. 3.12. Calorimeters of this type are capable of maintaining the temperature in the reaction vessel isothermal to $\pm 5 \times 10^{-6}$ (root mean square or standard deviation of the temperature signal at $\tau_T = 0.5$ s) for an indefinite time (2,8,42). The accuracy attainable by these instruments is illustrated by data given in Table 3.1 for the heat of neutralization of a strong acid by a strong base.

The form of the data obtained on a quantitative reaction using an isothermal, continuous titration calorimeter is illustrated in Fig. 3.13. During the initial period, the data represent the rate of heat production from the control heater required to balance the heat effects from the cooler and heat effects in the reaction vessel arising from stirring, thermistor self-heating, etc. At time t_0, titrant addition is begun and an increased heat effect is seen due to the combined effects of addition of the titrant and the resulting reaction. At the end of the reaction region, the reaction is complete. The heat effects in the excess reagent region are the same as those present in the initial region, plus any effects of continued addition of titrant. At the end of the excess reagent region, the buret is turned off and the heat effects measured in the final region are the same as those present in the initial region except that the volume in the reaction vessel has increased due to the addition of titrant.

If the isothermal reaction vessel were able to respond instantaneously to the heat effects associated with addition of the titrant, then the data at the start of the reaction and excess reagent regions would follow the dashed lines. The deviation of the data from this ideal represents the period when the instrument is not in control and is dependent on the time constants of the instrument. Usually the dashed and solid lines are indistinguishable after about 3 or 4 min following the step change in the heat production rate that accompanies the start or completion of the reaction. Data obtained during this time interval are in error by the

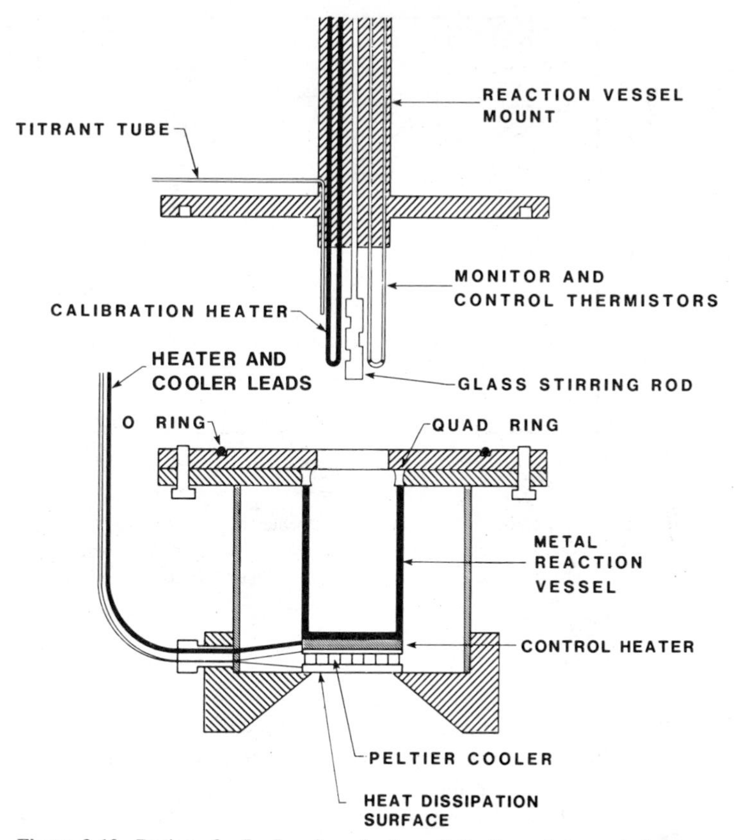

Figure 3.12. Design of a 5-mL-volume isothermal titration solution calorimeter.

difference between the areas under the dashed and solid lines. However, once the calorimeter is back in control, the total area under the solid and dashed lines will be the same. Techniques have been described to deconvolute the data obtained to generate the input function represented by the dashed line using Laplace transformation techniques (43). However, in practice the start of data analysis is usually delayed until the calorimeter is in isothermal control.

The rate of heat input from the control heater during the initial and final regions, dq_I/dt and dq_F/dt, respectively, differ slightly from each other since the increased volume in the final region, as compared to the initial region, results in a changed heat effect due to stirring. The same effect is also responsible for the linear change in the baseline in the reaction and excess reagent regions (dotted line in Fig. 3.13).

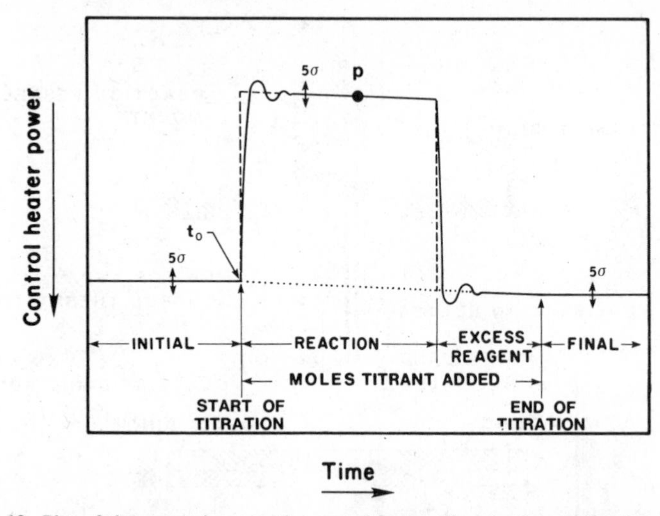

Figure 3.13. Plot of the power input to the control heater of an isothermal titration calorimeter versus time during a continuous titration experiment involving a quantitative reaction. The vertical double-ended arrows are the peak-to-peak noise in the indicated measurement. The peak-to-peak noise is 5σ.

Isothermal flow calorimeters capable of collecting data on reacting systems at high temperatures and pressures have been reported (10, 11). The high accuracy and precision achieved with these instruments is illustrated by the data in Table 3.1. A diagram showing an isothermal flow reaction vessel is given in Fig. 3.14. Auxiliary equipment needed to operate the isothermal flow calorimeter are a fluid-flow circuit consisting of two precision pumps and a flow-control programmer, and a constant-temperature environment for the reaction vessel. A reaction is initiated by starting the pumps and letting the reactants flow, either at a constant rate or constantly changing rate, as controlled by the programmer, through thecoil on the isothermal plate. A Peltier cooler (10, 11) or a controlled temperature differential across a thermal barrier (44, 45) may be used to remove energy from the coil and plate at a constant rate and discharge it to the surrounding bath. A controlled heater compensates for the energy liberated or absorbed by the reaction and maintains the coil and plate at a constant temperature. As was the case with isothermal solution calorimeters, the difference in the energy supplied by the control heater before and during the reaction is a direct measure of the energy of the interaction. Calibration of the instrument and data analysis are identical to calibration of an isothermal solution calorimeter as described previously. Flow rates for the instruments, which have been described in the literature (10, 11), are usually between 0.2 and 3.0 mL min^{-1} total flow.

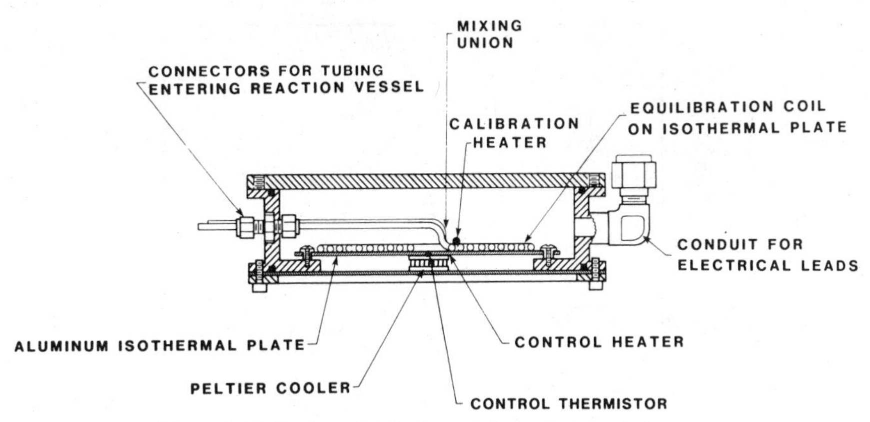

Figure 3.14. Design of an isothermal flow solution calorimeter.

3.4.2. Calibration and Data Analysis

3.4.2a. Calibration of the Control Heater

In an isothermal batch, titration, or flow calorimeter, the rate of heat input by the control heater is measured as a function of time. The heater may be powered with either direct current, in which case the power is measured by measuring voltage and current, or by a square-wave alternating current, in which case the power is proportional to the frequency. The latter case has significant advantages in electronic design and accuracy for automatic control. Usually the data are presented as a time-averaged frequency η over a given time interval Δt. The energy equivalent of each pulse, or square wave, is most accurately measured by introducing a constant current through the calibration heater. The electrical calibration is conducted until the calorimeter is well past the transient period and a stable control heater frequency is established as illustrated in Fig. 3.15. Since the volume of solution in the reaction vessel does not vary over the calibration period, the initial and final base lines, η_I and η_F, should be identical. The energy equivalent of the pulsed control heater ε_η, in energy/pulse, is given by

$$\varepsilon_\eta = \frac{dq_E/dt}{\eta_H - (\eta_I + \eta_F)/2} \tag{3.21}$$

$$\varepsilon_\eta = \frac{(dq_E/dt)t_H}{\left(\sum_{i=1}^{p} [\eta_i - (\eta_I + \eta_F)/2][t_i - t_{i-1}]\right)} \tag{3.22}$$

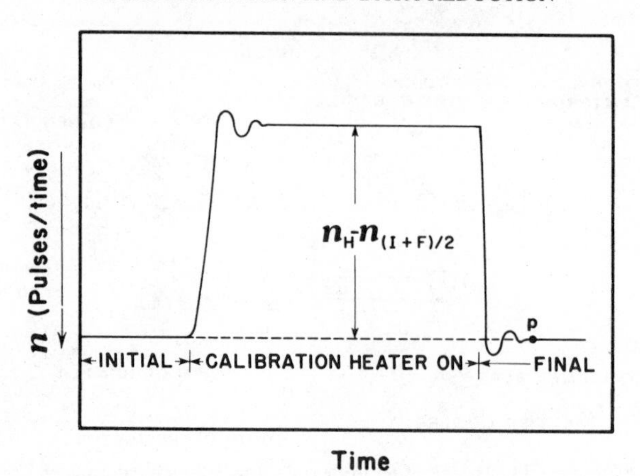

Figure 3.15. Plot of the rate of heat input to the control heater of an isothermal calorimeter versus time during an electrical calibration.

In Eq. (3.22), t_H is the time interval between "heater on" and "heater off" and point p must be after the transient at the beginning of the final region. The value of ε_η is usually independent of the volume or type of liquid in the reaction vessel.

3.4.2b. Analysis of Isothermal Calorimetric Data

The total heat produced in the reaction vessel from the start of the titration or flow experiment, point t_0 (or any other beginning point for data analysis), to any data point p (see Fig. 3.13) is given by

$$q_p = \sum_{i=1}^{p} [\eta_i - \eta_I + (\eta_F - \eta_I)(t_i + t_{i-1})/2t_{\text{tot}}]\, \varepsilon_\eta (t_i - t_{i-1}) \quad (3.23)$$

In Eq. (3.23), the sum is over p data points with η_i being the average frequency in the control heater during the time of the ith data point, η_I and η_F are the average baseline frequencies in the heater at the start and end of the titration (measured during the initial and final periods, respectively), t_i is the total time from the start of the titration to the end of the ith data interval, and t_{tot} is the total time between the start and the end of the titration.

The data produced by a batch isothermal calorimetric experiment are illustrated in Fig. 3.16. The experiment is initiated at time t_0 by mixing the total volumes of the two reagents in the calorimeter reaction vessel. The resulting heat effect causes the deviation from the baseline as shown in the reaction region. The exact

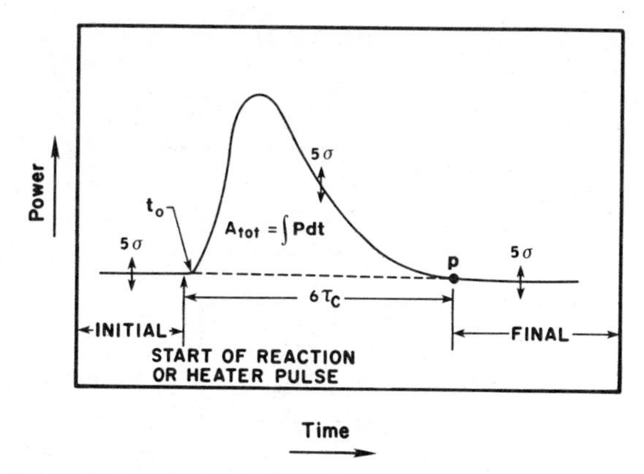

Figure 3.16. Calorimetric data obtained during a batch isothermal or heat-conduction calorimetric experiment. The vertical double-ended arrows are the peak-to-peak noise in the indicated measurement. The peak-to-peak noise is 5σ.

shape of the curve in this region will depend on the kinetics of the reaction, on the magnitude of the heat effect measured, and on the time constants for the isothermal calorimeter. For instruments similar to those described in the literature (8), the time from t_0 to the final region is about 5 min. The heat produced in a batch experiment q_p is calculated with Eq. (3.23), where the time from t_0 to point p is greater than $6\tau_C$.

3.4.2c. Calculation of Non-Chemical Energy Correction Terms

The only non-chemical contribution to the heat measured in an isothermal solution or flow reaction vessel is caused by the temperature difference between the titrant and titrate or between the incoming flow streams and the reaction vessel control plate. The technique for handling this effect has been described in Section 3.3.6.

3.5. HEAT-CONDUCTION SOLUTION CALORIMETRY

3.5.1. Equipment Description

In a heat-conduction calorimeter, the heat produced within the reaction cell results in an emf response of a thermoelectric device as heat flows through the device to a constant-temperature heat sink. The design of a flow calorimeter is illustrated in Fig. 3.17. The same design can be used for batch and titration experiments by replacing the flow cells with appropriate reaction vessels. For either a thermopile or a thermoelectric device the voltage $\mathbf{E}_C$ and the heat flow dq/dt are

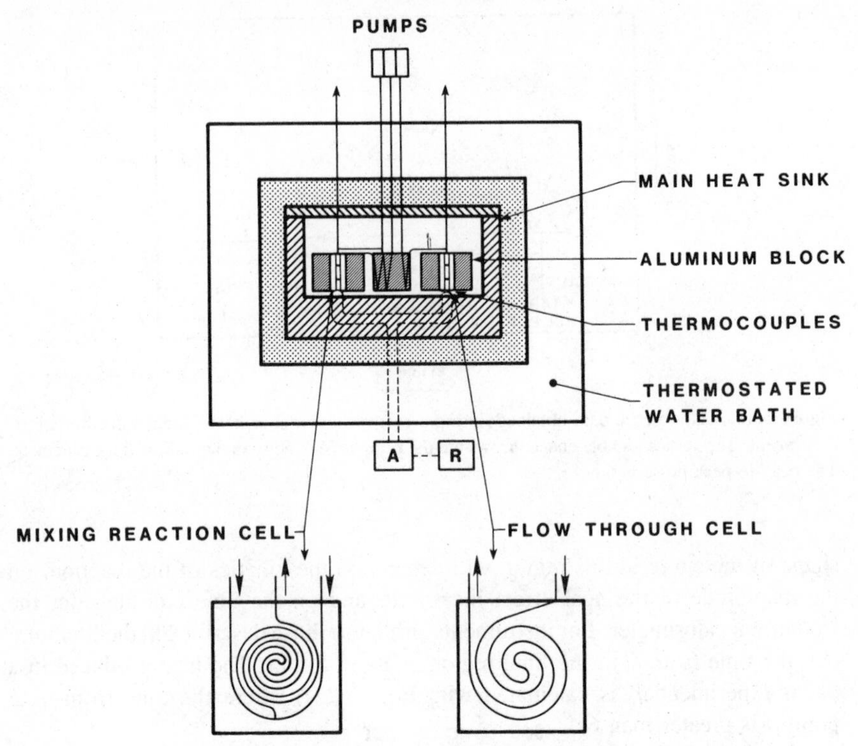

Figure 3.17. Schematic of a heat-conduction flow calorimeter.

proportional to the temperature difference between the reaction vessel wall and the heat sink (5, 46). Therefore, $\mathbf{E}_C = \varepsilon_C(dq/dt)$. Calibration and analysis of the data from a heat-conduction batch or flow calorimeter (5, 46) are comparable to analysis of the data from an isothermal batch or flow calorimeter. However, the minimum time constant of the measurement is that of the thermoelectric device. Typically this is of the order of minutes. The factors that control the time response of a heat-conduction calorimeter have been discussed (5, 46). The large time constant introduces a delay of several minutes between the start of a reaction and the attainment of a baseline or steady-state condition in a batch or a flow cell. Techniques to deconvolute data from a heat-conduction calorimeter have been described (18–21). The precision and accuracy that can be obtained with either a batch (3) or a flow (14) heat-conduction calorimeter are illustrated by the data for the protonation of THAM and the dilution of sucrose. There has been one report in the literature of the development of a continuous titration heat-conduction calorimeter (9), but no details were given. The usefulness of such an instrument will probably be limited by the long time constant.

3.5.2. Calibration and Data Analysis

3.5.2a. Calibration of Heat-Conduction Calorimeters

A heat-conduction calorimeter may be calibrated electrically either by addition of a known pulse of electrical energy (Fig. 3.16) or by addition of a constant rate of electrical energy until a stable output signal is obtained (Fig. 3.8). The method of choice is that which best reproduces the output obtained from the chemical reactions to be studied.

The data produced by a batch (or an incremental titration or batch injection) heat-conduction calorimeter during a chemical reaction are illustrated in Fig. 3.16. Similar data would be obtained by calibration of any heat-conduction calorimeter with a single electrical pulse. The value of $6\tau_C$ for most heat-conduction calorimeters is about 20 min. The calibration constant for heat-conduction calorimeters ε_C is calculated from the heat introduced by the electrical pulse q_E and the total integrated area A_{tot} shown in Fig. 3.16:

$$\varepsilon_C = \frac{q_E}{A_{tot}} \tag{3.24}$$

The heat produced q_p during an experiment is then calculated from

$$q_p = \varepsilon_C A_{tot} \tag{3.25}$$

The calibration constant ε_C is independent of the reactants and volumes used in the calorimeter reaction cell.

If a heat-conduction calorimeter is used as a flow calorimeter, then a plot of the calorimeter output as a function of time for either an electrical calibration or chemical reaction will be of the form shown in Fig. 3.8. The calibration constant of a flow heat-conduction calorimeter is most accurately measured electrically after the calorimeter is well past the transient period and a stable output relative to the baseline is established. The calibration constant is then given by

$$\varepsilon_C = \frac{dq_E/dt}{T_p - T_0} \tag{3.26}$$

The value of ε_C should be tested to determine if it is independent of the type of liquid in the flow reaction vessel and the flow rate through the reaction vessel (14, 47). Electrical calibration of the instrument at different flow rates can serve as a check on the thermal equilibration of the fluid prior to exiting the reaction vessel. The rate of heat production during a flow experiment is found from

$$\frac{dq_p}{dt} = \varepsilon_C(T_p - T_0) \tag{3.27}$$

The total heat associated with flow over a time interval Δt is then calculated from

$$q_p = \left(\frac{dq_p}{dt}\right) \Delta t \tag{3.28}$$

3.5.2b. Baseline Stability of Heat-Conduction Calorimeters

The time between initiation of heat production from either an electrical calibration run or a chemical experiment and the establishment of a steady-state condition with a heat-conduction calorimeter will be of the order of tens of minutes. During this transient period, the baseline of the instrument may drift from that measured during the initial period. The accuracy of data obtained with a heat-conduction calorimeter is limited by baseline drift. The effects of baseline drift may be minimized by improving the design of the calorimeter (48, 49), by referencing to a twin calorimeter cell (3), and by using some type of data deconvolution to make use of the data obtained during the transient period (18–21). Data deconvolution is much simpler in this case than for an isothermal calorimeter since it involves only one time constant and the well-known equation for exponential decay.

The baseline drift of a heat-conduction calorimeter is very dependent on the temperature stability of the heat sink enclosing the reaction cell and on the effect of temperature variations on the electrical signal amplification equipment (48, 49). It has been shown that in order to achieve a peak-to-peak baseline noise and long-term drift of less than 100 nW in a heat-conduction calorimeter using thermoelectric modules, the bath surrounding the air chamber containing the heat-sink block and reaction cell must be thermostated to about $\pm 1 \times 10^{-4}$ K.

3.5.2c. Calculation of Non-Chemical Energy Correction Terms

The only non-chemical contribution to the heat measured in a heat-conduction calorimeter is caused by the temperature difference between the reaction vessel and incoming solutions. The calculation for correcting for this effect is described in Section 3.3.6.

3.6. COMPARISON OF INSTRUMENTS

3.6.1. General

The choice of a calorimetric method is not a simple matter. Factors such as availability, price, time required to acquire and set up the instrument, and ease of operation, data acquisition, and analysis may be important aspects of the decision. In addition to these extrinsic considerations, the suitability of a given calorimeter for the desired measurement must always be considered. The physical properties of the material (e.g., viscosity, stability, melting point, etc.) may dictate the choice of a method. For example, if the reactants to be studied are stable and suitable for use in any calorimeter, but the products of the reaction are not stable and slowly decompose or precipitate, then it may be necessary to choose a calorimeter with a rapid response time. In this case, an isoperibol calorimeter may be superior to either an isothermal or a heat-conduction calorimeter. It may also be desirable to run batch, as opposed to titration, calorimetric experiments. However, if the reaction to be studied is a very slow process, then isoperibol calorimetry may not be suitable because of the problems associated with making heat-leak corrections over a long time period. These factors must be considered specifically for each chemical system and will not be considered in further detail here.

If the factors previously outlined do not dictate the choice of a calorimeter, it may be desirable to choose an instrument that has the best detection limit or optimizes reagent use for a given system. These factors become more important if the reagents to be used are limited in quantity. This is frequently the case in the study of rare-element chemistry or in the study of biochemicals. The factors which may then influence the choice of a calorimetric system are outlined in this section. The general problem is to maximize the amount of information obtained per unit of reagent.

3.6.2. Calorimeter Specifications

The difficulty in choosing a method from the nine possible combinations of calorimetric measurement and reactant mixing comes in making direct comparison among the various types of calorimetric experiments. This is complicated since the primary measurements for an isoperibol, isothermal, or heat-conduction calorimeter are of different quantities and the amount of reactant used per experiment is time dependent in some cases and not in others. It is the purpose of this section to review ways in which meaningful comparisons of different calorimeters can be made. Details of the material presented here have been recently published (42). It is not our purpose to attempt to make comparisons of commercially available instruments since such comparisons would have only a short term validity. However, we shall use data comparable to current best technology for illustrative purposes. Table 3.2 gives the necessary "state-of-the-art" specifications we have chosen (42) to represent each calorimetric method. The combination of heat-conduction calorimetry and continuous titration of one reactant has not been reported in detail, so we have simply estimated the best set of specifications that we believe could be achieved.

3.6.3. Theory

Figures 3.6–3.8, 3.13, and 3.16 show the form of the data collected by each type of calorimeter as well as defining the uncertainties associated with the measurements. The standard deviation in the measurement of the temperature or power, σ, is approximately equal to one-fifth of the peak-to-peak noise. It is assumed that the temperature or power measurement has a constant peak-to-peak noise envelope, 5σ, associated with it throughout the measurement, including the baseline periods. Also, as shown in Figs. 3.8 and 3.16, we assume that the baselines are at the same value of the power in the initial and final regions for an isothermal or a heat-conduction calorimeter.

3.6.4. Calorimeter Capabilities

One quantity that can be calculated for all types of calorimeters by use of the above assumptions and the data in Table 3.2 is the minimum detectable total heat effect, which is proportional to the smallest observable shift in the baseline temperature or power for the isoperibol batch and flow, isothermal titration and flow, and heat-conduction titration and flow calorimeters. Neglecting the effect of the extrapolation of the non-zero and changing slope over the $6\tau_C$ seconds in the case of the isoperibol batch calorimeter, the standard deviation of any difference on the measurement axis is given by

$$\sigma_{\text{diff}} = (2\sigma_i^2)^{0.5} \tag{3.29}$$

The units of σ_{diff} are kelvins for the isoperibol batch calorimeter and watts for the isothermal titration and flow and the heat-conduction flow and titration calorimeters. In the case of the isoperibol batch calorimeter, multiplying σ_{diff} by the thermal equivalent, -6.3 J/K for a 1.5-mL-volume reaction vessel, converts the result to joules. The results for the other calorimeters must be multiplied by $6\tau_C$, the minimum time in which a measurement could be made in order to obtain the result in joules. The results of these calculations are given in Table 3.2.

The problem is slightly different for the isoperibol titration calorimeter in that it is the change in slope that must be detectable instead of a baseline shift. The

TABLE 3.2. Microcalorimeter Specifications and Resulting Capabilities

Calorimeter Type	Peak-to-Peak Noise (5σ)	Time Constant, τ_C (s)	V and dV/dt[a]	Detection Limit[b] (μJ)	Volume/Data Point[c]	C[d]
Isoperibol batch (Fig. 3.6)	25[e]	0.5	1.5, —	90[f]	—	—
Isoperibol titration (Fig. 3.7)	25[e] (50[g])	0.5	1.5, 0.01	50[f]	0.5	833
Isoperibol flow (Fig. 3.8)	25[e] (0.3[g])	1.0	—, 0.2[h]	1.2[i]	10	1
Isothermal batch (Fig. 3.16)	30[g]	30[j]	3.0, —	3000	—	—
Isothermal titration (Fig. 3.13)	30[g]	30[j]	3.0, 0.01	3000	30	833
Isothermal flow (Fig. 3.8)	15[g]	15[j]	—, 0.2[h]	750	150	42
Heat-conduction batch (Fig. 3.16)	0.1[g]	200	0.1, —	70	—	—
Heat-conduction titration (Fig. 3.13)	10[g]	20	0.5, 0.01	780	20	325
Heat-conduction flow (Fig. 3.8)	1[g]	200	— 0.20[h]	780	2000	3

[a]Reaction vessel volume, mL, and flow rate, mL min^{-1}.
[b]Minimum detectable total heat effect in μJ.
[c]Volume of reactant solution delivered in $6\tau_C$ in μL.
[d]Relative concentration of reactant required to obtain the minimum detectable total heat effect in $6\tau_C$. Normalized to lowest value = 1.
[e]Peak-to-peak noise, μK. Equal to 5σ.
[f]Directly proportional to reaction vessel volume.
[g]Peak-to-peak noise, μW. Equal to 5σ.
[h]Total flow for both reactant lines, that is, the flow rate is 0.1 mL min^{-1} in each line.
[i]Directly proportional to flow rate.
[j]An actively controlled system actually has a minimum of three time constants. We have chosen to simplify this by using only the time constant that describes a critically damped system, that is, one-sixth of the settling time.

uncertainty in the lead slope can be very small because the number of data points, N, can be made very large. The uncertainty in the reaction region slope is

$$\sigma_{\text{slope}} = \{N\sigma^2/\tau_C^2[N\Sigma i^2 - (\Sigma i)^2]\}^{0.5} \tag{3.30}$$

if data points are taken at τ_C intervals (42). Because the minimum time over which a meaningful measurement could be made is $6\tau_C$, we choose $N = 6$ and obtain $\sigma_{\text{slope}} = 8$ μK/s. Multiplying by the thermal equivalent, ~6.3 J/K, and $6\tau_C$ gives the value of the minimum detectable total heat effect given in Table 3.2.

The calculation of the minimum detectable total heat effect for isothermal and heat-conduction batch calorimeters is again a different problem since the calculation of a total heat effect requires integration of the measurement signal. The desired result is thus given by the uncertainty in the area under the curve, which must be at least $6\tau_C$ in length. The areal uncertainty for small signals is approximated by

$$\sigma_{\text{area}} = (2)^{0.5}\,(6\tau_C\sigma) \tag{3.31}$$

The results are shown in Table 3.2.

3.6.5. Summary and Conclusions

Nine different kinds of calorimeters have been described, and the calculations required to convert the raw data to q_p values have been outlined. Which calorimeter is best for a given study depends on the results desired. A comparison of instrument operation characteristics and the parameters of response time and peak-to-peak noise that determine the minimum detectable heat change for each instrument have been summarized in Table 3.2. The best instrument to use to measure a single q_p value is either the heat conduction or isoperibol batch calorimeter. With these instruments, the heat can be measured to 0.1 mJ. Since the calorimetric measurement for a batch heat-conduction calorimeter is independent of the heat capacity of the reaction vessel and its contents, the detection limit is independent of the volume used in the reaction cell. The same is not true for the batch isoperibol calorimeter. The detection limit is directly proportional to the calorimeter reaction vessel volume for this instrument. The batch instrument with the poorest detection limit is the isothermal solution calorimeter. If a large amount of data is required from a single set of solutions, the instrument of choice is the isoperibol titration calorimeter. For example, a complete titration can be made with 1.5 mL of titrate in the isoperibol reaction vessel and a data point recorded every $6\tau_C$, that is, 3 s, in a 20-min continuous titration with a detection limit of 0.05 mJ for each point. On the other hand, only eight data points can be obtained with the same amount of solution in a heat-conduction flow calorimeter and the uncertainty in each point will be an order of magnitude higher. Twenty minutes will be required to obtain each point in the flow instrument. In contrast, there is increasing uncertainty in heat-leak corrections for the small-volume isoperibol titration calorimeter as titration times become longer (Fig. 3.5) (7). Consequently, 20 min is the longest run that can be made with reasonable accuracy in a small-volume isoperibol titration calorimeter. However, more than 100 data points can be obtained during this 20-min time interval. The reason that the isoperibol titration calorimeter has not been used more in biological studies is probably related to the extensive calibrations and calculations required for proper data reduction. The high sensitivity of this instrument compared to either the isothermal titration or the heat-conduction flow calorimeter is due to its short response time. It is possible to improve on the detection limits of the heat-flow measuring instruments by inverse Laplace transformation or similar data analysis techniques to reduce the effective time constant. A study that illustrates the best features of heat conduction and isothermal calorimeters for biological studies has recently been published (50). For very small heat production rates from a bacterial culture, a heat-conduction calorimeter was found to work very well. The heat production rates seen in this part of the study were below the detection limit of the isothermal calorimeter. At higher heat production rates the heat-conduction calorimeter became unsuitable because of depletion of oxygen in the growth medium, a situation for which the stirred reaction vessel on the isothermal calorimeter proved to be ideal.

REFERENCES

1. S. Sunner and I. Wadso, *Acta Chem. Scand.* **13**, 97 (1959).
2. D. J. Eatough, J. J. Christensen, and R. M. Izatt, *J. Chem. Thermodyn.* **7**, 417 (1975).
3. I. Wadso, *Acta Chem. Scand.* **22**, 927 (1968).
4. J. Jordan, J. K. Grime, D. H. Waugh, C. D. Miller, H. M. Cullis and D. Lohr, *Anal. Chem.* **48**, 427A (1976).
5. C. H. Spink and I. Wadso, *Methods in Biochemical Analysis,* D. Glick, Ed., Wiley, New York, 1975.
6. J. J. Christensen, R. M. Izatt, and L. D. Hansen, *Rev. Sci. Instrum.* **36**, 779 (1965).
7. L. D. Hansen, T. E. Jensen, S. Mayne, D. J. Eatough, R. M. Izatt, and J. J. Christensen, *J. Chem. Thermodyn.* **7**, 919 (1975).
8. J. J. Christensen, J. W. Gardner, D. J. Eatough, R. M. Izatt, P. J. Watts, and R. M. Hart, *Rev. Sci. Instrum.* **44**, 481 (1973).
9. P. Paoletti, *Gazz. Chim. Ital.* **112**, 135 (1982).

10. J. J. Christensen, L. D. Hansen, D. J. Eatough, R. M. Izatt, and R. M. Hart, *Rev. Sci. Instrum.* **47**, 730 (1976).
11. J. J. Christensen, L. D. Hansen, R. M. Izatt, D. J. Eatough, and R. M. Hart, *Rev. Sci. Instrum.* **52**, 1226 (1981).
12. P. Picker, C. Jolicoeur, and J. E. Desnoyers, *J. Chem. Thermodyn.* **1**, 469 (1969).
13. P. Picker, P.-A. Leduc, P. R. Phillip, and J. E. Desnoyers, *J. Chem. Thermodyn.* **3**, 631 (1971).
14. P. Monk and I. Wadso, *Acta Chem. Scand.* **22**, 1842 (1968).
15. R. S. Schifreen, C. S. Miller, and P. W. Carr, *Anal. Chem.* **51**, 278 (1979).
16. E. B. Smith, C. S. Barnes, and P. W. Carr, *Anal. Chem.* **44**, 1663 (1972).
17. G. Steinberg, *CHEMTECH* **11**, 730 (1981).
18. R. L. Berger and N. Davids, "True Data Reconstructed by Computer Simulation as Applied to Heat Conduction Microcalorimeters," Presentation at the 32nd Annual Calorimetry Conference, Sherbrooke, Quebec, Canada, 1972, NIH Technical Report, 1974.
19. E. Cesari, J. Vinals, and V. Torra, *Thermochim. Acta* **63**, 341 (1983).
20. S. L. Randzio and J. Suurkuusk, "Interpretation of Calorimetric Thermograms and Their Dynamic Corrections," in *Biological Calorimetry*, A. E. Beezer, Ed., Academic Press, New York, 1980, pp. 311–341.
21. J. R. Rodriguez, C. Rey, V. P. Villar, V. Torra, J. Ortin, and J. Vinals, *Thermochim. Acta* **63**, 331 (1983).
22. L. D. Hansen, R. M. Izatt, D. J. Eatough, T. E. Jensen, and J. J. Christensen, "Recent Analytical Applications of Titration Calorimetry," in *Analytical Calorimetry*, R. S. Porter and J. F. Johnson, Eds., Plenum Press, New York, 1974, Vol. 3, pp. 7–16.
23. C. J. Martin and M. A. Marini, *Critical Reviews in Analytical Chemistry*, CRC Press, Cleveland, 1979, Vol. 8, pp. 221–286.
24. C. J. Martin, B. R. Sreenathan, R. L. Berger, and M. A. Marini, Presentation at the 32nd Annual Calorimetry Conference, Sherbrooke, Quebec, Canada, 1977.
25. R. L. Berger, W. S. Friauf, and H. E. Cascio, *Clin. Chem.* **20**, 1009 (1974).
26. L. D. Bowers and P. W. Carr, *Thermochim. Acta* **10**, 129 (1974).
27. A. E. Van Til and D. C. Johnson, *Thermochim. Acta* **23**, 1 (1978).
28. D. J. Eatough, J. J. Christensen, and R. M. Izatt, *Thermochim. Acta* **3**, 219 (1972).
29. H. A. Skinner, *Experimental Thermochemistry*, Interscience, New York, 1962, Vol. 2, p. 169.
30. H. A. Skinner, in Ref. 29, pp. 179–182.
31. M. V. Kilday, *J. Res. Natl. Bur. Stand.* **85**, 449 (1980).
32. V. B. Parker, *Thermal Properties of Aqueous Uni-Univalent Electrolytes*, National Bureau of Standards, NSRDS-NBS 2, U.S. Government Printing Office, Washington, D.C., 1965.
33. J. D. Hale, R. M. Izatt, and J. J. Christensen, *J. Phys. Chem.* **67**, 2605 (1963).
34. C. E. Vanderzee and J. A. Swanson, *J. Phys. Chem.* **67**, 2608 (1963).
35. J. O. Hill, G. Ojelund, and I. Wadso, *J. Chem. Thermodyn.* **1**, 111 (1969).
36. L. D. Hansen and E. A. Lewis, *J. Chem. Thermodyn.* **3**, 35 (1971).

37. C. E. Vanderzee, D. H. Waugh, and N. C. Haas, *J. Chem. Thermodyn.* **13**, 1 (1981).
38. F. D. Rossini, "Introduction: General Principles of Modern Thermochemistry," in *Experimental Thermochemistry,* F. D. Rossini, Ed., Interscience, New York, 1956, Vol. 1, pp. 16–18.
39. J. Coops, R. S. Jessup, and K. Van Nes, "Calibration of Calorimeters for Reactions in a Bomb at Constant Volume," in *Experimental Thermochemistry,* F. D. Rossini, Ed., Interscience, New York, 1956, Vol. 1, pp. 28–35.
40. H. D. Johnston, Ph.D. Dissertation, Brigham Young University, Provo, Utah, 1968.
41. D. Smith-Magowan and R. H. Wood, *J. Chem. Thermodyn.* **13**, 1047 (1981).
42. L. D. Hansen and D. J. Eatough, *Thermochim. Acta* **70**, 257 (1983).
43. R. M. Hart and L. D. Hansen, Presentation at the 32nd Annual Calorimetry Conference, Sherbrooke, Quebec, Canada, 1977.
44. L. D. Hansen, R. M. Hart, D. M. Chen, and H. F. Gibbard, *Rev. Sci. Instrum.* **53**, 503 (1982).
45. J. J. Christensen and R. M. Izatt, *Thermochim. Acta* **73**, 117 (1984).
46. R. L. Biltonen and N. L. Langerman, *Methods in Enzymology,* Academic Press, New York, 1979, Vol. 61, pp. 287–318.
47. V. M. Poore and A. E. Beezer, *Thermochim. Acta* **63**, 133 (1983).
48. L. D. Hansen and R. M. Hart, *J. Electrochem. Soc.* **125**, 842 (1978).
49. C. Mudd, R. L. Berger, H. P. Hopkins, W. S. Friauf, and C. Gibson, *J. Biochem. Biophys. Meth.* **6**, 179 (1982).
50. A. S. Gordon, F. J. Millero, and S. M. Gerchakov, *Appl. Environ. Microbiol.* **44**, 1102 (1982).

CHAPTER

4

FLOW ENTHALPIMETRY (ISOPERIBOL FLOW-INJECTION CALORIMETRY)

R. S. SCHIFREEN

E. I. DuPont de Nemours & Co.
Biomedical Products Department
Wilmington, Delaware

The term flow enthalpimetry is used in this chapter to describe a small family of methods that measure a short-lived change in temperature induced in a flowing liquid by a chemical reaction designed to measure the concentration of an analyte of interest introduced as a discrete liquid sample. In terms of the nomenclature system proposed in Chapter 1, the definitive classification of the technique is isoperibol flow-injection calorimetry. The technique can be thought of as a marriage of flow-injection analysis with direct-injection enthalpimetry, although that concept does not follow the historical development of the methodology. This chapter will be devoted to developing a historical perspective and theoretical understanding of flow enthalpimetry. As stated in Section 7.9.1. the nomenclature used to describe this area is quite confused, and it is important to distinguish between the various types of thermal instrumentation that utilize thermistors, enzymes, and continuously flowing reagent streams. Flow enthalpimetry only includes those methodologies that analyze a *discrete sample* introduced into a *flowing stream* by measurement of a temperature change in an *adiabatic reactor*. This definition specifically excludes closely related systems such as heat-conduction calorimetry and thermal enzyme probes. In heat-conduction calorimetry the reaction chamber is essentially isothermal with the analytical signal being provided by heat flow through a thermopile detector. Although the flow system is similar to that used in flow enthalpimetry, the fundamental mechanism of measurement is quite different, as discussed in Chapter 3. Thermal enzyme probes are also based on a fundamentally different measurement principle than the flow enthalpimeter as described in Section 7.9.5. The major difference between these devices is in the presence or absence of a flow system for sample delivery and in the diffusion of both sample and heat within the analytical reactor.

Calorimetric instrumentation used for continuous process monitoring is described in Section 7.9.2 and may be similar to that used in flow enthalpimetry except for the use of continuous rather than discrete sampling. Although inferences regarding the fundamental characteristics of some of these systems can be made from an understanding of flow enthalpimetry, this must be done with caution as much of the theory developed for the latter is based on the measurement of discrete thermal peaks.

Flow enthalpimetry, like calorimetric analysis in general, has the advantage of being potentially applicable to any endo- or exothermic chemical system. Applications have been developed using a broad cross section of reaction chemistries. Sensitivity and specificity depends on the chemistry chosen to measure the analyte of interest and is totally under the control of the analyst. Whereas classical methods of calorimetric analysis are sometimes slow and tedious, flow enthalpimetry is designed for routine operation and rapid throughput. Current systems can analyze 60–80 samples per hour with sample volumes ranging from microliters to milliliters. Commercial instrumentation to perform flow enthalpimetry is not currently available, but can be easily constructed by an interested investigator with access to machine shop and glass-blowing facilities.

The purpose of this chapter is to familiarize the interested reader with the technique of flow enthalpimetry and impart an understanding of how and why it works. To date, most applications of the technology have been chosen to demonstrate its unique capabilities and to probe its fundamental mechanism. This pioneering work has progressed sufficiently so that the methodology has reached a degree of maturity where it must prove its ability to solve real analytical problems. It is hoped that this discussion will help to catalyze the movement of this technology from the research to the applications laboratory.

4.1. HISTORICAL DEVELOPMENT OF INSTRUMENTATION

At first glance, the devices classified as being historically related to flow enthalpimetry may appear quite diverse; however, they all share a number of structural and functional similarities. Evolution of the technology can be followed by improvements in the structure and organization of the various components that enhance the capabilities of the methodology.

A block diagram of an idealized flow enthalpimeter is shown as Fig. 4.1. Similarities to flow-injection techniques are immediately obvious. A discrete sample is injected into a flowing stream, mixed with a reagent, allowed to react, and the resulting change in temperature measured. A thermally controlled environment must be provided, since even small changes in normal ambient temperature are quite large compared to those generated by the analytical reaction.

Major improvements in the performance of these devices have been achieved through combination of the functions of two or more discrete components into a single unit. The importance of these changes will be reflected both in the development of improved instrumentation and in greater understanding of fundamental concepts.

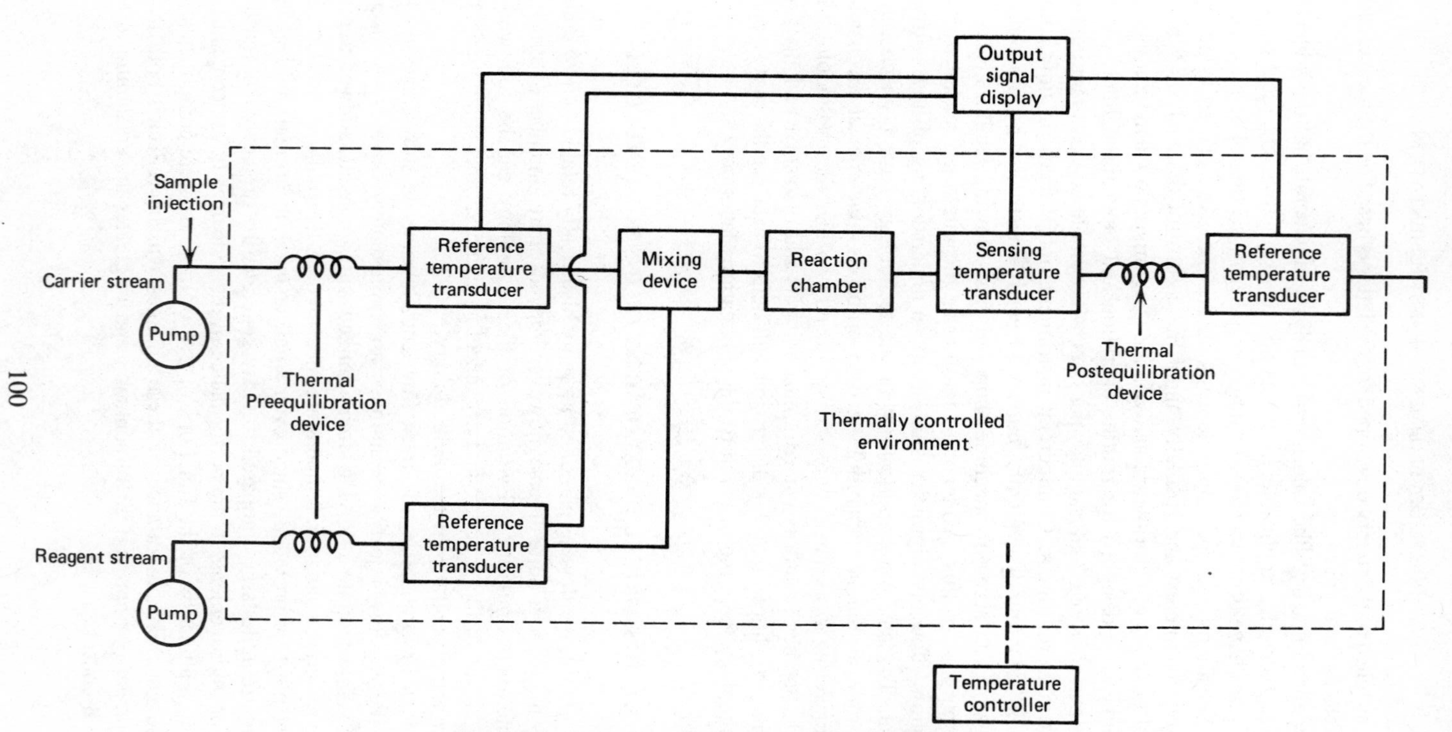

Figure 4.1. Block diagram of an idealized flow enthalpimeter.

4.1.1. Continuous-Flow Enthalpimetry

The first device which can be classified as a "flow enthalpimeter" was reported by Priestley et al. of Kodak Ltd. in 1965 (1). Their instrument consisted of a Perspex cylinder drilled so that opposing sample and reagent streams each passed through a small orifice into a reaction chamber as is shown in Fig. 4.2. This arrangement allowed for efficient mixing at delivery rates of 15 mL min^{-1} for each stream provided by a single multichannel peristaltic pump. Thermal measurement was achieved with a set of three thermistors monitoring the sample, reagent, and final reaction streams. Any two thermistors could be monitored differentially, which helped compensate for the lack of a temperature-controlled environment. All analyses were performed at steady state (see Section 4.4.5) with a minimum of 10 mL of sample.

A number of important concepts were described by these workers and incorporated into the design of their device for "continuous flow enthalpimetry." They recognized that the sample and reagent stream flow rates required careful and precise matching. The need to achieve efficient mixing was identified as was the importance of constructing the reaction chamber to have low thermal capacity and conductivity. This latter concept was confirmed over a decade later as one of the key determinants of instrument performance and one of the most difficult to achieve (Section 4.2.2.). Finally, the need to compensate for changes in ambient temperature was described, and the effectiveness of differential measurement in performing that function was demonstrated.

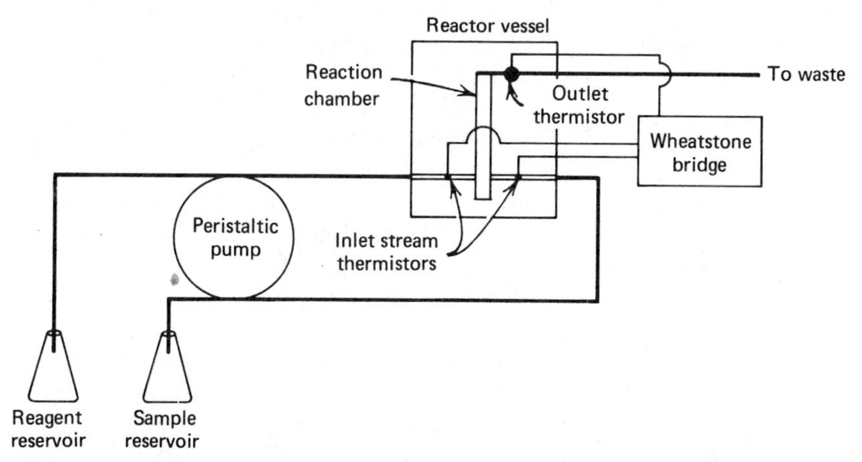

Figure 4.2. Schematic diagram of the instrument described by Priestley et al. (1).

4.1.2. Peak Enthalpimetry

Almost a decade later, Censullo and co-workers described the technique of "peak enthalpimetry" (2). Their device utilized a reaction chamber loosely packed with glass beads and agitated by an electromagnetic vibrator. The sample and reagent streams were introduced through a 2 m length of coaxial Teflon tubing, and the temperature of the final reaction stream was monitored by a single thermistor. The concentric sample and reagent tubes and the reaction chamber were all submerged in a water bath whose temperature was precisely controlled. Hydrostatic pressure was used to provide pulseless flow, with control being achieved by adjusting the height of the solution reservoirs. Sample introduction could either be continuous, by filling the sample reservoir with the analyte solution or discrete, by injection of a sample into an inert flowing stream. The advantages of simultaneous electronic integration of the thermistor-bridge output were recognized and utilized for quantitative measurement.

These investigators noted the importance of efficient mixing and of controlling and matching the flow rates in both solution channels. They also identified the system as a concentration-sensitive detector and the corresponding integrated output as a measure of total heat evolved or absorbed. The use of a controlled water bath both provided a stable environment and allowed for preequilibration of the solution temperatures prior to reaction.

The analysis of chloride in 0.1 mL samples of simulated serum was demonstrated with the "peak enthalpimeter." Chloride concentrations from 50 to 170 mmol L^{-1} yielded measurable peaks with baseline widths of approximately 2 min.

The "peak enthalpimeter" represented a transition from steady state to discrete sample measurement. It also marked a change from compensating for variations in ambient temperature to controlling the environment by instrumental means. While an important demonstration of capabilities, the thermal instrument could not match the sample throughput and flexibility of continuous flow spectrophotometric systems available at the time.

4.1.3. The Thermometric Analyzer

A somewhat different approach to continuous flow thermal analysis was taken by Peuschel, Hagedorn, and co-workers (3–5) and marketed by Technicon Instruments Corporation as the Thermometric Analyzer. This instrument combined the standard Technicon air-segmented sampling system with an equilibration coil and a 1 mL Teflon reaction cell immersed in a temperature-controlled bath. Mixing was achieved by a magnetic stirrer in the cell and the reaction temperature monitored by a single thermistor encased in the reaction vessel cover. Throughput was 20–40 samples per hour with resolution of temperature changes as small as 5 m°C. Key advantages of this system were its compatibility with the popular air-segmented flow system and the fact that it could be employed to measure analytes using reactions based on precipitaton. This would quickly clog all other instruments that perform flow enthalpimetry. The analyzer was designed to be especially useful for determining medium to high concentrations of analyte in complex sample matrices.

Conceptually, this instrument is similar to that described by Censullo and co-workers in that the reaction is monitored by a single thermistor and the thermal environment for both solution preequilibration and the reaction itself is electronically controlled. The flow systems are also quite similar except for the substitution of peristaltic pumps for hydrostatic flow and the use of air segmentation in the flow stream of the Thermometric Analyzer. Although the reactor vessels appear quite different, they were both designed to optimize mixing efficiency.

4.1.4. The Enzyme Thermistor

A key problem affecting the development of improved instrumentation to perform flow enthalpimetry was the need to achieve adequate mixing of the sample and reagent streams. This increased the complexity of the reaction vessel and required a trade-off against achieving optimal thermal characteristics. Mosbach and co-workers recognized that a reactor packed with enzyme immobilized to a solid support could substitute for the reagent stream and eliminate the need for mixing (6). Early designs of the "enzyme thermistor" incorporated a single flow channel with a sample injector leading to a flow-through immobilized-enzyme reactor and a single thermistor to detect the final temperature of the stream leaving the reactor. As with earlier designs, a preequilibration coil and the reaction vessel itself were immersed or embedded into a thermostatted environment. Later improvements included better thermal contact between the thermistor and reactant stream and the use of both differential thermistors and columns to further reduce noise due to ambient temperature fluctuations (7).

These changes produced a simpler and more reliable instrument. The use of a single sample stream avoided the need for both mixing and careful matching of dual reagent flow rates. Applications were limited, however, to reactions catalyzed by enzymes and choice of the enzyme support used in the reactor was an additional critical variable. Recently, the use of the "enzyme thermistor" has been extended to analyses utilizing solid-phase immunochemical reactors (8). A more-detailed discussion of the design and application of "enzyme thermistors" is presented in Section 7.9.1.

4.1.5. Thermal Enzyme Probe (Flow Configuration)

An interesting and different approach to continuous flow thermal measurement was taken by Weaver and co-workers (9). Their device incorporated a pair of differential thermistors, one of which was coated with a 100 μm layer of immobilized enzyme. The enzyme was immobilized to lyophilized polyacrylamide gel and glued to the thermistor with contact cement. Both thermistors were mounted in a stainless-steel flow-through cell and immersed in an oil bath controlled to 1 m°C.

The instrument was designed to minimize both thermal and flow-generated noise. Buffer flow was initiated and controlled by pressurizing a 3 L stainless-steel vessel with nitrogen. Bubble nucleation was prevented by placement of a vacuum degassing membrane interface designed for a mass spectrometer prior to the detection cell. Flow rates were optimized to yield a quiet, laminar liquid flow.

Thermal noise was reduced by employing matched thermistors in the differential detector cell. The two thermistors were placed directly opposite each other to ensure sensing the same instantaneous flow, but far enough apart to prevent heat generated at the enzyme-coated sensing thermistor from interfering with the reference thermistor. The thermal enzyme flow-through probe was capable of resolving temperature changes of 2–3 μ°C corresponding to a sample of 10^{-4} mol L^{-1} urea. Calibration of analyte concentration with temperature change was non-linear, and analyses typically required 1–3 min per sample. Predictably, application of "thermal enzyme probes" has been confined to biochemical and clinical analyses. A more-detailed discussion is provided in Section 7.9.5.

4.1.6. The Flow Enthalpimeter

Carr and co-workers also designed an instrument that avoided mixing by use of an immobilized-enzyme reactor (10). The original flow enthalpimeter was similar to Mosbach's "enzyme thermistor" except for important differences in the design of the thermal reactor. This reactor, which will be described in detail in the next section, was a thin-glass tube surrounded by a vacuum jacket. It was designed to have minimal thermal capacity and conductivity as first recommended by Priestley (1). Other related improvements were the use of controlled pore glass as the enzyme support and the location of the sensing thermistor in direct proximity to the reactor bed.

A second-generation instrument was developed which allowed for the elimination of the temperature-controlled water bath (11, 12). It was designed so that a stable and uniform thermal environment could be achieved by stirring and by careful insulation of a cylindrical water bath. This eliminated both the expense and inconvenience of using electronic temperature control and substituted a slow drift in temperature for cyclical thermal fluctuations characteristic of electronic temperature controllers. This slow drift was compensated for by use of a second reference thermistor located in the flow stream as originally described by Priestley.

The most recent design of the immobilized enzyme flow enthalpimeter introduced a number of modifications for convenient use and maintenance and further improvements to the reactor to enhance its thermal characteristics (13). A schematic illustration of this instrument is shown in Fig. 4.3.

An instrument capable of utilizing two fluid streams was also designed and constructed (14). This device incorporated a reactor column packed with inert glass beads to allow mixing of the reactants initiated in a tee fitting just prior to the column. The reference thermistor was moved as shown in Fig. 4.4, so that both thermistors were exposed to exactly equal flow rates. In addition, switching of the instrument between the immobilized-enzyme and soluble reagent configurations was facilitated by the use of a modular design and simple fittings.

This third-generation instrument allows for a throughput of up to 60–80 samples per hour with a 120μL sample volume. It is capable of measuring temperature changes as small as 0.5 m°C generated by a wide variety of different reaction mechanisms and will form the basis for the remainder of this discussion.

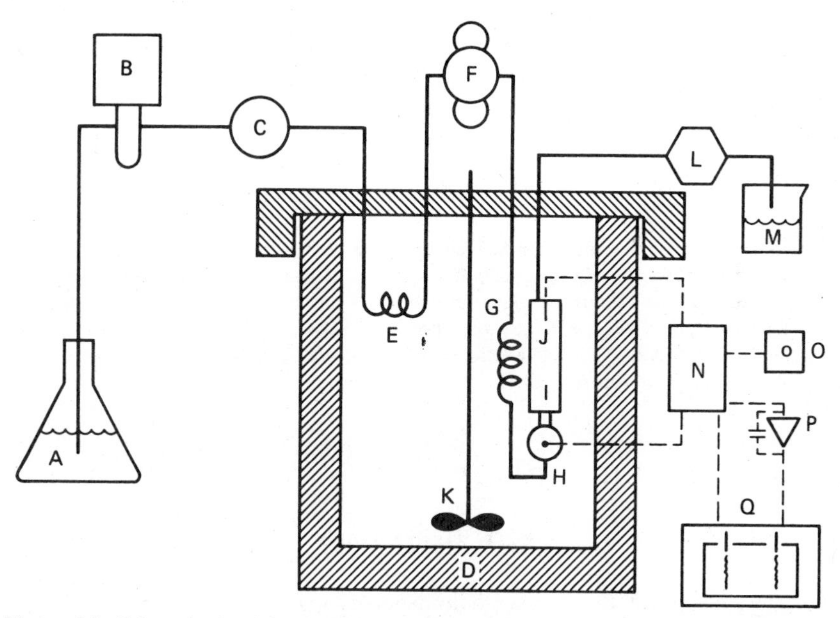

Figure 4.3. Schematic drawing of the immobilized enzyme flow enthalpimeter: (A) reservoir, (B) pump, (C) pulse damper–pressure gage, (D) insulated water bath, (E) preequilibration coil, (F) sample injection valve, (G) equilibration coil, (H) reference thermistor, (I) adiabatic column, (J) reference thermistor, (K) stirrer, (L) flowmeter, (M) waste, (N) bridge electronics, (O) oscilloscope, (P) intergrator, (Q) recorder. (Reprinted with permission from Ref. 13. Copyright 1977 American Chemical Society.)

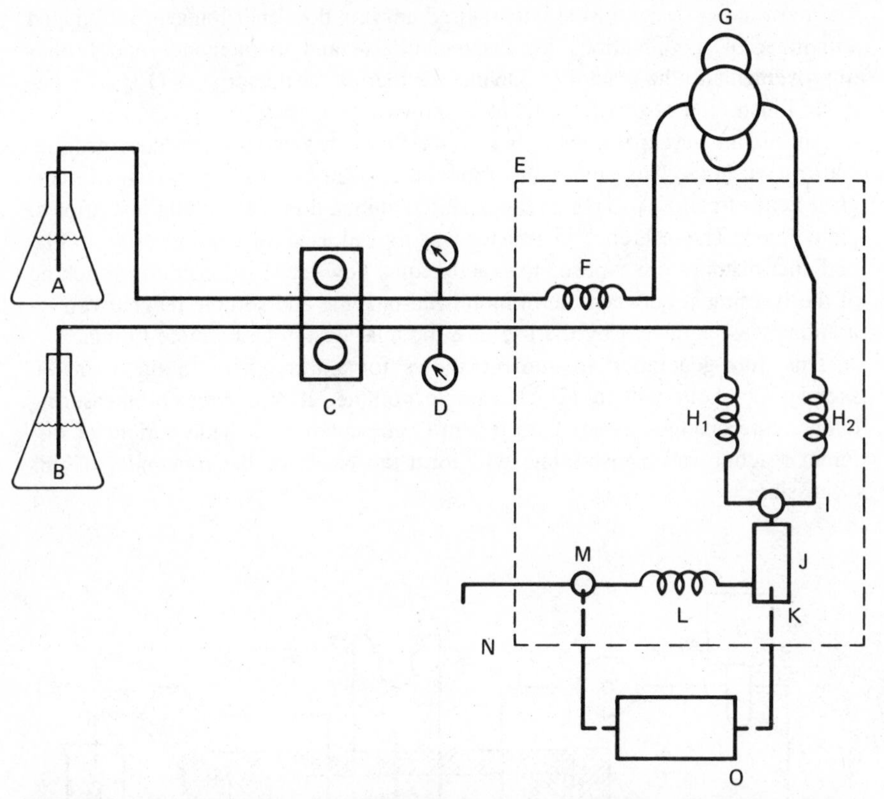

Figure 4.4. Schematic drawing of the flow enthalpimeter designed for use with soluble reagents: (A) sample buffer reservoir, (B) reagent buffer reservoir, (C) pump, (D) pulse damper–pressure gage, (E) insulated water bath, (F) preequilibration coil, (G) sample injection valve, (H_1, H_2) matched equilibration coils, (I) mixing tee, (J) reactor column, (K) sensing thermistor, (L) equilibration coil, (M) reference thermistor, (N) bridge electronics, (O) oscillator. (Reprinted with permission from Ref. 14. Copyright 1979 American Chemical Society.)

4.2. FLOW ENTHALPIMETERS—DESIGN AND CONSTRUCTION

4.2.1. Design Goals

The latest design of the flow enthalpimeter evolved as various practical and theoretical needs were recognized and addressed. Most fundamental is the requirement that the analytical system provide sufficient sensitivity, reproducibility, and throughput to determine a variety of analytes on a routine basis. Fortunately, all of the above are complementary and are dependent on the achievement of optimum thermal and flow characteristics. In short, one must endeavor to detect as much of the heat evolved or absorbed by the reaction as possible by preventing heat transfer in the reactor either through its walls or by diffusion or dispersion through the flow stream. Ultimately, this will result in large, narrow analytical peaks similar to those sought in chromatography, flow-injection analysis, segmented flow analysis, and other continuous flow systems.

A second concern is practicality. It is an unfortunate fact that all instrumentation requires maintenance and repair, especially when it is used on a routine basis. In the flow enthalpimeter servicing is facilitated by the use of robust, modular components. Since virtually all of the working components are submerged in a water bath, there is an obvious challenge in controlling corrosion. This design also mandates that as much testing of critical components as possible be accomplished without dismantling the unit in order to avoid disturbing the thermal equilibrium which must develop for efficient operation.

A third consideration is that the flow enthalpimeter be readily convertible between the immobilized-enzyme reactor and soluble reagent configurations. Detailed descriptions of the instruments' construction have been published and should be consulted by the interested investigator (13, 14). Schematic drawings of the instrument in both configurations appear as Figs. 4.3 and 4.4. Important considerations relating to the construction and operation of the instrument are summarized in Table 4.1.

4.2.2. The Thermal Reactor

The unique adiabatic vacuum-jacketed glass thermal reactor shown in Fig. 4.5 is the heart of the flow enthalpimeter (13). Reactors are typically 3–12 cm long with 3–4 mm internal diameters and volumes of 0.4–0.8 mL. They must be packed with a support material such as 200–400 mesh controlled pore glass for use with immobilized enzymes or inert 80–120 mesh solid glass beads for use with mixed reagent streams. Non-compressible supports are preferred, because of their superior flow characteristics. Compressible supports such as cellulose and agarose exhibit a variable resistance to flow, which results in increased back pressure and baseline noise.

Adiabaticity is a key determinant of reactor performance. Thermal interactions with the environment reduce the temperature change induced in the flowing stream by the analytical reaction and, consequently, decrease sensitivity. Effective insulation is provided by the vacuum jacket surrounding the reactor. Early designs incorporated a silvered outer wall, but this was subsequently found to be unnecessary. At optimum flow rates the reactor retains approximately 66% of the heat generated within its walls. In the soluble reagent mode heat is also generated within the precolumn mixing tee. This results in an additional 20%

TABLE 4.1. Instrumental Parameters for Optimal Performance of the Flow Enthalpimeter

	Optimum	Comments
Thermal reactor (Section 4.2.2)		
Length	3–12 cm	Must be sufficient for reaction to reach completion.
Internal diameter	3–4 mm	Minimize dispersion due to interactions with the reactor wall (see Section 4.5.4).
Internal wall thickness	1 mm	Reduce heat loss and peak tailing due to thermal mass.
Insulation	Vacuum	Effective insulation.
Water bath (Section 4.2.3)		
Volume	10L	Provide a stable thermal environment.
Stirrer	600 rpm with interrupted shaft	Maintain thermal homogeneity.
Thermal equilibration coils (Section 4.5.4)		
Length	1 m	Provide adequate thermal equilibration with minimal dispersion (see Section 4.5.4).
Internal diameter	0.5 mm	
Internal volume	200 μL	
Sample volume	100–750 μL	Yield adequate sensitivity and minimum peak width (see Section 4.4.4).
Flow rate	1–4 mL min^{-1}	Balance the effects of heat loss and dispersion (see Section 4.5.2).

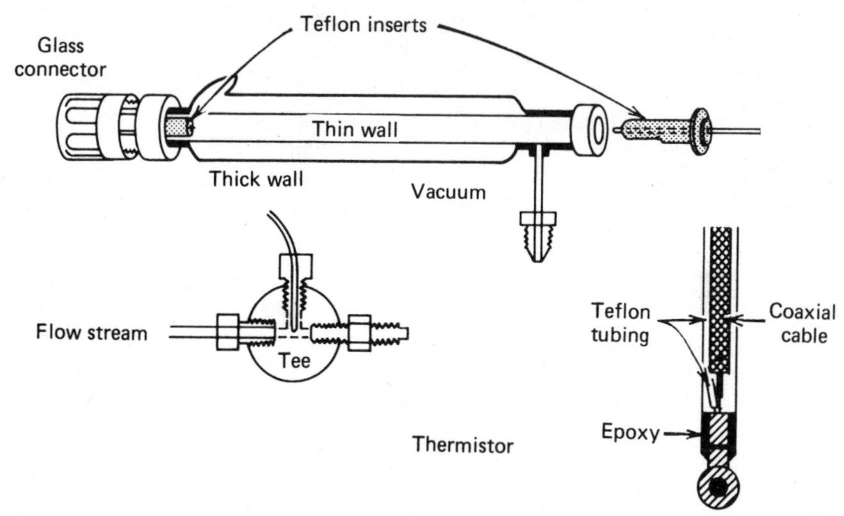

Figure 4.5. Column reactor. (A) Thin-walled vacuum-jacketed column, (B) reference thermistor assembly, (C) sensing thermistor insert, (D) details of thermistor probe assembly. (Reprinted with permission from Ref. 13. Copyright 1977 American Chemical Society.)

loss of the reaction heat despite a residence time in the tee and connection fittings of less than 1 s. Considering the very small volume contained in these reactors, both configurations are quite efficient (15).

The importance of minimizing thermal mass is less obvious, but equally important. Consider a heated segment of the flowing stream in thermal contact with a segment of the reactor wall for a short period of time. During that time heat will diffuse into the reactor wall. When the heated segment has passed, heat will diffuse from the wall back into the flowing stream and the water bath resulting in both thermal peak tailing and reduced sensitivity. Conceptually, this condition is identical to the classical case of a thermal conductor connecting two bodies with different temperatures. A thermal gradient forms through the conductor, and heat will flow from the warmer to cooler body until the temperatures are equalized or the mismatch is removed by an external perturbation of the system. Likewise, as long as a heated segment of the flow stream is in thermal contact with a segment of the cooler reactor wall, heat will flow from the flow stream to the reactor vessel. In this case the heated flow segment moves past the reactor wall before complete equilibration can be achieved. The warmer reactor wall is then in thermal contact with both the now cooler flowing reagent stream and the surrounding water bath and will lose heat to each. Thermal energy diffusing from the reactor into the water bath is lost permanently from the measurement system, whereas that heat diffusing back into the flowing stream increases its temperature and is detected as thermal peak tailing. These processes are shown in Fig. 4.6.

The thermal mass of the reactor wall determines its capacity to store heat. The magnitude of the thermal perturbations to the measurement system described above are proportional to the amount of heat initially transferred to the reactor wall while it is in thermal contact with the heated sample stream segment. These undesirable effects can, therefore, be reduced by minimizing the thermal mass or capacity of the reactor wall and reducing the time a sample stream segment and the reactor wall are in thermal contact.

Thermal mass in the reactor is minimized by constructing the inner walls of the adiabatic reactor of thin (1 mm) glass tubing and by placement of the sensing thermistor directly against the base of the reactor column. Surrounding the inner reactor wall by a vacuum allows for both efficient insulation and low thermal capacity. The time during which there is thermal contact between the heated stream and reactor is minimized by proper choice of flow rate as described in Section 4.5.2.

The importance of this aspect of reactor design is evident in comparing the performance of an instrument in which the heated sample stream was in contact with several millimeters of plastic fittings before reaching the sensing thermistor

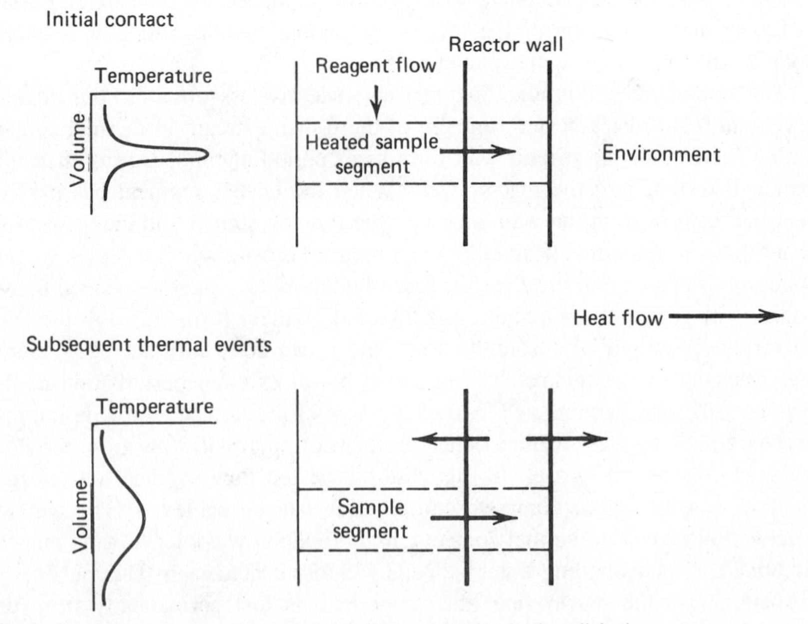

Figure 4.6. Model for heat loss and peak tailing originating in the adiabatic reactor.

(11, 12). Changing to the most recent reactor design resulted in a threefold enhancement in sensitivity and halving of the peak-width-at-half-maximum height.

4.2.3. The Water Bath

The water bath is designed to provide a homogeneous thermal environment with minimal short-term fluctuations in temperature. Once steady state is achieved, the temperature will drift slowly due to thermal interactions with the environment and heat generated by stirring. This slow drift is compensated by the use of the differential measurement system.

Four factors are critical to the function of the water bath: a tight seal, effective insulation, efficient stirring, and adequate thermal mass. In addition, the bath should be constructed to facilitate disassembly and provide proper electrical shielding.

The bath housing is constructed in two sections. The upper section contains the stirring mechanism and a suspended plastic cage which supports the components of the flow system. This section can be lifted off and out of the bath itself and will stand on the plastic cage for servicing. The lower section contains a 10 L stainless-steel container, which is filled with water. Both the upper and lower sections are insulated with at least 10 cm of styrofoam on all sides and covered by aluminum shells. A 600 rpm synchronous motor drives a three-dimensional Teflon-coated steel stirrer. It is important that the stirrer shaft be interrupted by a plastic coupling device above the water line in order to prevent conduction of both heat and electrical current to or from the bath along the steel rod. The upper and lower sections are sealed by a rubber gasket and an aluminum lip extending down from the upper section.

Both the stainless-steel bath container and aluminum outer instrument shells are connected to the common of the measurement electronics. The synchronous motor and its aluminum housing are electrically isolated from the rest of the instrument and connected directly to earthen ground.

The water bath usually requires about 24 h to reach steady state after being filled with deionized or distilled water at room temperature. Once steady state is reached, typical short-term thermal noise is 800 μ°C and drift is 9m°C h^{-1} as measured differentially by two thermistors on opposite sides of the stirred bath. Short-term noise is further dampened by the differential design of the flow system.

4.2.4. Thermal Equilibration Coils

It is critical that the temperature of the flowing stream be precisely matched to that of the water bath. The usual approach of allowing a long equilibration time is not appropriate for flow enthalpimetry owing to mass dispersion in the flow stream. Equilibration time is reduced by the use of 1m lengths of 0.5-mm-ID stainless-steel hypodermic needle tubing formed into a 1 cm coil. Adequate thermal equilibration is achieved with an internal volume of approximately 200 μL and an associated residence time of only 12 s at a flow rate of 1 mL min^{-1}

4.2.5. Temperature Measurement

As stated above, temperature measurement is accomplished by a pair of thermistors arranged in a differential configuration. Glass-probe-encapsulated thermistors are sealed into the end of a length of 1.6-mm-OD Teflon tubing and connected to miniature coaxial cable (13). This thermal probe is waterproof and has an expected lifetime of about 6 months. Thermistors need not be matched beyond what is necessary to balance the bridge, although it is suggested that pairs be assembled from the same production lot.

The sensing thermistor is placed within the adiabatic region of the thermal reactor against the end of the reactor bed. This arrangement improves sensitivity and isolates the sensor from thermal fluctuations of the bath. Thermal isolation of the reference thermistor is accomplished by its placement in a plastic tee fitting, the thermal mass of the fitting acting to insulate the probe from short-term temperature fluctuations. This arrangement results in a peak-to-peak baseline noise of 50 μ°C and drift of 0.5 m°C h^{-1}, only approximately 6% of the noise in the bath itself.

The temperature difference as measured by the thermistors is converted to a unbalanced voltage by an ac phase-lock bridge driven at 1.0 V peak-to-peak with a 2 kHz sine wave (16). The need to provide adequate shielding from the environment would make use of a dc bridge extremely difficult. BNC connectors between the probes and detector electronics allow for *in situ* testing of the thermistors.

It is important that both thermistors are exposed to exactly the same flow rate. This need arises due to heat generated from the small currents that flow through the devices. With the immobilized-enzyme configuration the reference thermistor can be located in front of the reactor, since only a single reagent stream is present. In the soluble reagent configuration the flow rate at the reactor outlet is the sum of the individual inlet flow rates. Placement of the reference thermistor in any one of the inlet streams is undesirable, and it therefore must be placed in the column effluent after it has passed through an additional thermal equilibration coil.

4.2.6. Flow Components

The flow enthalpimeter routinely operates at flow rates of 0.5–3.0 mL min^{-1} with back pressures of 5–20 psi. These requirements are easily met by any of a number of peristaltic or metering pumps. Multichannel peristaltic pumps are especially useful with the soluble reagent configuration. Flow pulsations, which would be detected by flow-sensitive thermistors, are effectively dampened by adding a tee connector with the third port connected to a coiled, sealed 1.5 m length of Teflon tubing followed in series by a bellows-type pressure gage. It would also be possible to use pumps designed for HPLC or pressurized-reservoir vessels; however, these approaches would not improve performance and would be considerably more expensive or inconvenient.

The instrument flow system is constructed of 0.5-mm-ID Teflon tubing and zero dead volume, chemically inert, fittings. The number of connections and the length of tubing are kept to a minimum to reduce dispersion. Narrower, 0.3-mm-ID tubing was found not to significantly reduce dispersion and to clog when used with serum samples.

Both slider valves and septum injectors were determined to be suitable for sample introduction. The slider-type sampling valves yielded better precision and lent themselves more readily to automation. The newer rotary sampling valves would also be acceptable.

4.3. METHODOLOGY

4.3.1. Operation

Thermal equilibration of the flow enthalpimeter requires starting the pump and stirring motor 5 min prior to performing analyses. Diluted samples are injected through the sampling valve and thermal peaks are detected 0.5–2 min later, depending on conditions. A typical recorder tracing appears as Fig. 4.7. The signal could also be processed for automatic calculation by analog or digital means. Routine maintenance is limited to keeping the system clean and flushing the lines at the conclusion of each day.

It is critical that pH and ionic strength perturbations not be introduced into the system. Changes of this type perturb the equilibrium between the flow buffer and reactor bed and result in extraneous thermal events. This can be avoided by preparing flow buffers with sufficient buffer capacity and by diluting three parts sample with one part buffer concentrate. By preparing the flow buffer as a fourfold dilution of the same buffer concentrate precise matching is ensured. To avoid viscosity-related flow noise the sample and reagent buffers should be as dilute as possible.

A different approach to this problem was taken by Mosbach and co-workers (7). They designed a system with two separate, but identical flow channels. One reactor column contained an inert support and the other enzyme immobilized to

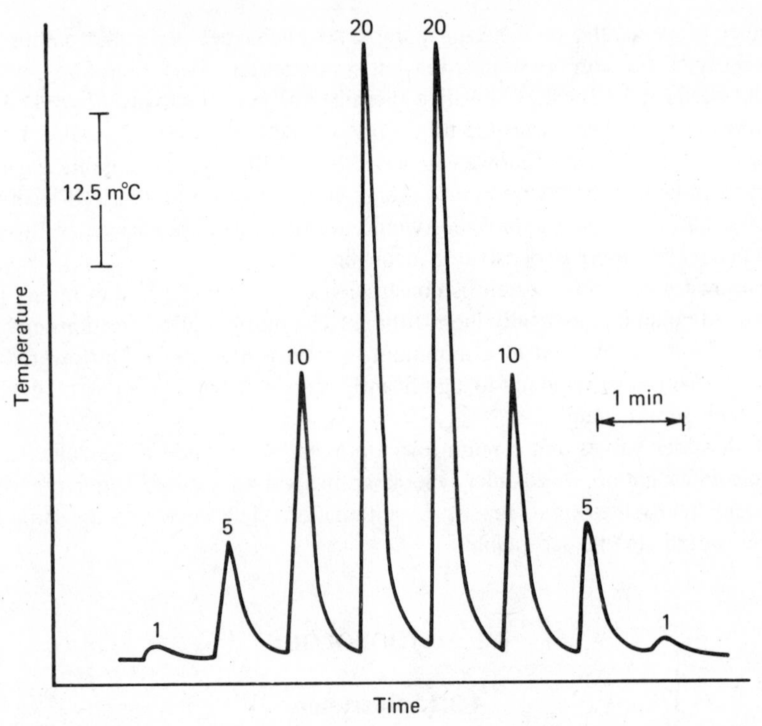

Figure 4.7. Recorder tracing for the neutralization of hydrochloric acid by THAM, 120-μL samples, flow rate = 2.5 mL/min, concentrations in mmol/L. (Reprinted with permission from Ref. 14. Copyright 1979 American Chemical Society.)

that same support. This differential arrangement allowed for compensation of the non-specific heats associated with a buffer mismatch between the sample and flow stream (see Section 7.9.1).

4.3.2. Applications

The published applications of flow enthalpimetry are reviewed in Chapters 6 and 7. In general, all that is required to perform an analysis by flow enthalpimetry is that the analytical reaction system be in aqueous phase with either soluble reactants or immobilized enzymes and that sufficient heat be evolved or absorbed to yield a peak height of 1–5 m°C. Thermal indicator reactions are sometimes useful in obtaining the needed sensitivity. The use of reusable immobilized enzymes can result in considerable savings in cost over the equivalent solution chemistry in the long term.

The importance of choosing conditions that allow the reaction to reach completion in the analytical reactor is often not appreciated. Carr has shown that, for a given analytical system, critical variables such as temperature and pH have a relatively greater influence on the rate constant than on the associated equilibrium constant (17). A reaction that proceeds to completion can therefore be expected to yield better precision and more reproducible assays over long periods of time and changing batches of reagents. In addition, sensitivity is obviously proportional to the extent of reaction.

Flow enthalpimetry is ideally suited to samples with poor optical qualities. In some cases spectrophotometric reaction systems can be simplified to eliminate coupling to a chromophore, reducing both cost and susceptibility to interferences. These classical advantages of thermal analysis, coupled with the convenience and speed of flow enthalpimetry, make it a tool that should be considered for the routine analytical laboratory.

4.4. REACTOR MODELS

The theory describing flow enthalpimetry is a combination of concepts which have evolved for both flow and calorimetric analysis (13). An exact closed-form mathematical model for this system has not been developed and, therefore, certain assumptions must be made. The use of these simplifications makes it possible to provide working theoretical descriptions of the flow enthalpimeter, but can also introduce error if not used with care.

The fundamental assumption which defines the theoretical model of the flow enthalpimeter is that conversion of sample to product occurs instantaneously. In fact, this assumption is not unreasonable. The chemical reactions such as acid–base neutralization, complexation, and oxidation–reduction which have been adapted to the flow enthalpimeter are relatively fast. It is also not unusual for immobilized-enzyme reactors to contain 1000 IU of enzymatic activity (see Section 7.9) which will convert a concentration of substrate equal to the enzymatic K_m in less than 12 s. Slow reactions would be expected to alter the peak shape and change the performance of the instrument.

For the sake of simplicity it is also assumed that the reactor bed equilibrates instantaneously to the temperature of the flowing stream and that the detector also responds very rapidly to the temperature of the fluid. Both of these assumptions are justified by the actual performance of the system.

The theoretical derivation only assumes the use of a concentration-sensitive detector, not necessarily a thermal transducer. Analogous equations can easily be written to describe spectrophotometric, refractive index, electrochemical, and other concentration-dependent detection devices.

4.4.1. Pure Plug Flow

The pure plug flow reactor represents the ideal case where there is no axial mixing of the reactant plug with the fluid around it. The sample flows through the column or packed bed with an infinitely sharp leading and trailing edge. Since there is no mixing, there is no dilution to consider. The extent of reaction is identical to that which would occur in a batch reactor under the same conditions.

Assuming perfect thermal characteristics, the behavior of an analytical pure plug flow reactor with thermal detection is shown in Fig. 4.8*a*.

The height of the temperature pulse, $\Delta T_{\max}$, in Celcius is given by

$$\Delta T_{\max} = \frac{\Delta H_R[\mathrm{S}]}{C^*_{P,mp}} \tag{4.1}$$

where [S], ΔH_R, and $C^*_{P,mp}$ designate sample concentration, reaction enthalpy, and the heat capacity of the mobile phase fluid per unit volume. This derivation assumes that the heat capacities of the sample and mobile phases are identical, which is true in most cases.

It is interesting to note that the predicted temperature change is dependent only on concentration and reaction chemistry and is independent of sample volume and the volume of the reactor. The volume of the reactor does, however, influence the residence time. Residence time t_R is a function of flow rate F and the reactor volume V_R:

$$t_R = \frac{V_R}{F} \tag{4.2}$$

The sample volume V_S determines the peak width in time $t' - t_R$ according to

$$t' - t_R = \frac{V_S}{F} \tag{4.3}$$

4.4.2. Continuously Stirred Reactor

The other extreme limit of behavior is the perfectly well-mixed reactor. In this reactor it is assumed that any sample or fluid entering the reactor is instantaneously mixed with all of the fluid resident in the reactor. Physically this is analogous to an exponential dilution flask. Assuming instantaneous conversion of reactant to products the reactor functions as shown in Fig. 4.8*b*. The instantaneous temperature difference from the baseline ΔT_t is

$$\Delta T_t = \Delta T_0 \exp - \left[\frac{FC^*_{P,mp}}{\overline{C}^*_P V_{\text{tot}}}\right] t \tag{4.4}$$

where ΔT_0 is the initial magnitude of the temperature peak, $\overline{C}^*_P$ is the mean heat capacity of the reactor and contents, and V_{tot} is the total volume. The initial temperature rise ΔT_0 is given as

$$\Delta T_0 = \frac{\Delta H_R[\text{S}]V_S}{\overline{C}^*_P V_{\text{tot}}} \tag{4.5}$$

Unlike the pure plug flow reactor, the maximum temperature change occurs instantaneously upon introduction of the sample. Peak width is a function of the reactor constant τ_R calculated as

$$\tau_R = \frac{V_R}{F} \tag{4.6}$$

A critical difference between these models is that the peak width for the continuously stirred reactor is determined by the reactor volume whereas it is determined by sample volume in pure plug flow.

Figure 4.8*c* illustrates the potential effect of changing the reaction kinetics on the theoretical performance of the reactor. In this case the reaction proceeds to completion by a zero-order mechanism at a finite rate expressed by k_0. This model has the unique property of relating peak width to initial sample concentration $[\text{S}_0]$. The peak-width-at-half-maximum height $\Delta t_{1/2}$ in time units is given by

$$\Delta t_{1/2} = \tau_R \ln\left(2 + \frac{[\text{S}_0]}{k_0 \tau_R}\right) \tag{4.7}$$

These models are important in defining the theoretical limits of performance for any given system. For example, the pure plug flow model represents the theoretical maximum limit of sensitivity and the continuously stirred reactor model represents the limit of sample dilution. A real system must fall somewhere in between.

4.4.3. Axially Dispersed Plug Flow

One model commonly used for describing chromatographic systems assumes axial dispersion following a Gaussian distribution as a sample moves through the flow reactor. Figure 4.8*d* depicts this system subject to the assumptions of

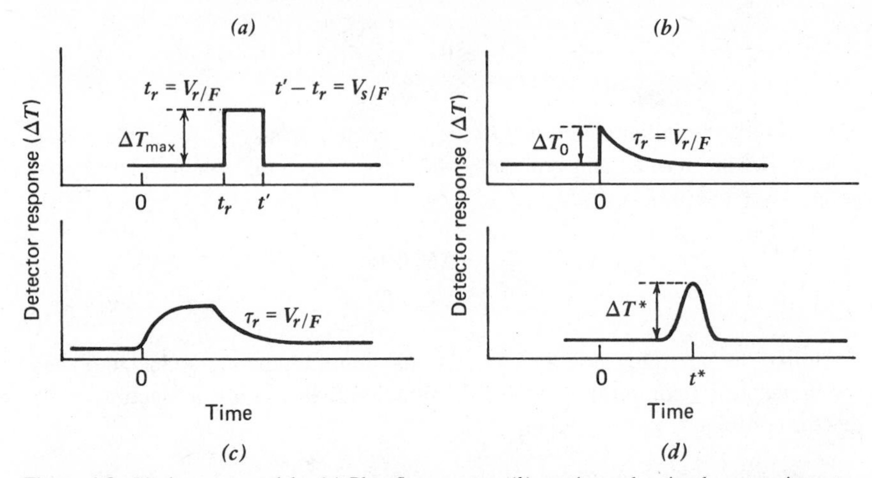

Figure 4.8. Ideal reactor models. (*a*) Plug flow reactor; (*b*) continuously stirred reactor, instantaneous conversion; (*c*) continuously stirred reactor, zero-order conversion; (*d*) axially dispersed plug flow. Sample is injected at time = 0. (Reprinted with permission from Ref. 13. Copyright 1977 American Chemical Society.)

instantaneous sample conversion to product, ideal thermal characteristics, radial homogeneity, and the usual approximations involved in considering the plate model developed for chromatography. For an infinitely narrow sample function of finite volume:

$$\Delta T_t = \Delta T^* \exp - \left(\frac{\mathcal{N}(t - t^*)^2}{2\, t^{*2}} \right) \tag{4.8}$$

and

$$\Delta T^* = \frac{\Delta H_R [\mathrm{S}] V_S}{C^*_{P,mp} V_{el} \sqrt{\mathcal{N}/2\pi}} \tag{4.9}$$

where ΔT^* is the maximum deviation from baseline that occurs at time t^* which is analogous to the retention time of a chromatographic peak, $\mathcal{N}$ is the number of "thermal" plates in the reactor, and V_{el} corresponds to the volume of fluid required to "elute" the heat peak through the column.

It is important to note that although these equations are written for a thermal detector, they could be defined in terms of any concentration-sensitive detector. If the system behavior were dictated by mass- rather than heat-transfer rates, then both $\mathcal{N}$ and V_{el} would refer to the corresponding chemical species rather than heat. The thermal diffusivity of water is 0.0014 $cm^2\ s^{-1}$, about 100 times as fast as a small molecule. Axial dispersion therefore includes components related to both thermal- and mass-transfer processes.

Using this model it can be shown that $W_{1/2,cl}$ the peak-width-at-half-maximum height due to dispersion in the column is related to the column variance in volume units, $\sigma^2_{V,cl}$, by

$$W_{1/2,cl} = \sqrt{8 \ln 2}\, \sigma_{V,cl} \tag{4.10}$$

and that

$$\Delta T^* = 2\sqrt{\ln 2/\pi}\, \frac{\Delta H_R[\mathrm{S}]V_S}{C^*_{P,mp} W_{1/2,cl}} \tag{4.11}$$

This equation predicts that the peak height will increase indefinitely in proportion to the added volume of sample. Clearly, this relationship is only true for a limited range of sample volumes.

4.4.4. Convoluted Axially Dispersed Plug Flow Model for Finite Sample Volumes

The apparent contradiction inherent in Eq. (4.11) arises from the assumption of an infinitely narrow sample distribution of finite volume. Mathematically, this is expressed as an impulse function, which is defined as having unit area. In reality, the width of the sample input must be proportional to volume and can only approximate an impulse function at low sample volumes.

This problem was addressed by Sternberg (18) who concluded that when the input function becomes sufficiently wide the system output can no longer be considered Gaussian. Qualitatively, as the input distribution is made wider, due to increasing sample volume, the peak height increases to some maximum value, a plateau develops in the output curve, and the peak width gradually increases. Alternatively, the system can be viewed as attaining a steady state. This is illustrated in Fig. 4.9 and has been shown to provide an excellent fit to experimental observations.

Sternberg's equation, translated in terms of temperature and time, may be written

$$\Delta T_t = \frac{\Delta T_{\max}}{2}\left[\operatorname{erf}\left(\frac{t - t^*}{\sqrt{2}\,\sigma_{t,cl}}\right) - \operatorname{erf}\left(\frac{(t - t^*)\,\Theta}{\sqrt{2}\,\sigma_{t,cl}}\right)\right] \tag{4.12}$$

where $\sigma^2_{t,cl}$ refers to the variance in time units due solely to the Gaussian column processes and Θ is the input plug width, in time units, defined as

$$\Theta = \frac{V_S}{F} \tag{4.13}$$

As sample volume decreases, that is, as Θ approaches zero, one obtains Eq. (4.8) representing pure Gaussian peaks.

The peak height $\Delta T'_{\text{peak}}$ is

$$\Delta T'_{\text{peak}} = \Delta T_{\text{max}} \operatorname{erf}\left(\frac{\Theta}{2\sqrt{2}\,\sigma_{t,cl}}\right) \tag{4.14}$$

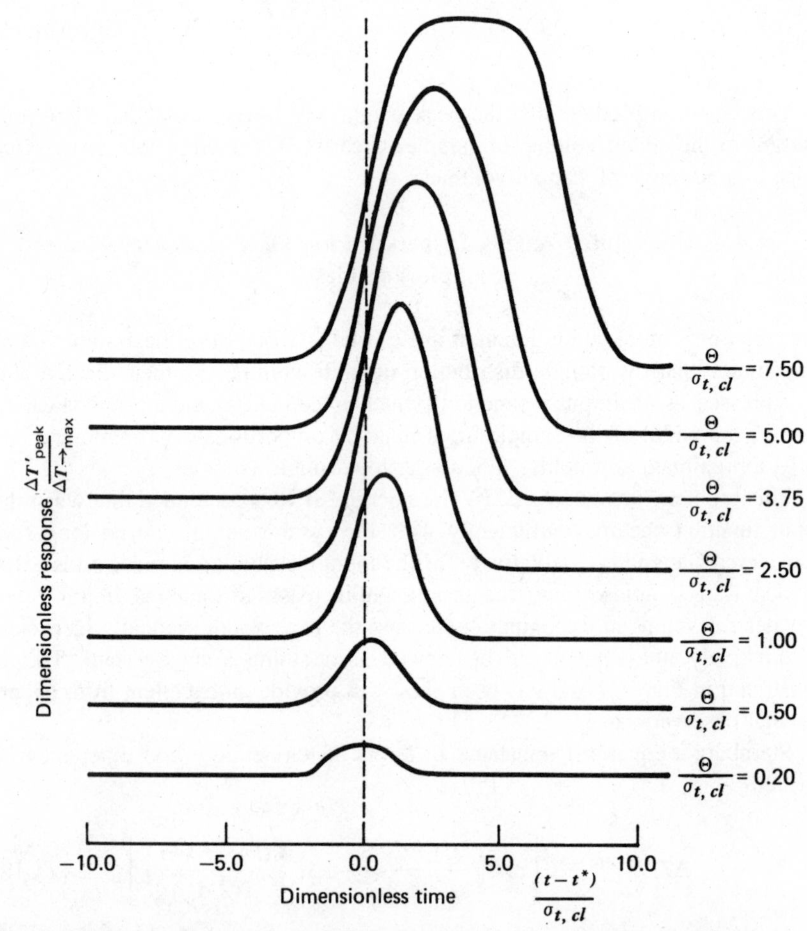

Figure 4.9. Theoretical signal-time curves and their dependence on sample volume. All curves computed from Eq. (4.13) with $\sigma_{t,cl}$ = 0.20 min and t^* = 1.0 min. (Reprinted with permission from Ref. 13. Copyright 1977 American Chemical Society.)

Sternberg has shown that the peak is symmetric about its maximum, which will occur when

$$t = t^* + \frac{\Theta}{2} \tag{4.15}$$

and that the total peak variance is given by

$$\sigma^2_{V_{tot},cl} = \sigma^2_{V,cl} + \frac{V_S^{\,2}}{12} \tag{4.16}$$

For very large sample volumes the peak half-width becomes equal to V_S or

$$W_{1/2,V} \cong V_S; \quad \text{when } V_S > 3\sigma_{V,cl} \tag{4.17}$$

A numerical analysis of the relationship defined by Eq. (4.16) and (4.17) is shown in Fig. 4.10.

Analysis of these models leads to important fundamental concepts that set the theoretical limits to performance in all continuous flow systems. Peak height can be increased by increasing sample volume only until steady state is achieved and then the peaks become flat-topped. Further increases in sample volume will decrease throughput by increasing peak width without any improvement in sensitivity. Likewise, throughput can only be improved by reducing sample volume until the limit of the Gaussian reactor model is reached. Further decreases in sample volume will continue to decrease sensitivity, but will not change peak width.

4.4.5. Equilibrium and Steady State

Equilibrium and steady state are simple concepts despite the confusion that currently exists in the continuous flow analysis literature. Equilibrium should be understood strictly as a thermodynamic concept. A flow system in which the reaction goes to completion is an equilibrium system, regardless of the shape of the peaks that appear at the detector. If the reaction does not go to completion, the system is kinetic and can be further classified according to the scheme recommended by Pardue (19).

Steady state is a term which defines a condition where the response of interest is no longer changing. It generally follows a period where the response either increases or decreases, designated as the transient state. The dividing line between

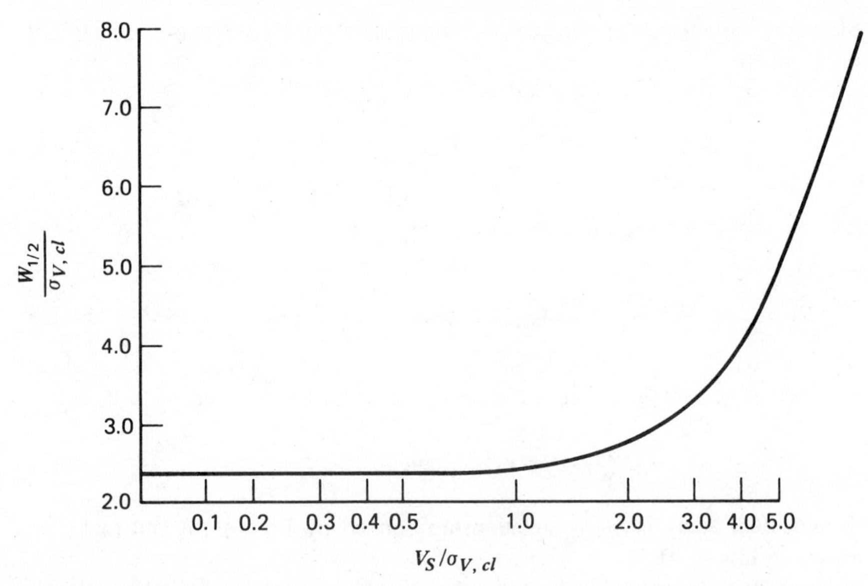

Figure 4.10. Theoretical dependence of peak half-width on sample volume. Computed with Eq. (4.13) with $\sigma_{V,cl}$ = 1.0 mL.

transient and steady state is often not completely clear and must be defined by convention. For example, steady state is generally agreed to be achieved after five time constants for an exponentially changing system such as a charging capacitor. In flow analysis, steady state is characterized by an analytical signal which is "flat-topped" or exhibits a plateau. A transient-state signal appears as the classical peak seen in gas and liquid chromatography.

The chemical behavior of a system, that is, equilibrium or kinetic, does not determine whether the analytical signal is transient or steady state. In fact, all four possible combinations can be found in the literature for various types of continuous flow systems. Equilibrium reaction chemistries have been used with the flow enthalpimeter with both steady- and transient-state analytical signals. What is often confusing is that the same variables may affect both the thermodynamic and signal characteristics of the system. For example, increasing the flow rate may decrease the extent of reaction by reducing reaction time and change the peak shape by increasing mass dispersion.

4.5. DISPERSION

Dispersion in the flow enthalpimeter is related to both thermal and mass transfer. Mass transfer is the sole contributor in the sample valve, equilibration coil, and other tubing and fittings until the sample reaches the reaction column. Once in the column, heat transfer becomes progressively more important as the reaction reaches completion. The principles employed to reduce non-ideal peak spreading due to thermal interactions are described in Section 4.2.2.

4.5.1. Analysis of Contribution from Instrument Components

Table 4.2 shows the contribution of the various components of the immobilized-enzyme reactor instrument to sample dispersion as measured with a refractive index detector. Comparison of the final refractive index and thermal peaks shows that mass dispersion predominates in this system and that the refractive-index detector is a useful tool for studying the behavior of the individual components of the thermal instrument.

Clearly, the sample valve, thermal equilibration coil, and the column reactor are the major contributors to sample dispersion. The latter two components should be carefully optimized, as described in Sections 4.2.4 and 4.2.2, to minimize dispersion. The spreading of the sample within the sample valve is difficult to measure since it must also reflect contributions from the detector itself. Nevertheless, the almost identical peak-widths-at-half-maximum height seen for both 40-μL and 120-μL samples indicates that reductions in dispersion are possible through improvements in this component.

In HPLC, dispersion is reduced through the use of stationary phases with smaller particle diameters, on the order of 5–10 μm. The 200–400 mesh

TABLE 4.2. Estimated Sources of Sample Dispersion[a]

System Component	Geometric Volume (mL)	Calculated Component Half-Widths (mL)	
		40-μL Sample	120-μL Sample
Sample valve[b]	—	0.14	0.15
Connection tubing	0.03	~0	~0
Thermal equilibration coil	0.20	0.19	0.18
Reference thermistor junction	0.08	~0	~0
Column reactor[c]	0.42	0.24	0.23
Total	—	0.34	0.33
Thermal system[d]	—	0.31	0.32

[a]Measured with a low-dead-volume refractive-index detector.
[b]Includes detector contribution.
[c]Packed with 200–400 mesh porous glass.
[d]Same conditions as with refractive-index detector.

(37–75 μm) porous or solid glass used in the flow enthalpimeter is much larger and cannot approach the performance seen in HPLC columns. Use of smaller particles in the flow enthalpimeter is not possible since they would vastly increase the pressure drop across the column and introduce additional thermal noise.

4.5.2. Flow Rate

The relationship between flow rate and dispersion has been well documented for chromatographic systems. In liquid chromatography, increasing flow rate results in increased dispersion reflected by decreased peak height and increased peak width. This same relationship can be seen in Fig. 4.11 for an immobilized-enzyme reactor. At flow rates greater than 1.5 mL min^{-1}, peak height is inversely proportional to flow rate. Identical behavior is seen with a refractive index detector. Peak height is independent of dispersion at steady state, as shown in Fig. 4.11. The behavior of the thermal reactor at flow rates less than 1.5 mL min^{-1} is due to heat loss and will be discussed in Section 4.6.1.

Dispersion due to mass transfer is a concern in flow enthalpimetry as it is in liquid chromatography and other forms of continuous flow analysis. It can be reduced by utilizing low-dead-volume components in the flow system, minimizing particle size in the reactor, and selecting the lowest value for flow rate that does not result in increased heat loss from the system. The relationship between dispersion and flow rate serves to define the limit of precision achievable with the flow enthalpimeter and is a function of the ability to supply a constant and pulseless fluid flow.

4.5.3. Peak Area

Quantitation by peak area compensates to some extent for changes in peak shape related to dispersion. Peak area, unlike peak height, does not decrease with increased dispersion. This is illustrated by comparing the relationships between corrected peak area and flow rate in Fig. 4.12 with their counterparts in Fig. 4.11. Peak area, when measured with a concentration-sensitive detector such as a thermistor or photometer, is proportional to flow rate. Errors related to imprecision in controlling flow rate, therefore, will not be reduced by measuring peak area. In addition, no improvement in analysis time is associated with measurement of peak area as the peak width still remains the same.

In Fig. 4.12 the measurement of peak area has been corrected by a factor proportional to flow rate. If this correction were ideal the plots of the corrected peak area versus flow rate would be flat at high flow rate. Since it is very difficult to accurately measure flow rate during an experiment the correction employed was only an approximation. Thus, it was only possible to show a significant reduction in the negative slope of corrected peak area versus flow rate at high

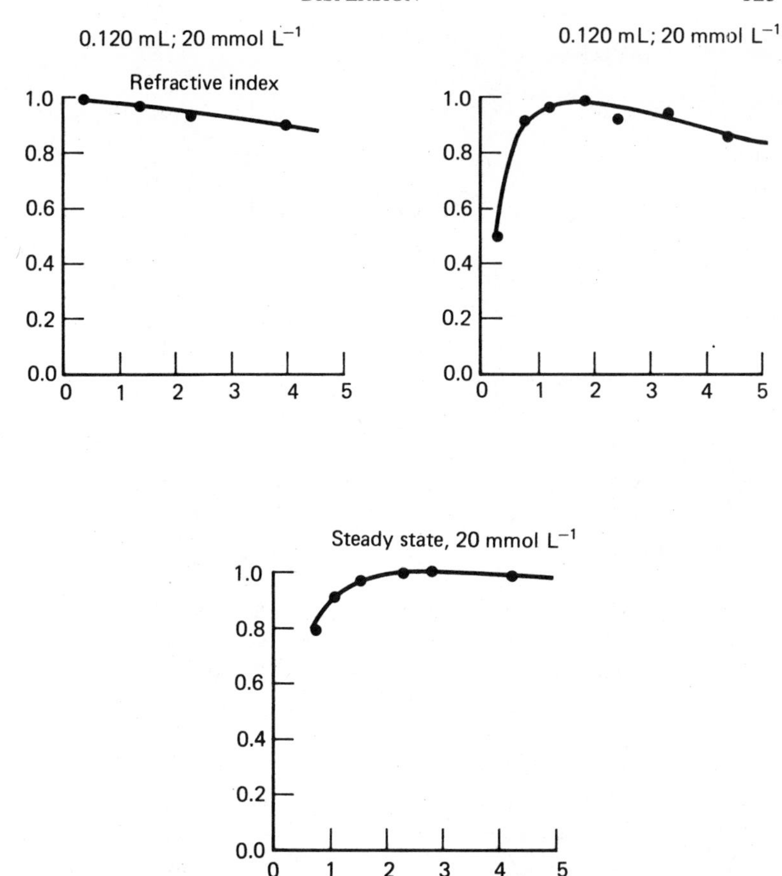

Figure 4.11. Dependence of normalized peak height on flow rate with an immobilized urease reactor. Sample volume and urea concentration as indicated. (Reprinted with permission from Ref. 13. Copyright 1977 American Chemical Society.)

flow rates when compared to the measurement of peak height shown in Fig. 4.11.

4.5.4. Fundamental Considerations

Dispersion is a complex topic and a detailed consideration is beyond the scope of this chapter. The interested reader should consult the excellent treatment given the subject by Carr and Bowers (20), which provided the basis for this discussion.

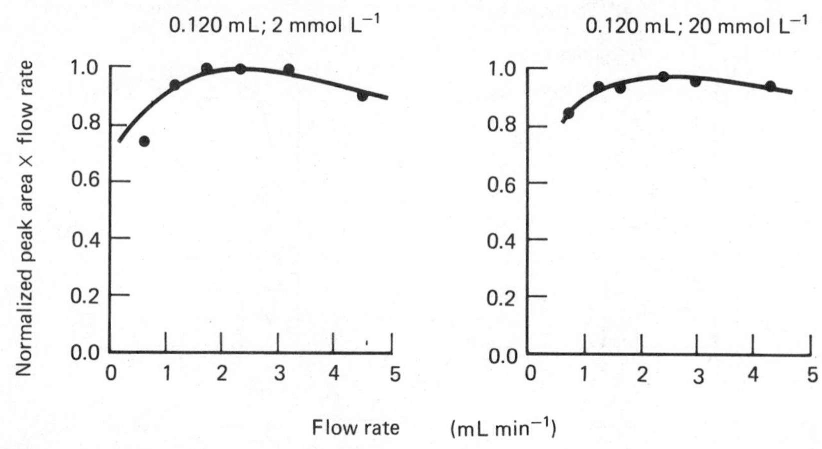

Figure 4.12. Dependence of normalized peak area on flow rate with an immobilized urease reactor. Sample volume and urea concentration as indicated.

There are, however, some considerations that should be understood in order to appreciate the operation of the flow enthalpimeter.

The simplest case of dispersion is that which occurs in the open tubing that connects the various components of the flow enthalpimeter. Under normal operating conditions the liquid flow is laminar in nature, which tends to disperse the sample parallel to the direction of flow due to the decrease in velocity of the liquid stream lines as they near the walls of the tubing. This process is opposed by the formation of a radial concentration gradient, which results in the sample spreading perpendicular to the direction of flow. In general, dispersion in an open tube will increase proportionately with the square of the internal diameter and with the length of the tubing. It is therefore important to minimize the internal diameter of the open tubing even at the expense of increasing the overall length.

Dispersion in the equilibration coils is somewhat more complex. In addition to the effects of laminar flow described above there is also a tendency for the molecules closer to the inner side of the tubing to stay near that wall and, consequently, follow a shorter path than those near the outside wall of the coil. This is similar to the advantage that a horse on the inside of an oval race track has and is referred to as the "race track effect." In addition, a set of secondary radial flows is also established by the centrifugal field due to the fluid motion. This process tends to augment radial mixing and diminish dispersion. In general, these latter effects do not exert a major influence on dispersion at low flow rates.

Thermal equilibration is enhanced by the formation of secondary radial flows in the thermal equilibration coil. These forces act to mix the fluid so that radial heat transfer is faster than it would be by diffusion alone. This process is favored by making a tighter coil, but at the expense of an increased resistance to flow.

The analysis of dispersion in the packed-bed reactor is quite complex owing to the necessity of considering a number of additional factors. Eddy diffusion does not exist in an open tube and arises in a packed bed because of the multitude of paths that molecules can follow through the bed are not equal in length. The column packing also interrupts the laminar flow profile of the open tube and establishes a number of new flow channels between particles and along the column wall. This process is termed mobile-phase resistance to mass transfer and acts in concert with eddy diffusion to disperse the sample along the direction of fluid flow. Dispersion due to the column wall can be virtually eliminated if the column internal diameter is wide enough so that the sample molecules never reach the wall; usually, an internal diameter equal to three times the product of particle diameter and column length is sufficient.

A final determinant of dispersion is diffusional resistance to mass transfer. This determines the rate at which molecules can move between the flowing mobile phase and stagnant liquid associated with porous, pellicular, or protein-coated particles.

The factors that determine dispersion in a packed bed are related to the nature, size, and geometry of the particles which constitute the packed bed, the reactor geometry, and flow rate. In general, small regular particles closely packed in a smooth wide column and operated at low flow rates is optimal. There are, of course, practical limitations to all of the above. Additionally, in the flow enthalpimeter, too low a flow rate will result in unacceptable heat loss from the reactor, as described in Section 4.6.1.

4.6. EFFECTS RELATED TO THE MEASUREMENT OF TEMPERATURE

4.6.1. Heat Loss

The sensitivity of the flow enthalpimeter at low flow rates, as shown in Fig. 4.11, is determined by heat loss. Since diffusion of heat across a boundary is an exponential function of time, it is not surprising that peak height decreases rapidly with increasing residence time of the thermal peak in the reactor. Figure 4.13 shows that heat loss can be reduced by improving the adiabaticity of the column. Calculating peak area, as shown in Fig. 4.12, does not improve sensitivity as the signal has been lost, not displaced as with mass dispersion.

4.6.2. Peak Spreading

The sources of thermal peak spreading have been described along with the design considerations necessary to reduce their impact on analytical performance (see Section 4.2.2). Thermal effects on peak shape tend to be more of a problem with the soluble reagent reactor than the immobilized enzyme-system because

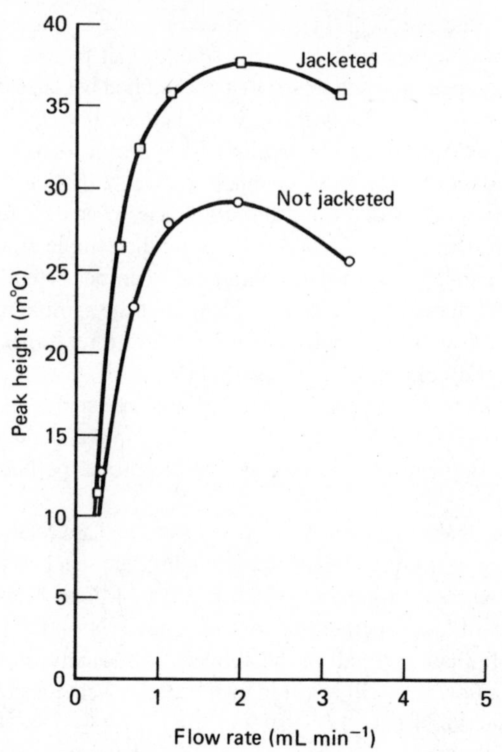

Figure 4.13. Effect of improved adiabaticity on peak height as a function of flow rate with an immobilized urease reactor: column without Teflon jacket (○); column with Teflon jacket (□). (Reprinted with permission from Ref. 13. Copyright 1977 American Chemical Society.)

part of the reaction occurs in the plastic mixing tee. The relatively high thermal capacity of this component results in significant peak tailing.

Equation (4.18) provides a tool for measuring the effect of thermal peak tailing on the analytical performance of the system. It simply states that the peak-width-at-half-maximum height at steady state is exactly equal to the sample volume. For the soluble reagent system this must be corrected for the dilution inherent in having separate sample and reagent streams. Assuming matched flow rates and dilute aqueous solutions, the slope of a plot of peak-width-at-half-maximum height versus sample volume should be exactly 2.00 for single sample and reagent streams. A slope of less than 2 reflects incomplete mixing, and a slope greater than 2 indicates overmixing or tailing. This calculated slope is referred to as the "mixing factor."

Figure 4.14 illustrates the calculation of the "mixing factor" for the neutralization of HCl with THAM in the soluble reagent reactor. This reaction proceeds

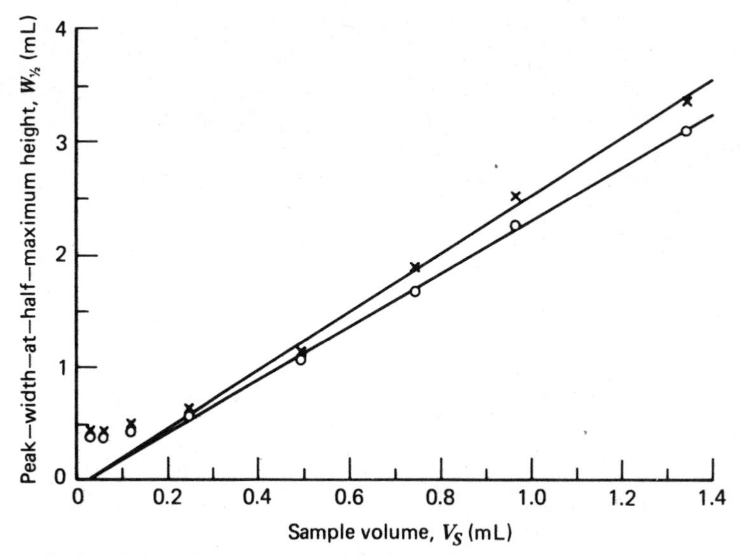

Figure 4.14. Mixing factor: Neutralization of hydrochloric acid with (tris-hydroxymethyl)aminomethane. (○) precolumn mixing volume = 30 μL, "mixing factor" = 2.35 ± 0.05; (×) precolumn mixing volume = 75 μL, "mixing factor" = 2.56 ± 0.09. (Reprinted with permission from Ref. 13. Copyright 1977 American Chemical Society.)

very rapidly and probably represents the worst case in which all of the heat of reaction is generated in the plastic mixing tee. As expected, minimizing the precolumn mixing volume improved the "mixing factor" by reducing peak tailing owing to thermal contact with the mixing fitting.

4.6.3. Thermistor Contributions

For the most part, the thermistors chosen for use in the flow enthalpimeter do not limit its performance. Further improvements in reducing mass dispersion or flow rate fluctuations might make selection of appropriate devices more critical. Key parameters are the temperature coefficient, dissipation constant, and response time (21). The temperature coefficient defines the change in resistance with temperature, and the dissipation constant defines the internal heating characteristics of the device. Together, they determine the sensitivity and signal-to-noise characteristics of the transducer. As a general rule, the response time constant should be no greater than 20% of the peak-width-at-half-maximum height.

The major cause of thermistor failure in the flow enthalpimeter is degradation of the epoxy seal that protects the connection between the transducer and cable from the aqueous environment. The presence of moisture around these connections produces stray capacitance, which results in unacceptable apparent thermal noise and drift. Extreme cases are detected as an inability to balance the capacitance of the ac bridge. Earlier detection is possible by testing for capacitance across the thermistor by measuring the rate of discharge of a dc voltage.

A final consideration in the use of thermistors is that their response to temperature is inherently exponential, not linear (21). In general, resistance at a given temperature R_{TH} is related to temperature T by

$$R_{TH} = \exp\left(A_{TH} + \frac{B_{TH}}{T + C_{TH}}\right) \tag{4.19}$$

where A_{TH}, B_{TH}, and C_{TH} are constants for a given thermistor. The temperature coefficient B_{TH} defines the percentage change in resistance with temperature. B_{TH} is negative for thermistors and is a function of temperature as defined by the first derivative of Equation (4.19):

$$B_{TH} \equiv \frac{100}{R_{TH}} \frac{dR_{TH}}{dT} \tag{4.20}$$

In a typical case, non-linearity was calculated as 0.24% over an 0.1°C interval and 2.4% over a 1°C interval.

The general effect of non-linear measurement systems on Gaussian peak shape and analyte has been described by Carr (22, 23). Detector non-linearity will change peak height and properties related to even moments such as peak area and peak width but will not alter properties related to odd moments such as retention time and peak symmetry. These effects are shown in Fig. 4.15. Other factors that influence peak shape such as non-linear adsorption isotherms can be distinguished from detector non-linearity by their tendency to distort peak symmetry and change retention time.

The concept of peak moments is derived from the general equations that define a Gaussian distribution. In chromatographic applications they are taken to be synonymous with physical characteristics of the Gaussian peaks. For example, the zero moment is peak area, the first moment is retention time, the second moment is peak variance, and the third moment is peak symmetry. A mathematical description of peak moments and the relationships among them can be found in most advanced chromatography textbooks.

It is evident that a given pair of thermistors will begin to yield non-linear calibration curves and increased peak-width-at-half-maximum height as the measured temperature change increases. This effect cannot be corrected by measuring

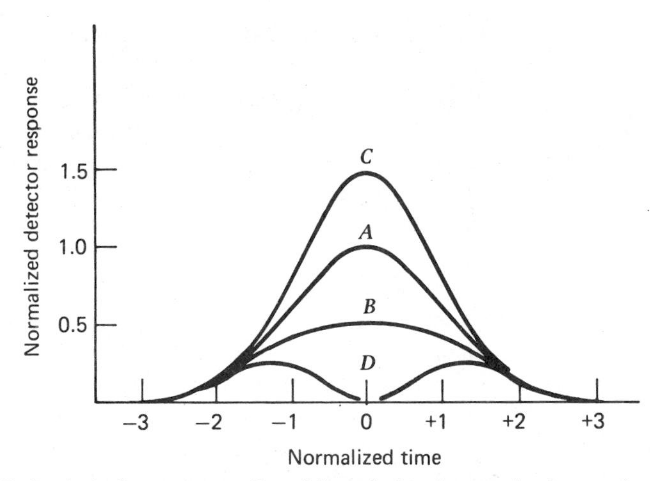

Figure 4.15. Synthetic chromatograms. Plot of dimensionless detector signal versus time from peak maximum. Peak *A* is peak seen with a linear detector. Peaks *B* and *D* represent increasingly negative detector non-linearity. Peak *C* simulates a positive non-linear detector. (Reprinted with permission from Ref. 22. Copyright 1980 American Chemical Society.)

peak area. As with other analytical techniques, samples containing large concentrations of analyte either must be diluted or the sample volume adjusted accordingly.

The behavior simulated by Fig. 4.15 would be expected to occur with any concentration-sensitive detector where the measured signal is not a linear function of sample concentration. In the case of a photometric chromatography detector, for example, it would be associated with samples having high absorbances when significant stray light is present. It may also be associated with non-linear reaction kinetics in photometric flow injection systems. For either flow enthalpimetry or chromatography, errors due to this phenomenon would be difficult to detect since changes in retention time and peak symmetry, two common indicators of instrumental problems, are not present. Calibration by either peak height or peak area would both be non-linear and peak width would change with concentration. Of even greater interest is the performance predicted for an extreme case of detector non-linearity where a peak associated with a single component is split into two symmetric peaks as shown in Fig. 4.15, peak *D*. The expected retention time of the peak with a linear detector would fall exactly midway between the two peaks observed with the non-linear detector. Certainly this would confuse the unwary analyst who would then begin a fruitless search for the identity of the non-existent extra sample component.

4.7. REACTOR KINETICS

As stated in Section 4.4, the theory developed to describe the flow enthalpimeter is based on the assumption of instantaneous conversion of sample to products. This ideal case is rarely encountered, even for fast reactions. In addition to the chemical rate constant, physical processes such as mixing and external and internal diffusion must also be considered as contributors to the overall reaction rate.

The attainment of an instantaneous reaction is actually not desirable for the reactor configuration where the reagents are mixed in a tee fitting external to the adiabatic column. For example, the reaction between HCl and THAM in the soluble reagent reactor is very fast and, consequently, results in undesirable thermal effects resulting from heat transfer to the precolumn mixing fittings. An instantaneous reaction would, however, be optimal if it occurs completely within the reactor as with the immobilized-enzyme configuration.

4.7.1. First-Order Reactions

Most of the chemistries utilized in the flow enthalpimeter have involved either true or pseudo-first-order reactions. If first-order kinetics prevail, the rate of depletion of reactant is proportional to the instantaneous concentration of sample $[S_t]$:

$$\frac{d[S_t]}{dt} = -k_1[S_t] \tag{4.21}$$

Solution of this equation yields the familiar relationship for the extent of reaction [see Eq. (7.92)] assuming 1:1 stoichiometry between reactant and product:

$$\frac{[S_0] - [S_t]}{[S_0]} = 1 - \exp(-k_1 t) \tag{4.22}$$

where $[S_0]$ is the concentration of sample at the beginning of the reaction.

The extent of reaction is determined only by the rate constant k_1 and the reaction time and not by sample concentration. For a first-order reactor the extent of reaction is always the same for all samples analyzed under the same conditions. There should be no observable effect on the peak shape or calibration curve related to sample concentration as long as the reaction remains first order. In general, deviations from the ideal instantaneous reaction model due to the differences between mass and thermal diffusion rates are too small to be observed experimentally with existing instrumentation. Other effects such as sensitivity to reaction temperature and the presence of inhibitors directly determine the extent of reaction (17).

4.7.2. Zero-Order Reactions

Zero-order reactions are observed with the flow enthalpimeter when the analyte is an enzymatic substrate and its concentration exceeds the enzymatic K_m. In this case the reaction rate is a constant or

$$\frac{d\,[S_t]}{dt} = -k_0 \tag{4.23}$$

and the extent of conversion, making the same assumptions as for Eq. (4.22), is

$$\frac{[S_0] - [S_t]}{[S_0]} = \frac{k_0 t}{[S_0]} \tag{4.24}$$

With a zero-order model the extent of conversion is a function of sample concentration. When analyzed by the pure plug flow model (see Section 4.4.1), any samples where $[S_0] > k_0 t$ should yield peaks having the same height and area regardless of sample concentration. Analysis of the same system by the continuously stirred reactor model (see Section 4.4.2) predicts that peak width will increase with sample concentration as described by Eqs. (4.7) and (4.8). The increase in peak-width-at-half-maximum height will become significant when $[S_0]$ is greater than $1.3k_0\tau_R$.

The interpretation of changes in peak width is complicated by the need to consider several alternative mechanisms. As sample concentration increases above the enzymatic K_m, the reaction kinetics become zero order and the peak width should increase as described above. However, a larger temperature change will also be measured and the recorded peaks will be affected by the non-linearity of the thermistors described in Section 4.6.3, which also affects both peak height and width. An additional complication is that a varying extent of reaction following cessation of first-order kinetics may also be present. To date a satisfactory determination of the relative importance of these processes in determining peak shape has not been reported.

4.7.3. Michaelis–Menten Kinetics

A discussion of the use of the Michaelis–Menten model in describing the extent of reaction in an immobilized-enzyme reactor can be found in Section 7.9.1a.

In many cases, the substrate concentration is either sufficiently less than or greater than the enzymatic K_m so that the reaction is either predominantly first or zero order. If this is true, then either the first-order model described in Section 4.7.1 or the zero-order model in 4.7.2 can be used to predict whether the reaction kinetics will affect peak shape or the linearity of the calibration curve. A model to describe the effect of the more complicated case where substrate concentration is near the enzymatic K_m and the kinetics are mixed first and zero order on peak shape has not yet been derived. In any case, the extent of reaction can be determined as described in Section 7.9.1a.

4.7.4. Catalyzed (Non-Enzymatic) Reactions

A study of the performance of the soluble reagent configuration of the flow enthalpimeter in measuring iodide by its catalytic effect on the reaction

$$2\ \text{Ce (IV)} + \text{As (III)} \rightarrow 2\ \text{Ce (III)} + \text{As (V)}$$

was conducted by Elvecrog and Carr (24). The first-order rate constant, and therefore the extent of reaction, was proportional to the iodide concentration in the sample as long as sufficient Ce (IV) was present. At high iodide concentrations the Ce (IV) was depleted and the reaction kinetics became first order in both iodide and Ce (IV).

The relationships between the change in kinetic mechanism and the performance of the flow enthalpimeter were quite complex. For example, peak-width-at-half-maximum height was a function of iodide concentration when the reaction was first order in both Ce (IV) and iodide, but not when the reaction was first order with respect to iodide alone. This effect was attributed to the non-linearity of the measured signal with respect to iodide concentration, which occurred when the reaction mechanism became dependent on both Ce (IV) and iodide at high iodide concentrations (see Section 4.6.3).

Dependence of peak height on flow rate also changed with the reaction kinetics, primarily as a result of the varying extent of reaction. The relationship between peak height and sample volume also appeared to be related to the reaction kinetics. This latter effect, however, could not be explained. Finally, the calculated "mixing factor" (see Section 4.6.2) was also inexplicably affected by the change in reaction kinetics.

The intricate relationships between reactor kinetics and detector non-linearity with peak height, area, width, and shape have only begun to be unraveled. These effects are not limited to the flow enthalpimeter or thermal systems, but require only the use of a flow system and a concentration-sensitive detector. They should be present and observable in other types of continuous flow systems such as flow-injection analysis and liquid chromatography.

REFERENCES

1. P. T. Priestley, W. S. Sebborn, and R. F. Selman, *Analyst (London)* **90**, 589 (1965).
2. A. C. Censullo, J. A. Lynch, D. H. Waugh, and J. Jordan, in *Analytical Calorimetry,* Plenum Press, New York, 1974, Vol. 3, pp. 217–235.
3. G. Peuschel and F. Hagedorn, *Fresnius Z. Anal. Chem.* **277**, 177 (1977).
4. F. Hagedorn, G. Peuschel, and R. Weber, *Analyst (London)* **100**, 810 (1975).
5. R. Weber, G. Blanc, G. Peuschel and F. Hagedorn, *Anal. Chim. Acta* **86**, 79 (1976).
6. K. Mosbach and B. Danielsson, *Biochim. Biophys. Acta* **364**, 140 (1974).
7. B. Mattiasson, B. Danielsson, and K. Mosbach, *Anal. Lett.* **9**, 867 (1976).
8. C. Borrebaeck, J. Borjeson, and B. Mattiasson, *Clin. Chim. Acta* **86**, 267 (1976).
9. S. P. Fulton, C. L. Cooney, and J. C. Weaver, *Anal. Chem.* **52**, 505 (1980).
10. L. M. Canning, Jr. and P. W. Carr, *Anal. Lett.* **8**, 359 (1975).
11. L. D. Bowers, L. M. Canning, Jr., C. N. Sayers, and P. W. Carr, *Clin. Chem. (Winston-Salem)* **22**, 1314 (1976).
12. L. D. Bowers, R. S. Schifreen, and P. W. Carr, *Clin. Chem. (Winston-Salem)* **22**, 1427, 2045 (1976).
13. R. S. Schifreen, D. A. Hanna, L. D. Bowers, and P. W. Carr, *Anal. Chem.* **49**, 1929 (1977).
14. R. S. Schifreen, C. S. Miller, and P. W. Carr, *Anal. Chem.* **51**, 278 (1979).
15. L. D. Hansen, T. E. Jensen, S. Mayne, D. J. Eatough, R. M. Izatt and J. J. Christensen, *J. Chem. Thermodyn.* **7**, 919 (1975).
16. E. B. Smith, C. S. Barnes, and P. W. Carr, *Anal. Chem.* **44**, 1663 (1972).
17. P. W. Carr, *Anal. Chem.* **50**, 1602 (1978).
18. J. C. Sternberg, in *Advances in Chromatography*, J. C. Giddings and R. A. Keller (eds.), Marcel Dekker, New York, 1966, Vol. 2, p. 205.
19. H. W. Pardue, *Clin. Chem.* **23**, 2189 (1977).
20. P. W. Carr and L. D. Bowers, *Immobilized Enzymes in Analytical and Clinical Chemistry,* Wiley, New York, 1980, pp. 332–347.
21. P. W. Carr, *CRC Crit. Rev. Anal. Chem.* **2**, 491 (1972).
22. P. W. Carr, *Anal. Chem.* **52**, 1746 (1980).
23. L. M. McDowell, W. E. Barber, and P. W. Carr, *Anal. Chem.* **53**, 1373 (1981).
24. J. M. Elvecrog and P. W. Carr, *Anal. Chim. Acta* **121**, 135 (1980).

CHAPTER

5

DETERMINATION OF ΔH_R AND K_{eq} VALUES

DELBERT J. EATOUGH, EDWIN A. LEWIS, and LEE D. HANSEN

Thermochemical Institute and the Department of Chemistry
Brigham Young University
Provo, Utah

5.1. DETERMINATION OF ΔH_R AND K_{eq} VALUES

In Chapter 3, the basic instrumental concepts of isoperibol, isothermal, and heat-conduction calorimetry were outlined. The methods used for reduction of the calorimetric measurement to a q_p value or array for the total heat produced during a calorimetric experiment were given. Equations for calculating the corrections for non-chemical effects, $q_{HL,p}$ and $q_{TC,p}$, were also given. From these values, $q_{C,p}$ is calculated by

$$q_{C,p} = q_p - q_{HL,p} - q_{TC,p} \tag{5.1}$$

After the data reduction outlined in Chapter 3 is completed, the $q_{C,p}$ data from any type of calorimetric equipment are equivalent, and the methods of data analysis used to calculate thermodynamic quantities for the reactions occurring during the calorimetric experiment will be the same. The methods for calculation of ΔH_R and K_{eq} values from the experimental $q_{C,p}$ values are outlined in this chapter. The procedures used to correct the $q_{C,p}$ values for all chemical processes other than the reaction(s) of interest are given. Least squares methods to calculate ΔH_R and K_{eq} values from the resulting corrected $q_{C,p}$ values are outlined. Finally, the application of these calculation techniques are illustrated by discussion of the determination of thermodynamic values for four examples of chemical systems.

5.1.1. Calculation of Corrections to $q_{C,p}$ Values

After the non-chemical corrections described in Chapter 3 have been made, the data from any calorimeter will have been reduced to an array of $q_{C,p}$ versus n_p values, where $q_{C,p}$ is the total heat produced from all chemical processes and n_p is the total moles titrated or mixed at data point p. The $q_{C,p}$ values will include the contributions from effects due to diluting and mixing the two reactant solutions and from reactions other than the ones for which values of thermodynamic constants are to be determined. In general, the corrections for dilution and extraneous reactions can be made using

$$q_{C,p} = q_{D,p} + \sum_i \Delta n_i \Delta H_i + \sum_j \Delta n_j \Delta H_j \tag{5.2}$$

where $q_{D,p}$ is the sum of the heat effects of diluting the titrant and titrate separately, the sum over i applies to all reactions for which ΔH_R and Δn are known, and the sum over j applies to all reactions for which ΔH_R and K_{eq} are to be calculated.

These corrections are detailed in the following sections.

5.1.1a. Dilution of Reactant Solutions

As the two reactant solutions are mixed in the calorimetric experiment, a heat effect caused by dilution and shifts in chemical equilibria will occur. The magnitude of the heat effect depends on the concentrations of species present in the two solutions. Because chemical reactions such as ion pairing, molecular association, and hydrolysis may occur, this heat effect will be present even if an inert electrolyte is used to maintain a constant ionic strength. The magnitude of the dilution heat effect must be determined by experimentation or from available data tables, for example, Refs. 1–3.

In a titration calorimetric experiment, the heat effects associated with dilution of the titrate solution are often negligible since the changes in concentration of the chemical species in the titrate are small. On the other hand, chemical species present in the titrant solution are diluted about an order of magnitude during the titration and the associated heat of dilution may be large. Similar changes will occur in batch or injection calorimetric experiments. Heat of dilution effects in a flow system are usually significant for both solutions. If the apparent partial molal heat contents Φ_L are known for the solutes in the two solutions as a function of concentration or ionic strength μ, then the heat effect due to dilution of each of the two mixed solutions, q_D, is given by

$$q_{D,p} = \sum_{j=1}^{2} (\Phi_{L,p,j} - \Phi_{L,i,j})\, n_{p,j} \tag{5.3}$$

where $\Phi_{L,p,j}$ refers to the hypothetical situation where the volume of the solutions has been changed to the total volume of the two solutions at point p but no chemical reaction between the two solutions has occurred; $\Phi_{L,i,j}$ refers to the initial condition; and $n_{p,j}$ is the number of moles of material added. If $\Phi_{L,p,j} - \Phi_{L,i,j}$ values are not available in the literature, they must be measured. However, it is not necessary to determine absolute Φ_L values since only the differences in $\Phi_{L,p,j} - \Phi_{L,i,j}$ need be known to calculate $q_{D,p}$. These differences are measured in separate experiments in which each solution is diluted into the appropriate solvent. A more laborious but equivalent approach is to correct all $q_{C,p}$ data to infinite dilution. The ΔH_R value calculated is valid at the ionic strength in the final solution or at infinite dilution depending on the method used. Both approaches are approximate in that heat effects due to mixing dissimilar solutes at a constant ionic strength or concentration are neglected.

5.1.1b. Heat Contributions from Other Reactions

If reactions other than those of interest occur in the calorimeter, corrections must be made for their enthalpy contributions to the $q_{C,p}$ values. Since hydrolysis of a ligand species is a commonly encountered reaction for which corrections must be made, this reaction will be used to illustrate the necessary calculations. If one of the reactants is a basic ligand B, then heat effects due to

$$B + H_2O = HB^+ + OH^- \tag{5.4}$$

will contribute to the total heat produced in the reaction vessel. The energy terms that arise from reaction (5.4) are those due to the formation of HB^+ and to the dissociation of water by the hydrolysis of B. In both cases the correction term will be the change in the number of moles of species involved times the appropriate ΔH_R value for the reaction. For example, the correction term associated with changes in the number of moles of the species HB^+, $q_{RB,p}$, will be

$$q_{RB,p} = (c_{HB,p}V_p - c_{HB,x}V_x)\Delta H_{HB} = (n_{HB,p} - n_{HB,x})\Delta H_{HB} \tag{5.5}$$

where ΔH_{HB} is the change in enthalpy for the reaction $H^+ + B = HB^+$ and the c_{HB} and n_{HB} terms refer to the concentration and number of moles of HB^+ species, respectively. Since the number of moles of water hydrolyzed is the same as the number of moles of OH^- formed, the correction for the formation of water, $q_{RW,p}$, is

$$q_{RW,p} = (c_{OH,p}V_p - c_{OH,x}V_x)(-\Delta H_W) = (n_{OH,p} - n_{OH,x})(-\Delta H_W) \tag{5.6}$$

where ΔH_W is the change in enthalpy for the reaction $H^+ + OH^- = H_2O$. The initial conditions (V_x, $c_{HB,x}$, and $c_{OH,x}$) may refer either to the initial solution or to a hypothetical solution diluted to the total volume at point p, but with no interaction between the two solutions. The latter case would be used if q_D values are calculated from Φ_L values reported in the literature. It is important that the data for the heat of dilution and for the correction for hydrolysis are consistent so that the correct heat effect is calculated.

Corrections for other reactions such as ion pair formation or dimerization, will involve terms of the form $(\Delta n_i)(\Delta H_i)$ where ΔH_i is the enthalpy change associated with the reaction and Δn_i is the change in the number of moles of the species involved from the start of the reaction (point 0) to the data point in question (point p).

5.1.1c. Calculation of Heat Released by Reactions for which ΔH_R Values are to be Determined

The heat change to point p due to the reaction(s) of interest is calculated by subtracting the previously described correction terms from the total chemical heat, $q_{C,p}$:

$$q_{R,p} = \sum_j n_{j,p}\Delta H_j = q_{C,p} - q_{D,p} - \sum_i n_{i,p}\Delta H_i \qquad (5.7)$$

The term $\Delta n_{i,p}\Delta H_i$ refers to those reactions occurring other than the ones for which the ΔH_R values are to be determined [e.g., see Eq. (5.5) and (5.6)]. The $q_{R,p}$ value is a function only of the K_{eq} and ΔH_R values for the reactions of interest.

5.1.2. Calculation of ΔH_R Values

The mixing of two liquid solutions or a solution and a solid in the calorimetric experiment produces one or more reactions where the extent of the reaction(s) and the heat produced are related by Eq. (5.7) to the corresponding equilibrium constant(s) and enthalpy change(s) for the reaction(s). In Eq. (5.7), $\Delta n_{j,p}$ is the change in moles of product j formed from point 0 to point p and is a function of the equilibrium constant for reaction j.

In general, the best values for ΔH_j are calculated by a least squares analysis of Eq. (5.7). The error square sum over the m data points is given by

$$U(\Delta H_j) = \sum_{p=1}^{m} \left(q_{R,p} - \sum_j \Delta n_{j,p}\Delta H_j \right)^2 \qquad (5.8)$$

where the subscript p is over all the data points and the subscript j is over all the reactions being studied. The best values for ΔH_j in a given run are those that minimize $U(\Delta H_j)$; that is, those values which satisfy

$$\frac{\partial U(\Delta H_j)}{\partial \Delta H_k} = 0 = \sum_{p=1}^{m} \left(q_{R,p}\Delta n_{k,p} - \Delta n_{k,p} \sum_j (\Delta n_{j,p}\Delta H_j) \right) \qquad (5.9)$$

where $k = 1, \ldots, j$. The j expressions given by Eq. (5.9) are all homogeneous first-order linear equations in the ΔH_j values and are easily solved.

5.1.3. Calculation of Equilibrium Constants

5.1.3a. General Description of Calculations

The equilibrium constant for a given reaction can be determined simultaneously with ΔH_R if the magnitudes of K_{eq} and ΔH_R for the overall reaction taking place in the reaction vessel are within certain limits. The family of curves presented in Fig. 5.1*a* shows that increased overall curvature of the thermogram is obtained with decreasing values of K_{eq} (ΔH_R and concentrations assumed constant); in other words, the shape of a given curve is a function of the K_{eq} value. The curves for systems with K_{eq} values greater than approximately 10^4 differ only slightly from one another; hence, it is difficult to make accurate calculations of K_{eq} in these cases. For reactions with K_{eq} values less than 10, very little reaction takes place and, hence, very little heat is evolved. In addition, the shape of the curve is little affected by incremental decreases in K_{eq} for K_{eq} values less than 10. The curves shown in Fig. 5.1*a* hold only for titration of 10^{-2} mol L^{-1} A with B. For example, if the concentration of A in Fig. 5.1*a* is 10^{-4} mol L^{-1} A, then the upper limit for measuring K_{eq} is about 10^6 and the lower limit is about 10^3.

The magnitude of the q_R value is directly proportional to ΔH_R, as illustrated in Fig. 5.1*b*. It follows that the lower the K_{eq} value, the higher ΔH_R must be in order to calculate K_{eq} and ΔH_R values with a given reliability. However, while the calculated value of ΔH_R is directly related to the magnitude of the measured q_R values, the calculated value of K_{eq} depends only on the curvature. For this reason, one can obtain accurate K_{eq} values even though ΔH_R is erroneous because of linear errors, for example, errors in calibration constants. More detailed information on error analysis and possible errors in the calorimetric determination of K_{eq} and ΔH_R have been published (4–8).

The successful application of the calorimetric method of determining equilibrium constants to a given system, therefore, depends on the equilibrium constant and concentrations of reacting species being within the bounds required to yield a sufficiently curved plot of q_R versus moles of titrant and the ΔH_R value for the reaction being large enough that the total measured heat is known with sufficient accuracy. It has been shown that by use of selective titrants the method can be extended to the determination of equilibrium constants for proton ionization (5) and metal–ligand interaction (8) of almost any magnitude. Specific examples illustrating the calculation of equilibrium constants from calorimetric data are given in Section 5.2.

5.1.3b. Approximate Calculation of K_{eq} Values

Accurate evaluation of K_{eq} and ΔH_R (Section 5.1.3*c*) usually requires sophisticated calorimetric equipment and complex computation. It is possible, however,

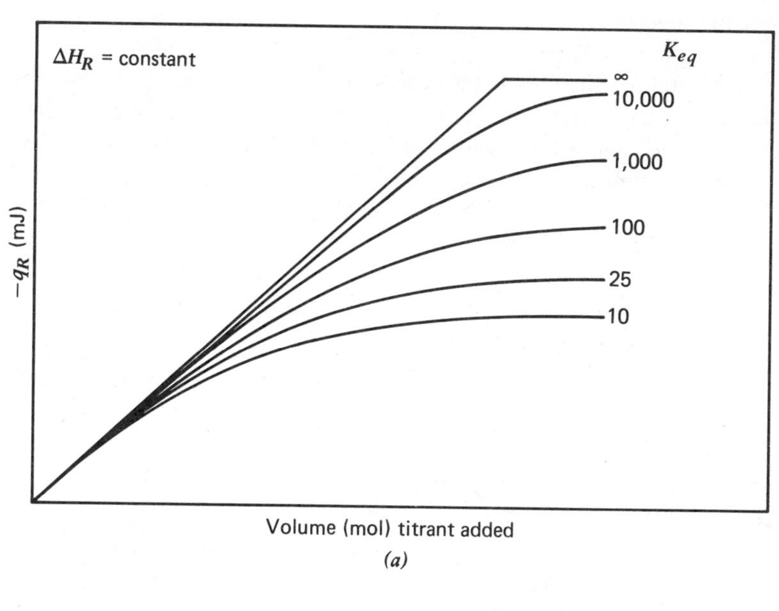

(a)

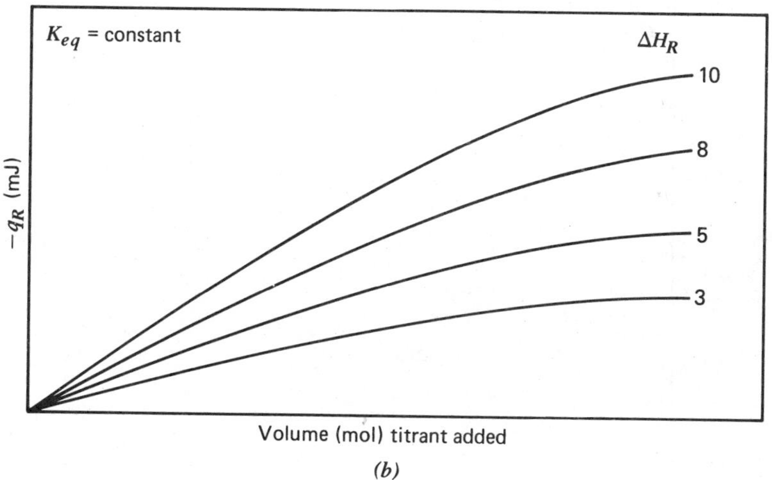

(b)

Figure 5.1. Dependence of calorimetric data on (*a*) K_{eq} and (*b*) ΔH_R for the titration of 0.01 mol L^{-1} *A* with *B*.

to obtain moderately accurate results using simple calorimeters and simplified calculation procedures (9, 10).

This section describes one of the many simplified techniques for the calculation of the equilibrium constant for a simple system with only one incomplete reaction occurring. The example is intended to illustrate calculation procedures further developed later in the least squares evaluation of ΔH_R and K_{eq} as outlined in Section 5.1.3c. Assume that the reaction producing the isoperibol calorimetric titration data in Fig. 5.2 is for the association of A and B to yield AB. The heat, corrected for all extraneous heat effects, due to the reaction from the start of the titration to any point p on the plot is $q_{R,p}$. This quantity is related by Eq. (5.10) to the number of moles of AB formed:

$$q_{R,p} = \Delta H_R(\Delta n_p) \tag{5.10}$$

where ΔH_R is the change in enthalpy for the reaction and Δn_p is the moles of AB formed from the start of the titration to point p. The quantity Δn_p can be

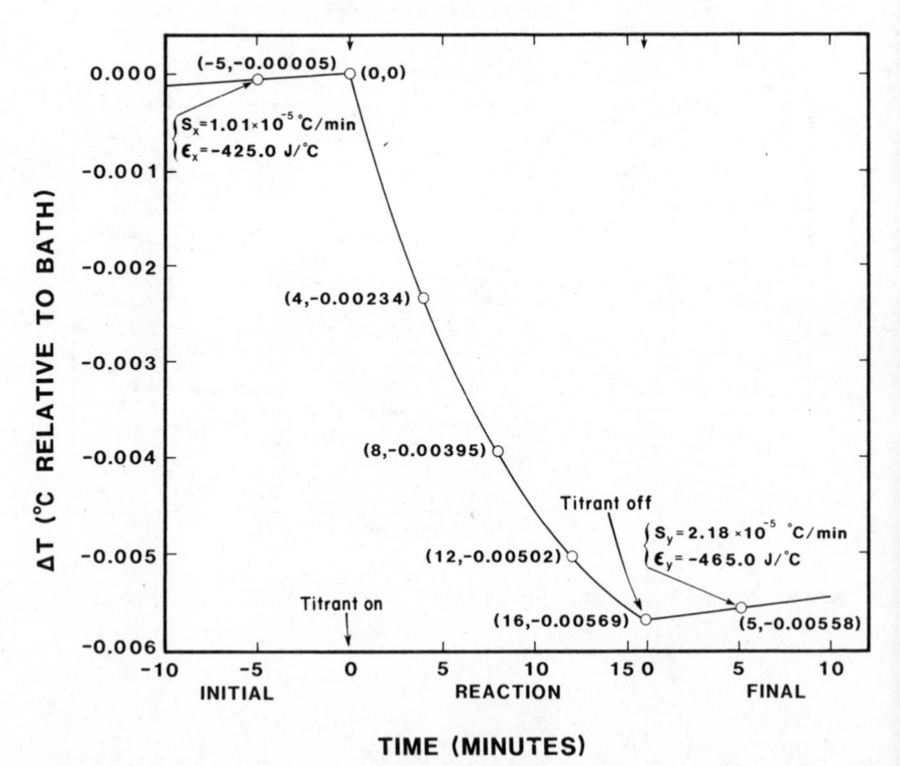

Figure 5.2. Sample isoperibol calorimetric data for the titration of 100.0 mL of 0.0100 mol L^{-1} B with 0.200 mol L^{-1} A.

calculated from the concentration of each species present in the reaction vessel at point p if the value of the equilibrium constant for the reaction is known. However, in this case, the K_{eq} value for the formation of AB is not known, and therefore Δn_p is not known. If an approximate value for K_{eq} is assumed, a corresponding approximate value of Δn_p can be calculated. A value for ΔH_R is then obtained at a given point p using Eq. (5.10). This operation is carried out for each of the chosen points—$q_{R,1}$, $q_{R,2}$, $q_{R,3}$, . . . , $q_{R,m}$—using the same assumed value of K_{eq}. Because ΔH_R and K_{eq} are constants at a constant ionic strength μ and temperature T (μ and T are assumed to vary insignificantly during the course of the reaction), it is expected that the calculated ΔH_R value will be the same for each point. If such is not the case, a new value of K_{eq} is chosen and ΔH_R values are calculated until a K_{eq} value is found that yields the same ΔH_R value at each point p throughout the run. When this occurs, the proper values for K_{eq} and ΔH_R will have been determined. The value for ΔG_R is determined by the relationship $\Delta G_R = -RT \ln K_{eq}$, and the change in entropy ΔS_R is determined by the relationship $\Delta S_R = (\Delta H_R - \Delta G_R)/T$.

Data previously reported (9) for the hypothetical experiment for the calorimetric study of the reaction A + B = AB by continuous titration calorimetry can be used to illustrate the above procedure. The data are illustrated in Fig. 5.2. A is the titrant, B is the titrate, and four data points taken at 4-min intervals are used in the calculations. The titrant, 0.200 mol L^{-1} A, is titrated into 100.00 mL of 0.0100 mol L^{-1} B at a constant rate of 0.600 mL min^{-1}, giving a total of 2.40, 4.80, 7.20, and 9.60 mL of titrant delivered, respectively, at the end of each interval. After the appropriate heat corrections are made, it is determined that the total heat of reaction at the end of each time interval is, respectively, $q_{R,1} = 1.04$ J, $q_{R,2} = 1.81$ J, $q_{R,3} = 2.37$ J, $q_{R,4} = 2.77$ J. The concentrations of the various species in the calorimeter at the end of each time interval can be calculated using

$$K_{eq} = \frac{[AB]_p}{[A]_p[B]_p} \tag{5.11}$$

$$[A_{tot}]_p = [A]_p + [AB]_p \tag{5.12}$$

$$[B_{tot}]_p = [B]_p + [AB]_p \tag{5.13}$$

where the bracketed quantities represent the concentrations of the indicated species and tot = total.

Substitution of Eq. (5.12) and (5.13) into Eq. (5.11) gives

$$K_{eq} = \frac{[AB]_p}{([A_{tot}]_p - [AB]_p)([B_{tot}]_p - [AB]_p)} \tag{5.14}$$

Rearranging gives

$$K_{eq}[AB]_p^2 - (K_{eq}[B_{tot}]_p + K_{eq}[A_{tot}]_p + 1)[AB]_p + K_{eq}[A_{tot}]_p[B_{tot}]_p = 0 \quad (5.15)$$

The initial concentrations of components A_{tot} and B_{tot} are known. As the titration proceeds, values for $[A_{tot}]_p$ and $[B_{tot}]_p$ can be calculated at the end of each time interval. For example, at the end of the first time interval 2.40 mL of titrant have been added to the reaction vessel, and $[A_{tot}]_1 = (0.200 \text{ mol L}^{-1})$ (2.40 mL/102.40 mL) $= 4.69 \times 10^{-3}$ mol L^{-1}, and $[B_{tot}]_1 = (0.0100 \text{ mol L}^{-1})$ (100.00 mL/102.40 mL) $= 9.77 \times 10^{-3}$ mol L^{-1}. With the appropriate values of $[A_{tot}]_p$ and $[B_{tot}]_p$ and an assumed value of K_{eq}, the values of $[AB]_p$, $[A]_p$, and $[B]_p$ are calcualted from Eq. (5.15), (5.12), and (5.13), respectively. In this example, only the concentration of $[AB]_p$ is necessary to complete the calculations. The moles of AB formed, Δn_p, can be calculated from the relation $\Delta n_p = [AB]_p V_p$, where V_p is the volume of solution in the reaction vessel at the end of the time interval in question. A value for ΔH_R is then calculated using Eq. (5.10). In this example, two values of K_{eq} were chosen, 200 and 100, and the resulting species concentrations and ΔH_R values at the end of each time interval are given in Table 5.1. Values of ΔH_R are constant when K_{eq} is chosen to be 100, but not when K_{eq} is taken to be 200. Therefore, for this hypothetical case, $K_{eq} = 100$ and $\Delta H_R = 5.0$ kJ mol^{-1}. An actual determination would require that activity coefficients be incorporated into the calculations, that many more data points be taken from the curve, and that smaller increments be used in estimating

TABLE 5.1. Calculation of K_{eq} and ΔH_R Values from Titration Calorimetric Data using Eqs. (5.10)–(5.15)

Quantity, Units		4 min (p = 1)	8 min (p = 2)	12 min (p = 3)	16 min (p = 4)
$[A_{tot}]_p$, mol L^{-1}		4.69×10^{-3}	9.16×10^{-3}	1.34×10^{-2}	1.75×10^{-2}
$[B_{tot}]_p$, mol L^{-1}		9.77×10^{-3}	9.54×10^{-3}	9.33×10^{-3}	9.12×10^{-3}
$[AB]_p$, mol L^{-1}	($K_{eq}{}^a$ = 200)	2.74×10^{-3}	4.57×10^{-3}	5.67×10^{-3}	6.31×10^{-3}
$[AB]_p$, mol L^{-1}	($K_{eq}{}^a$ = 100)	2.04×10^{-3}	3.46×10^{-3}	4.42×10^{-3}	5.06×10^{-3}
n_p, mol	($K_{eq}{}^a$ = 200)	2.80×10^{-4}	4.79×10^{-4}	6.08×10^{-4}	6.92×10^{-4}
n_p, mol	($K_{eq}{}^a$ = 100)	2.09×10^{-4}	3.63×10^{-4}	4.74×10^{-4}	5.55×10^{-4}
$q_{R,p}$, J		1.04	1.81	2.37	2.77
ΔH_R, J mol^{-1}	($K_{eq}{}^a$ = 200)	3729	3790	3897	4011
ΔH_R, J mol^{-1}	($K_{eq}{}^a$ = 100)	5000	5000	5000	5000

Source: Ref. 9.

[a]Correct values of K_{eq} and ΔH_R are 100 mol^{-1} L and 5.000 kJ mol^{-1}, respectively.

K_{eq}. {Incidentally, if K_{eq} is accurately known, the ΔH_R value for the reaction is determined by solving Eq. (5.15) for $[AB]_p$, converting $[AB]_p$ to Δn_p, and calculating ΔH_R from Eq. (5.10).}

It should be pointed out that the above procedure is a least squares technique if Eq. (5.8) is used to test for constancy of ΔH_R (see Section 5.1.3c.).

5.1.3c. Least Squares Calculation of K_{eq} Values

The method of calculating the heat $q_{R,p}$ [Eq. (5.7)] released by the reaction(s) for which the K_{eq} value(s) is to be determined has been described in Section 5.1.

The mathematical relationships among the heat produced, the equilibrium constant(s), and the enthalpy change(s) for the reaction(s) are given in Eq. (5.7). The term $\Sigma\ \Delta H_{i,p}\Delta n_i$ in Eq. (5.7) refers to all reactions other than the ones for which the K_{eq} and ΔH_R values are to be determined. A simple method of solving Eq. (5.7) for K_{eq} and ΔH_R was outlined in Section 5.1.3b. for the reaction $A + B = AB$ ($j = 1$).

In general, the best values for K_j and ΔH_j are those which minimize $U(K_j, \Delta H_j)$ [see Eq. (5.8)]; that is, those values which satisfy both

$$\frac{\partial U(K_j, \Delta H_j)}{\partial \Delta H_k} = 0 = \sum_{p=1}^{m} \left(q_{R,p}\Delta n_{k,p} - \Delta n_{k,p} \sum_{j} (\Delta n_{j,p}\Delta H_j) \right) \tag{5.16}$$

and

$$\frac{\partial U(K_j, \Delta H_j)}{\partial K_k} = 0 = \sum_{p=1}^{m} \left(q_{R,p} - \sum_{j} (\Delta n_{j,p}\Delta H_{j,p}) \right) \sum_{j} \frac{\partial n_{j,p}}{\partial K_k} \tag{5.17}$$

where $k = 1, \ldots, j$. The j equations given by Eq. (5.16) are all homogeneous first-order linear equations in the ΔH_j values and may be solved easily if K_j values, and therefore $\Delta n_{j,p}$ values, are known. The j equations given by Eq. 5.17 are non-linear in the K_j values and must be solved by trial and error or by some iterative technique. A complete and accurate solution of Eq. (5.16) and (5.17) involves five steps: (1) assumption of initial K_j values; (2) calculation of the concentration of each species in the reaction vessel at each data point using the assumed K_j values; (3) calculation of the best ΔH_j values corresponding to the K_j values chosen; (4) evaluation of the error square sum [Eq. (5.8)] using these K_j and ΔH_j values to establish how well they fit the experimental data; (5) recalculation of the quantities in steps 2, 3, and 4 using new K_j values until the best set of K_j and ΔH_j values is found. These five steps have been discussed in detail (8, 11, 12).

5.1.3d. Calculation of K_{eq} Values Using Calorimetric and Free Ligand Concentration Data

If the system involves several simultaneous or sequential reactions, the accuracy with which one or more of the desired K_{eq} values can be determined may be unacceptable due to interdependence of some of the variables in the complex expression represented by Eq. (5.7). The probability of accurately defining the thermodynamic values for such a system is improved if two independent data sets are combined. It is frequently the case that the concentration of uncomplexed ligand can be measured spectroscopically, the concentration of uncomplexed metal ion measured potentiometrically, or the hydrogen ion concentration measured with a glass electrode. Any of these or similar data sets may be combined with the calorimetric data to simplify the calculations required to obtain K_{eq} and ΔH_R values.

The combining of calorimetric and other data sets can be illustrated by a system involving the interaction of A with multiple binding sites on B. This could be a model for binding of ligands, metal ions, or protons to a protein or for metal ion complexation with sites on the surface of a solid or a lipid bilayer membrane. The equilibria involved are

$$\begin{array}{ll} A + B = AB, & K_1, \Delta H_1 \\ 2A + B = A_2B, & K_2, \Delta H_2 \\ \vdots & \vdots \\ jA + B = A_jB, & K_j, \Delta H_j \end{array} \tag{5.18}$$

The system is described by the equilibrium expressions, Eq. (5.18), the mass balance expressions, Eqs. (5.19) and (5.20), and the equation relating q_R to the reacting species, Eq. 5.21:

$$[A_{tot}]_p = [A]_p + [AB]_p + 2\,[A_2B]_p + \ldots + j\,[A_jB]_p \tag{5.19}$$

$$[B_{tot}]_p = [B]_p + [AB]_p + [A_2B]_p + \ldots + [A_jB]_p \tag{5.20}$$

$$\frac{q_{R,p}}{V_p} = [AB]_p\,\Delta H_1 + [A_2B]_p\Delta H_2 + \ldots + [A_jB]_p\Delta H_j \tag{5.21}$$

Equations (5.18)–(5.21) can be combined to give Eq. (5.22) where $n_{B_{tot},p} = [B_{tot}]_pV_p$ is the total number of moles of B:

$$\frac{q_{R,p}}{n_{B_{tot},p}}\left(1 + \sum_j [A]_p^j K_j\right) = \sum_j [A]_p^j \Delta H_j K_j \tag{5.22}$$

Equation (5.22) is a linear expression in the $2j$ unknowns K_j and $\Delta H_j K_j$. If more than $2j$ data points have been measured, then the set of equations represented by Eq. (5.22) can be solved by linear least squares analysis. For the special case where all the K_j values and all the ΔH_j values in a set of j reactions are equal, Eq. (5.22) reduces to

$$q_{R,p} = j\Delta H_R n_{\mathrm{B_{tot}},p} + \frac{q_{R,p}}{[\mathrm{A}]_p K_{eq}} \tag{5.23}$$

and a plot of $q_{R,p}$ versus $q_{R,p}/[\mathrm{A}]_p$ will be linear with K_{eq} being determined from the slope and $j\Delta H_R$ calculated from the intercept, $j\Delta H_R n_{\mathrm{B_{tot}},p}$.

5.2. ILLUSTRATIONS

In this section the concepts and calculation techniques developed in previous sections are applied to several chemical systems. The examples considered are chosen because of their relevance to understanding the calculation procedures used in analyzing calorimetric data. The general use of calorimetry in the study of chemistry is not covered here, but is discussed in Chapters 6 and 7.

The first example in this chapter involves the calculation of the equilibrium constant for the binding of adenosine diphosphate to bovine liver glutamate dehydrogenase. The calorimetric data for the formation of a 1:1 complex is analyzed by a simplified procedure and also by complete least squares analysis of the data. Determination of equilibrium constants for reactions with large K_{eq} values is illustrated for the binding of 1,10-phenanthroline to Cu^{2+}. The calorimetric data that allow calculation of the K_{eq} values were obtained by titration of a solution of $Cu(NO_3)_2$ and 1,10-phenanthroline with strong acid. Analysis of the data requires independent determination of the K_{eq} and ΔH_R values for proton ionization from protonated 1,10-phenanthroline. The least squares analysis of calorimetric data is also illustrated with data for the complexation of CN^- by $Hg(CN)_2(aq)$ and for the adsorption of aniline by Linde Molecular Sieve 13X. In these two examples, the use of least squares analysis procedures to determine activity coefficient values and to calculate adsorption capacities in addition to calculating free energy, enthalpy, and entropy changes for the reactions of interest is illustrated.

5.2.1. Binding of Adenosine Diphosphate to Bovine Liver Glutamate Dehydrogenase

This system involves the binding of the inhibitor adenosine diphosphate (ADP) with the enzyme bovine liver glutamate dehydrogenase (GDH) to form a complex. The calorimetric data (13) are illustrated in Fig. 5.3 and are given in Table

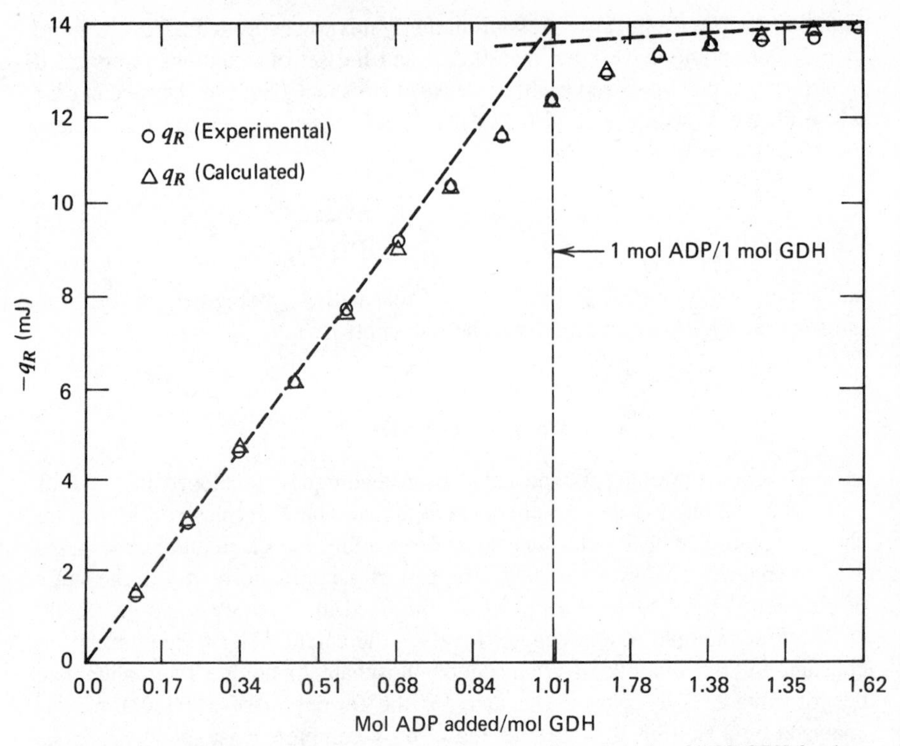

Figure 5.3. Experimental and calculated $q_{R,p}$ values versus the mole ratio of ADP/GDH for data given in Table 5.2. The calculated $q_{R,p}$ values were obtained using $K_{eq} = 2.70 \times 10^5$ and $\Delta H_R = -54.35$ kJ mol^{-1}.

5.2. The results suggest incomplete formation of a complex with 1 mol ADP:1 mol GDH. It is possible to calculate an equilibrium constant for the binding of ADP to GDH from the calorimetric data. However, Beaudette and Langerman (13) chose to use the K_{eq} value determined by Subramanian et al. (14), $K_{eq} = 2.6 \times 10^5$ mol^{-1} L, and calculated a ΔH_R value of -54.4 ± 2.9 kJ mol^{-1} by least squares analysis of the data. We will use this data set to illustrate the calculation of K_{eq} and ΔH_R values for the formation of the complex between ADP and GDH by two different approaches.

5.2.1a. Estimation of K_{eq} and ΔH_R for Enzyme-Inhibitor Binding

The binding of ADP to GDH is typical of many systems where the enzyme (E)–inhibitor (I) reaction can be represented by

$$\mathrm{E} + \mathrm{I} = \mathrm{EI}, \quad K_{eq} = \frac{[\mathrm{EI}]}{[\mathrm{E}][\mathrm{I}]} = K_{\mathrm{I}}^{-1} \tag{5.24}$$

Bjurulf et al. (15) have derived Eqs. (5.25) and (5.26), where $q_{R,p}$ is the total heat measured due to the reaction of n_{I} moles of I with n_{E} moles of E, q_{tot} is the total heat that would be measured for stoichiometric reaction of I with E ($q_{\mathrm{tot}} = n_{\mathrm{E_{tot}}} \Delta H_R$), $[\mathrm{I_{tot}}]$ and $[\mathrm{E_{tot}}]$ are the total concentrations of I and E at any data point p in the experiment, and [I] is the concentration of the uncomplexed inhibitor:

$$[\mathrm{I}]_p = [\mathrm{I_{tot}}]_p - \frac{[\mathrm{E_{tot}}]_p q_{R,p}}{q_{\mathrm{tot}}} \tag{5.25}$$

$$\frac{1}{q_{R,p}} = \frac{1}{q_{\mathrm{tot}}} + \frac{1/q_{\mathrm{tot}} K_{eq}}{[\mathrm{I}]_p} \tag{5.26}$$

It should be noted that Eq. (5.26) is identical to Eq. (5.23) ($j = 1$) used for the analysis of calorimetric data when the concentration of the uncomplexed species is known. Equation (5.26) has been used to derive K_{eq} and ΔH_R values for the binding of a wide variety of inhibitors to enzymes (see, for example, Refs. 15 and 16). This calculation technique involves assumption of a q_{tot} value to calculate K_{eq} and ΔH_R values for the complexation of I by E from a linear least squares analysis of the data using Eq. (5.25) and (5.26), recalculation of a q_{tot} value from the resulting ΔH_R value, and reiteration of the data set until a consistent result is obtained.

TABLE 5.2. Calorimetric Titration Data for the Addition of 6.17 mmol L^{-1} ADP to 2.00 mL of 0.1340 mmol L^{-1} GDH at pH 7.6 and 25°C

Titrant Volume (μL)	$-q_{R,p}$ (mJ)	Titrant Volume (μL)	$-q_{R,p}$ (mJ)
0.000	0.00	39.000	11.57
4.875	1.50	43.875	12.33
9.750	3.10	48.750	12.87
14.625	4.68	53.625	13.31
19.500	6.22	58.500	13.61
24.375	7.76	63.375	13.72
29.250	9.28	68.250	13.74
34.125	10.56	73.125	13.92

Source: Ref. 13.

The ΔH_R value (-54.35 kJ mol^{-1}) obtained for the titration of GDH with ADP by least squares analysis of the calorimetric data using the K_{eq} value determined by Subramanian et al. (14) gives a q_{tot} value of -14.566 mJ. The [GDH_{tot}] and [ADP_{tot}] values can be calculated for each data point to give the 1/[ADP] values [Eq. (5.25)] summarized in Table 5.2 and plotted versus $1/q_R$ in Fig. 5.4. Least squares fitting of the data to Eq. (5.26) should then give values of q_{tot} and K_{eq} for the binding of ADP to GDH. The curve in Fig. 5.4 shows obvious non-linearity for 1/[ADP] > 400 mmol^{-1} L. This non-linearity results from the fact that when ADP is first added to GDH, almost all the ADP is bound. As a result the $[I_{tot}]_p$ and $[E_{tot}]_p q_{R,p}/q_{tot}$ values in Eq. (5.25) are nearly equal. Thus, small errors in the calculated $q_{R,p}$ value will result in large errors in the calculated $[I]_p$ terms. The effect these errors will have on the calculated q_{tot} values is illustrated in Table 5.3. When the first three data points are used in the analysis, the resulting q_{tot} and K_{eq} values differ significantly from those obtained by least squares analysis of the remaining data. Although the effect

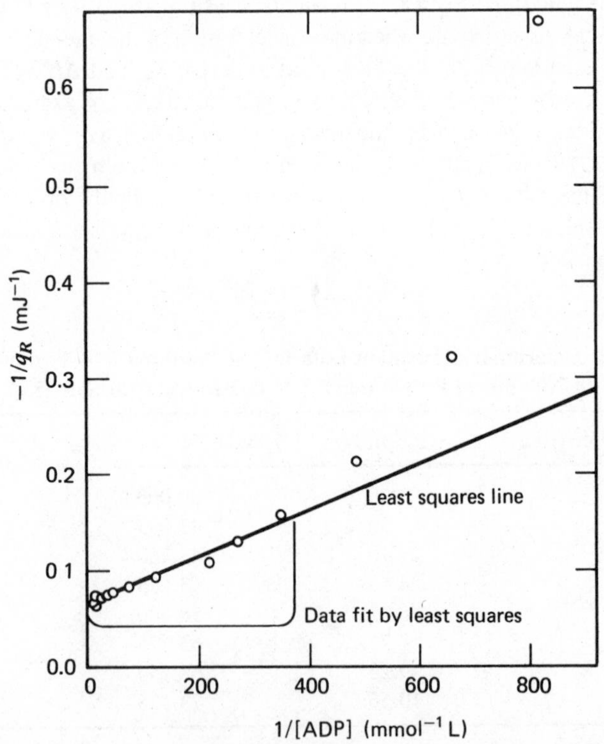

Figure 5.4. Plot of $1/q_{R,p}$ versus 1/[ADP] from the data in Table 5.2 for the binding of ADP to GDH.

TABLE 5.3. Calculated Values of q_{tot} and K_{eq} for the Data in Table 5.2

Data Points Analyzed	$-q_{tot}$ (mJ)	$K_{eq} \times 10^{-5}$ (mol^{-1} L)
1–15	26.85	0.66
2–15	16.70	1.72
3–15	15.16	2.39
4–15	14.62	2.85
5–15	14.27	3.40
6–15	14.09	3.87
7–15	14.23	3.44
4–12	14.77	2.79

will be less significant, errors can also be expected in the last few data points as the $[E_{tot}]_p q_{R,p}/q_{tot}$ values become small compared to the $[I_{tot}]_p$ terms, and the calculated $[I]$ values will be relatively insensitive to the measured $q_{R,p}$ values. It must thus be argued that the mid-data points should give the best result. Analysis of data points 4–12 gives $K_{eq} = 2.79 \times 10^5$ mol^{-1} L and $q_{tot} = -14.77$ mJ. These results illustrate that determination of K_{eq} and ΔH_R values for reactions of the type given in Eq. (5.24) by fitting that data to Eq. (5.26) requires accurate data near the 1:1 mol ratio of E and I. Errors can be expected in the fitted parameters distant from this 1:1 mol ratio. A weighted least square analysis technique has been recommended (16) to avoid these problems when analyzing data using Eq. (5.26). In contrast, complete least squares analysis of all the data as illustrated below is not subject to the same errors.

5.2.1b. Least Squares Calculation of K_{eq} and ΔH_R for Enzyme-Inhibitor Binding

The data given in Table 5.2 may also be analyzed by the least squares procedure outlined in Section 5.1.3c. We have analyzed these data by this procedure and calculated the values $K_{eq} = 2.70 \times 10^5$ and $\Delta H_R = -54.35$ kJ mol^{-1}. The $q_{R,p}$ values calculated from these K_{eq} and ΔH_R values are shown in Fig. 5.3. The plot of the error square sum versus log K_{eq} for the analysis is shown in Fig. 5.5. As illustrated, the data result in a well-defined minimum in the error square sum. The data are fitted with an average standard deviation of ± 0.071 mJ/data point, in agreement with the expected precision of the data (see Chapter 3). The log K_{eq} value determined by least squares analysis of the calorimetric data, 2.70×10^5, is in agreement with the value of 2.6×10^5 determined by a spectroscopic procedure (14). This value is also in agreement with the value of 2.79×10^5 determined by the simplified procedure given in Section 5.1.3b. However, the fit is more precise for the least squares analysis procedure using Eq. (5.8).

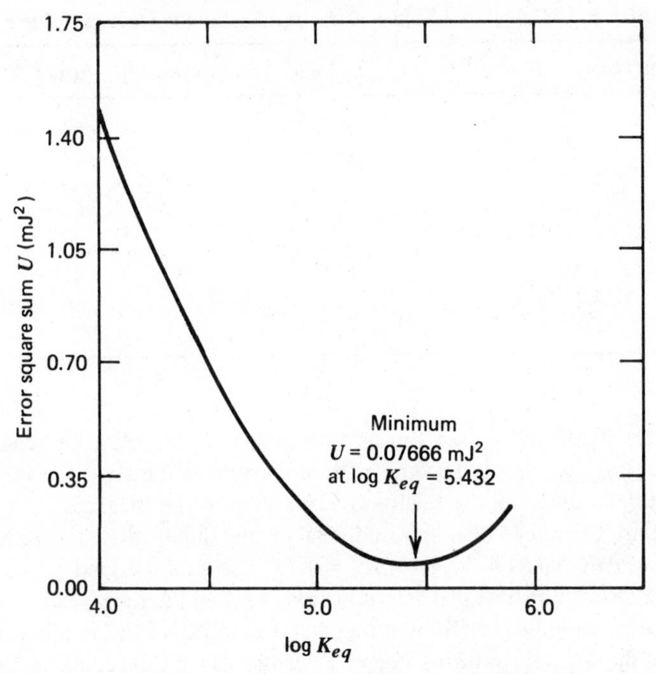

Figure 5.5. Error square sum U versus log K_{eq} for the data given in Table 5.2.

5.2.2. Complexation of 1, 10-Phenanthroline by Cu^{2+} in Aqueous Solution

Frequently, the equilibrium constant for the reaction(s) occurring in solution is large enough that addition of the two reactants A and B

$$A + B = AB, \quad K_B \tag{5.27}$$

will not result in the needed curvature for calculation of K_B from the calorimetric data. However, it may be possible to find a competing reaction

$$A + C = AC, \quad K_C \tag{5.28}$$

such that the equilibrium constant K_C/K_B,

$$AB + C = AC + B, \quad K_C/K_B \tag{5.29}$$

is within the bounds required for calculation from calorimetric data. If K_C is known, K_B may then be obtained. This technique has been used to determine the pK_a value for a weak acid by titration with another acid of similar pK_a (4, 17) or to determine the log K_{eq} values for metal complexes with a weak acid by titration of the complex with acid (8). The ΔH_R values for reactions (5.27) and (5.28) must be dissimilar. The procedure is well illustrated by the chemical system involving the reactions:

$$\begin{aligned} Cu^{2+} &+ P = CuP^{2+}, & K_1, \Delta H_1 \\ Cu^{2+} &+ 2P = CuP_2^{2+}, & K_2, \Delta H_2 \\ Cu^{2+} &+ 3P = CuP_3^{2+}, & K_3, \Delta H_3 \\ H^+ &+ P = HP^+, & K_{H^+}, \Delta H_{H^+} \end{aligned} \tag{5.30}$$

where P is 1, 10-phenanthroline.

The values of log K_1, log K_2, and log K_3 (9.1, 6.9, and 5.4, respectively) are large enough that the complexation is stoichiometric in a neutral solution when $[P_{tot}]_p/[Cu_{tot}^{2+}]_p = 3$. However, the log K_{H^+} value (4.9) is large enough that calorimetric titration of a solution of CuP_3^{2+} with H^+ results in the data shown in Fig. 5.6 with equilibrium between the Cu^{2+}, CuP^{2+}, CuP_2^{2+}, CuP_3^{2+}, HP^+, and H^+ species (Fig. 5.6) being such that K_1, K_2, and K_3 can be calculated from the data using literature values for K_{H^+} and ΔH_{H^+}. These experiments have been described in the literature (8) giving the results shown in Fig. 5.6.

5.2.3. Complexation of CN^- by $Hg(CN)_2$ in Aqueous Solution

$Hg(CN)_2(aq)$ will complex with CN^- anions to form $Hg(CN)_3^-$ and $Hg(CN)_4^{2-}$ ions. The formation of these complexes has been studied using titration calorimetry (18). The $q_{R,p}$ values for the addition of a NaCN solution to a $Hg(CN)_2$ solution are available (19). The data have been analyzed by the least square technique outlined in Sections 5.1.2. and 5.1.3 to give the log K_{eq} and ΔH_R values for reactions (5.31) and (5.32):

$$Hg(CN)_2(aq) + CN^- = Hg(CN)_3^-, \quad K_1, \Delta H_1 \tag{5.31}$$

$$Hg(CN)_3^- + CN^- = Hg(CN)_4^{2-}, \quad K_2, \Delta H_2 \tag{5.32}$$

The addition of the aqueous NaCN titrant solution to the aqueous $Hg(CN)_2$ will result in two heat effects for which correction terms must be calculated. These are the heat effect due to dilution of the titrant and titrate solutions, and also the heat resulting from changes in the concentration of both OH^- and HCN (aq) due to hydrolysis, reaction (5.33):

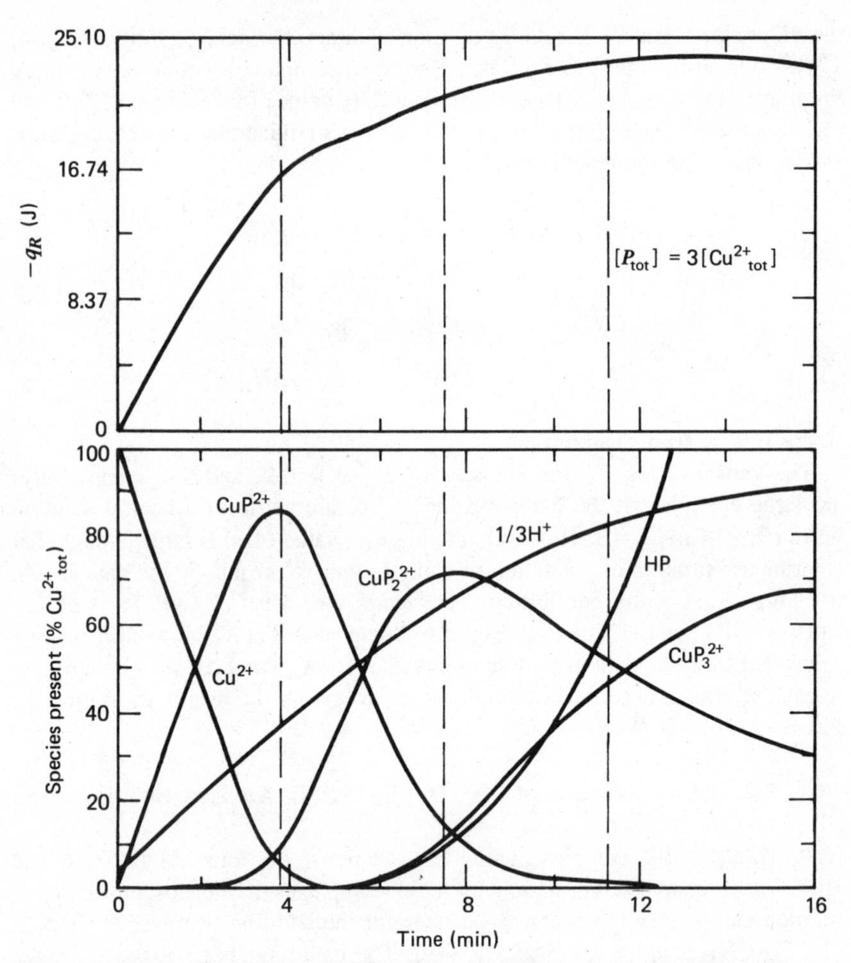

Figure 5.6. Titration of a $Cu(NO_3)_2$ − HNO_3 solution with P·HNO_3. (Reprinted with permission from Ref. 8. Copyright 1970 American Chemical Society.)

$$CN^- + H_2O = HCN(aq) + OH^- \tag{5.33}$$

Changes in the concentrations of CN^- and HCN(aq) during the titration result from the changes in the concentration of total cyanide in the solution as the titrant is diluted and from changes in the equilibria resulting from the dilution of titrant followed by the reaction of CN^- with $Hg(CN)_2$(aq). It was assumed in the treatment of the data that the heat due to dilution of 50 mL of the $Hg(CN)_2$(aq) solution by the addition of 5 mL of titrant was insignificant. Heat of dilution corrections for dilution of the NaCN solution were calculated from calorimetric data obtained on the addition of the NaCN titrant to water. The resulting $q_{D,p}$ values were subtracted from the measured $q_{C,p}$ values to correct for the heat of dilution. The corrections for the shift in equilibria due to hydrolysis [reaction (5.33)] were made by calculating the moles of OH^- and HCN(aq) present in the final solution of both titrant and titrate at each data point p, minus the moles calculated to be present in the titrant diluted to volume V_p but without addition of $Hg(CN)_2$. The correction term was then calculated using Eqs. (5.5) and (5.6). The log K_{eq} and ΔH_R values for reactions (5.34) and (5.35),

$$H^+ + CN^- = HCN(aq) \tag{5.34}$$

$$H^+ + OH^- = H_2O \tag{5.35}$$

are known (18). These values were used to calculate the correction term for the heat associated with the shift in equilibria of the hydrolysis, reaction (5.33). However, in the calculation of equilibrium constants for reactions (5.31) and (5.32), the correction to the measured $q_{C,p}$ values for the hydrolysis reaction will be dependent on the value of the equilibrium constants chosen for reactions (5.31) and (5.32). As a result, the hydrolysis correction term must be recalculated for each iteration of the log K_{eq} values.

The desired K_1 and K_2 values are thermodynamic constants valid at $\mu = 0$. The K_{eq} values in the literature for the formation of H_2O and HCN are also valid at $\mu = 0$. Activity coefficients of the species present in the reacting system must be known to calculate the equilibrium concentrations of the various species at each data point. In dilute aqueous solutions, the Debye–Hückel expression (20) given by

$$\log\gamma = -\mathcal{A}Z^2\,\mu^{1/2}/(1 + \mathcal{B}\mathring{a}\mu^{1/2}) + \mathcal{C}\mu \tag{5.36}$$

may be used to calculate activity coefficients. In Eq. (5.36), $\mathcal{A}$ and $\mathcal{B}$ are constants dependent on the temperature and solution dielectric constant, Z is the charge on each ion, μ is the ionic strength of the solution, and $\mathring{a}$ and $\mathcal{C}$ are empirically determined parameters. Earlier studies on the hydrolysis of HCN (18) have shown that the value of $\mathring{a} = 4$ Å correctly accounts for ionic strength effects at low μ values (where the $\mathcal{C}$ term is negligible). Therefore, it was assumed that this same value of $\mathring{a}$ would describe activity coefficients calculated from Eq. (5.36) for all the species present in the interaction of CN^- with $Hg(CN)_2$(aq). This value is also in reasonable agreement with expected values for an average "distance of closest approach." In a study of the $Hg(CN)_2(aq) - CN^-$ system by pH titration (19), it was found that consistent thermodynamic constants as a function of μ were obtained using a value of $\mathcal{C} = 0.55$. Since the error square sum given by Eq. (5.8) is a measure of how well the thermodynamic K_1

and K_2 values, corrected for activity coefficient effects, describe the data, the "correct" value of $\mathscr{C}$ would be that value which gives a minimum value of U. The effect of varying the value of $\mathscr{C}$ on the calculated error square sum, $U(K_j, \Delta H_j, \mathscr{C})$, is shown in Fig. 5.7. The results indicate that the system is best corrected for activity coefficient effects if the value $\mathscr{C} = 0.50$ is used in the Debye–Hückel expression. This value is in very good agreement with that ($\mathscr{C} = 0.55$) obtained in the pH study of the same system and indicates that the least square analysis of calorimetric data may be used to help in the empirical determination of activity coefficient corrections. The variation in $U(K_j, \Delta H_j, \mathscr{C})$ with varying $\mathscr{C}$ values is small as shown in Fig. 5.7. Such an approach should only be used if the magnitude of the expected detection limit for the measured heats is less than the error residuals in the error square sum at the minimum.

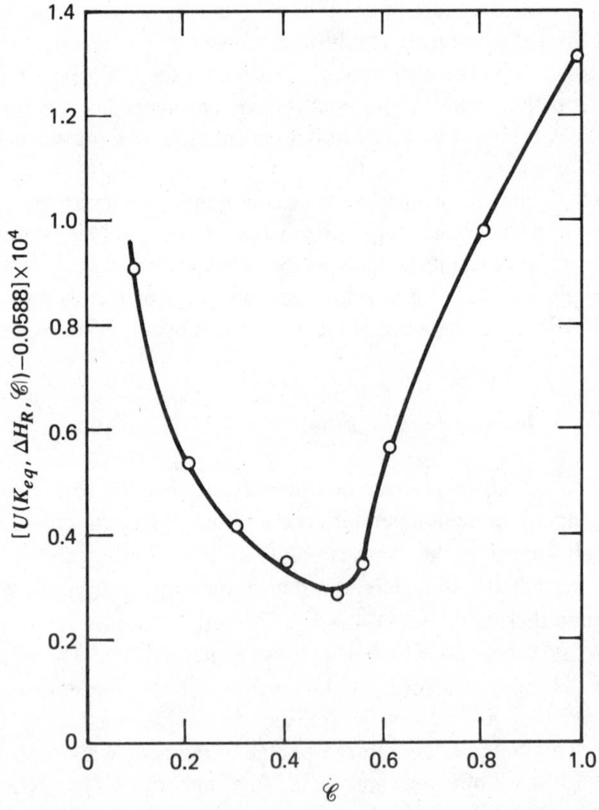

Figure 5.7. Error square sum U versus $\mathscr{C}$ in Eq. (5.36) for the titration of $Hg(CN)_2(aq)$ with NaCN(aq) (from Ref. 19).

5.2.4. Adsorption of Aniline by Linde Molecular Sieve 13X

Calorimetric data can be used to determine reaction stoichiometries (21) and adsorption capacities (22–24) for interactions at surfaces for both complete and incomplete reactions. For example, the calorimetric titration data obtained for the addition of 0.4 mol L^{-1} aniline in *n*-hexane to a suspension of 0.5 g Linde Molecular Sieve (LMS) 13X in 50 mL of *n*-hexane (22) is illustrated in Fig. 5.8. The data can be fit to a Langmuir isotherm by the least squares techniques outlined in Section 5.1.3c to determine the values for K_{eq}, ΔH_R, and the adsorption capacity *n* of the molecular sieve for aniline. Least squares treatment of triplicate runs similar to those illustrated in Fig. 5.8 gave the values (22) log $K_{eq} = 3.08 \pm 0.14$, $\Delta H_R = -26.0 \pm 0.4$ kJ mol^{-1}, and $n' = 1.64 \pm 0.10$ mmol g^{-1}. The log K_{eq} value and the adsorption capacity for aniline were determined independently by analysis of the concentration of aniline in equilibrium with that adsorbed by the zeolite as a function of total aniline added. In this non-calorimetric determination, the adsorption capacity and log K_{eq} value was also found from a plot of the amount of aniline adsorbed versus the total aniline added. The values obtained by this more laborious approach were log $K_{eq} = 3.05 \pm 0.03$ and $n' = 1.67 \pm 0.01$ mmol g^{-1}. The two independent approaches are in good agreement. These results illustrate that the mathematical approach outlined in Sections

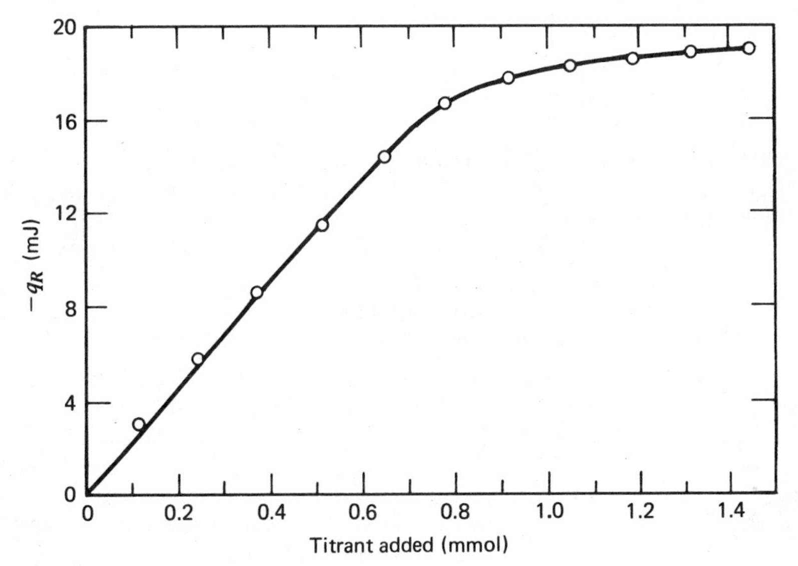

Figure 5.8. Calorimetric data for the addition of aniline in hexane to 0.5g of LMS 13X in hexane. (Reprinted with permission from Ref. 22. Copyright 1974 American Chemical Society.)

5.1.2 and 5.1.3 can be used to analyze calorimetry data and obtain reliable thermodynamic values for reactions involving adsorption on solids.

The data analysis for a heterogeneous system with many adsorption sites with differing ΔG_R, ΔH_R, and ΔS_R values would require more extensive modeling of the chemistry (23–26).

5.3. SUMMARY

The concepts and calculation procedures developed in Chapter 3 have been applied to the determination of K_{eq} and ΔH_R values from calorimetric data. The least squares analysis procedure is the best method for analysis of calorimetric data. This procedure has been illustrated for enzyme-inhibitor, metal–ligand, and sorbent–solid interactions.

In certain cases, it may be necessary to incorporate other data (e.g., measuring the concentration of uncomplexed species) with the calorimetric data to provide reliable thermodynamic values for complex systems and to provide a check on the assumed chemistry of the system. The use of data combinations of this type have been illustrated for a solid–sorbent reaction and an enzyme-inhibitor binding reaction.

Identifying and characterizing reactions occurring in complex chemical systems present a particular challenge to the scientist. Calorimetry provides an excellent means in many instances to probe such systems. In this chapter, examples of the use of calorimetry in several such instances have been given and it has been pointed out how the resulting data can be used to obtain definitive information concerning the systems involved. It is intended that this chapter together with Chapter 3 will be useful in delineating the procedures required to collect meaningful calorimetric data, to analyze these data, and to extract chemical information from them.

REFERENCES

1. D. D. Wagman, W. H. Evans, V. B. Parker, S. M. Bailey, and R. H. Schumm, *Selected Values of Chemical Thermodynamic Properties,* National Bureau of Standards, Technical Note 270-3, U.S. Government Printing Office, Washington, D.C., 1968.
2. V. B. Parker, *Thermal Properties of Aqueous Uni-Univalent Electrolytes,* National Bureau of Standards, NSRD-NBS 2, U.S. Government Printing Office, Washington, D.C., 1965.
3. F. D. Rossini, D. D. Wagman, W. H. Evans, S. Levine, and I. Jaafe, *Selected Values of Chemical Thermodynamic Properties,* National Bureau of Standards, Circular 500, U.S. Government Printing Office, Washington D.C., 1952.

4. J. J. Christensen, D. P. Wrathall, J. L. Oscarson, and R. M. Izatt, *Anal. Chem.* **40**, 1713 (1968).
5. J. J. Christensen, D. P. Wrathall, and R. M. Izatt, *Anal. Chem.* **40**, 175 (1968).
6. J. J. Christensen, J. H. Rytting, and R. M. Izatt, *J. Chem. Soc.* **A** 47 (1969).
7. R. M. Izatt, D. J. Eatough, R. L. Snow, and J. J. Christensen, *J. Phys. Chem.* **72**, 1208 (1968).
8. D. J. Eatough, *Anal. Chem.* **42**, 635 (1970).
9. J. J. Christensen, J. Ruckman, D. J. Eatough, and R. M. Izatt, *Thermochim. Acta* **3**, 203 (1972).
10. C. J. Martin and M. A. Marini, *CRC Crit. Rev. Anal. Chem.* **8**, 221 (1979); **8**, 407 (1979).
11. D. J. Eatough, J. J. Christensen, and R. M. Izatt, *Thermochim. Acta* **3**, 219 (1972).
12. D. J. Eatough, J. J. Christensen, and R. M. Izatt, *Thermochim. Acta* **3**, 233 (1972).
13. N. V. Beaudette and N. Langerman, *Anal. Biochem.* **90**, 693 (1978).
14. S. Subramanian, D. C. Stickel, and H. F. Fisher, *J. Biol. Chem.* **250**, 5885 (1975).
15. C. Bjurulf, J. Laynez, and I. Wadso, *Eur. J. Biochem,* **14**, 47 (1970).
16. R. L. Biltonen and N. Langerman, *Methods Enzymol.* **61**, 287 (1979).
17. L. D. Hansen, J. J. Christensen, and R. M. Izatt *Chem. Commun.* **3**, 36 (1965).
18. J. J. Christensen, R. M. Izatt, and D. J. Eatough, *Inorg. Chem.* **4**, 1278 (1965).
19. D. J. Eatough, PhD Dissertation, Brigham Young University, 1967.
20. H. S. Harned and B. B. Owen, *The Physical Chemistry of Electrolyte Solutions,* 3rd ed., Reinhold Publishing Corp., New York, 1958.
21. L. D. Hansen, R. M. Izatt, and J. J. Christensen, in *New Developments in Titrimetry,* J. Jordan, Ed., Marcel Dekker, New York, 1974, pp. 1–89.
22. D. J. Eatough, S. Salim, R. M. Izatt, J. J. Christensen, and L. D. Hansen, *Anal. Chem.* **46**, 126 (1974).
23. S. J. Rehfeld, L. D. Hansen, E. A. Lewis, and D. J. Eatough, *Biochim. Biophys. Acta* **691**, 1 (1982).
24. E. A. Lewis, "Binding of Surfactants to Clays," submitted, Brigham Young University (1984).
25. W. L. Gardner, D. P. Wrathall, and A. H. Herz, *Photograph. Sci. Eng.* **21**, 325 (1977).
26. D. P. Wrathall and W. Gardner, in *Temperature, Its Measurement and Control in Science and Industry,* H. Plumb, Ed., Instrument Society of America, Pittsburgh, 1972, Vol. 4, pp. 2223–2233.

CHAPTER

6

GENERAL ANALYTICAL APPLICATIONS OF SOLUTION CALORIMETRY

J. Keith Grime

Procter and Gamble Company
Ivorydale Technical Center
Cincinnati, Ohio

6.1. SCOPE

The application of calorimetric methods to analytical chemistry has been reviewed exhaustively in four books (1–4) and several reviews (5–11). More narrowly focused reviews have been published on organic compound analysis (12), inorganic solution thermochemistry (13), and catalyzed (non-enzymatic) reactions (14).

The published research prior to 1973 has been discussed in some detail in these texts. Accordingly, the emphasis in this review will be on work reported from 1973 until the present. Reports prior to that date will be cited only if new information has become available or for comparison purposes.

The instrumental bias is toward isoperibol technology, since, with few exceptions, instantaneous reactions form the basis of non-biochemical analytical calorimetry and, therefore, in general, few benefits accrue from the use of isothermal techniques. Extensive use has been made of tabulated information to maximize literature coverage. For ease of reference, each application area is preceded by a comprehensive but not exhaustive list of citations. Further discussion of a particular report in the text is denoted accordingly in each table.

The primary classification of applications is by compound type (anion, cation, organic, etc.) or by field of study (soils, pharmaceuticals, environmental, etc). Separate sections have been included on the specialized techniques, catalyzed indicator reactions, and gas enthalpimetry, which have developed considerably since the last major publication on analytical solution calorimetry.

6.2. ANIONS (SINGLE COMPONENT DETERMINATIONS)

A summary of the application of calorimetry to the determination of individual anionic species is presented in Table 6.1.

6.2.1. Azides—Protonation

Both soluble (sodium) and insoluble (silver, lead, copper, and mercury) azides can be determined directly by titration of the solution or suspension with hydrochloric acid (15). The enthalpy change is -15.06 kJ mol^{-1}. Endpoint location can be improved by addition of sulfate to induce the endothermic conversion to bisulfate as the titration reaction per se comes to its conclusion.

6.2.2. Chromate—Polymerization and Protonation

As part of a series of studies on the formation of polyanions in the course of the titration of neutral solutions of oxyanions (16, 18–21), a method has been developed for the determination of chromate by titration with perchloric acid (16). The titration curve contains three breaks at mole ratios (acid:chromate) of 0.1, 0.25, and 1.0. While the first two breaks are unexplained, the latter is presumed to be due to the formation of dichromate, that is, $2H^+ + 2CrO_4^- = Cr_2O_7^{2-} + H_2O$ and is used as the analytical endpoint. Obviously this procedure is severely limited in analytical application, since it would be subject to interference from other oxyanions or bases.

6.2.3. Molybdate (Molybic Acid and Molybdenum Trioxide)—Polymerization, Depolymerization, and Protonation

In the presence of a five-fold excess of a neutral salt, for example, Cs^+, Rb^+, K^+, NH_4^+, Na^+, Mg^{2+}, and Li^+, the stoichiometry of molybdate protonation has been shown to be $8H^+$:$7MoO_4^{2-}$ (18), which is consistent with the reaction

$$8H^+ + 7MoO_4^{2-} = Mo_7O_{24}^{6-} + 4H_2O \qquad (6.1)$$

Under these conditions, the assay of 0.005–0.5 mol L^{-1} orthomolybdate can be achieved with an accuracy and precision (rsd) better than 1.8%. In the same report, it was noted that the depolymerization of ammonium paramolybdate by hydroxyl ion can also be used as an analytical reaction. The enthalpogram exhibits two inflection points attributed to the reactions

$$(NH_4)_6Mo_7O_{24} + 8NaOH = 3(NH_4)_2MoO_4 + 4Na_2MoO_4 + 4H_2O \qquad (6.2)$$

and

TABLE 6.1. Application of Calorimetry to the Determination of Anions: Single Component Determinations

Analyte	Reagent	Mode[a]	Solvent	Detection Limit (DL), Range (R)	Reference	Text Section Reference
Azide	H_3O^+	T	Water	—	15	6.2.1
Chromate	H_3O^+	T	Water	—	16	6.2.2
Iodide	$Cr_2O_7^{2-}$	F	Water	1×10^{-2}–1×10^{-1} mol L^{-1} (R)	17	—
Molybdate	H_3O^+/OH^-	T	Water	5×10^{-3}–5×10^{-1} mol L^{-1} (R)	16, 18–21	6.2.3
	H_3O^+/OH^-	T	Water	1×10^{-2}–5×10^{-1} mol L^{-1} (R)	22	6.2.3
Nitrite	$Cr_2O_7^{2-}$	F	Water	1×10^{-2}–2 mol L^{-1} (R)	17	—
	NH_2SO_3H	F	Water	1×10^{-5} mol L^{-1} (DL)	23	6.2.4
Sulfate	Ba^{2+}	T	Water	1–40 g L^{-1} Na_2SO_4 (R)	24	6.2.5a
	CrO_4^{2-}/I^-	T	Water	1×10^{-4} mol L^{-1} (DL)	25, 26	6.2.5b
	Ba^{2+}	F	Water	1×10^{-2}–1 mol L^{-1} (R)	17	—
Sulfide	$Na_2S_2O_5$	T	Water	20–30 mg Na_2S (R)	27	6.2.6
Sulfite	$Cr_2O_7^{2-}$	F	Water	5×10^{-3}–1 mol L^{-1} (R)	17	—
Tungstate	H_3O^+	T	Water	5×10^{-3}–2×10^{-1} 1×10^{-4}–1×10^{-3} mol L^{-1} (R)	16 28	6.2.7
Vanadate	H_3O^+	T	Water	5×10^{-3}–5×10^{-1} mol L^{-1} (R)	16	6.2.8

[a]T = titration; I = injection; F = flow.

$$(NH_4)_2MoO_4 + 2NaOH = Na_2MoO_4 + 2NH_4OH \quad (6.3)$$

Although the quality of analytical data based on the first endpoint was found to be independent of molybdate concentration, data based on the second inflection deteriorated at low sample levels. Molybdic acid and molybdenum trioxide can also be determined by titration with hydroxide (18).

Subsequent detailed study has revealed the complex nature of molybdenum oxyanion proton-induced polymerization reactions (16). The shape of the enthalpogram and the number of inflection points are, in fact, dependent on the initial concentration of molybdate. At 5×10^{-3} mol L^{-1} only one inflection point is observed at an acid-to-molybdate mole ratio of 1.5. This can be rationalized in terms of the formation of the octomolybdate,

$$8MoO_4^{2-} + 12H^+ = Mo_8O_{26}^{4-} + 6H_2O \quad (6.4)$$

In contrast, the titration of 2×10^{-1} mol L^{-1} molybdate solutions produced an enthalpogram with three inflection points at acid-to-molybdate mole ratios of 1.14, 1.43, and 2.0 as shown in Fig. 6.1. The first break was assigned to the formation of heptamolybdate as in Eq. (6.1). The second change in slope was attributed to the reaction

$$Mo_7O_{24}^{6-} + 2H^+ = H_2Mo_7O_{24}^{4-} \quad (6.5)$$

and the third to

$$8H_2Mo_7O_{24}^{4-} + 32H^+ = 7H_4Mo_8O_{26} + 10H_2O \quad (6.6)$$

The salt effect observed in an earlier report (18) was later rationalized by invoking the formation of an ion pair between the heptamolybdate and the cations in

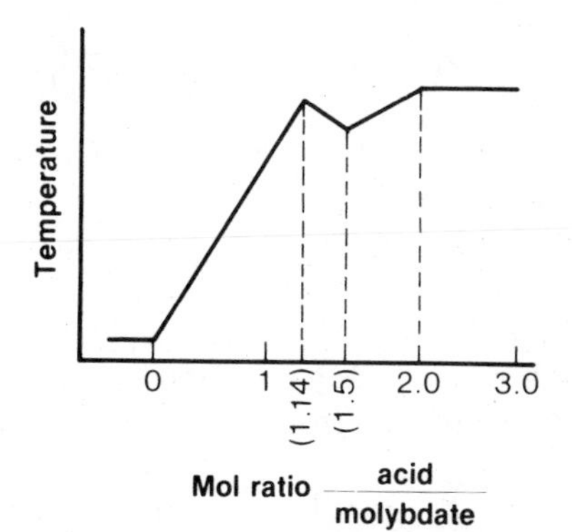

Figure 6.1. Enthalpogram for the titration of sodium molybdate (2.25×10^{-1} mol L^{-1}) with perchloric acid (16). (By permission of Pergamon Press.)

question (21), which inhibits further polymerization. The amount of salt necessary to achieve the maximum effect was found to vary linearly with the Stokes radius of the cation. Similar insights into the mechanism of depolymerization of molybdic acid and acidified molybdate solutions have been documented (20) by interpretation of potentiometric, thermometric, and spectrophotometric data of hydroxyl ion titrations. The initial concentration of molybdate proved once again to be a significant factor in determining the nature of the depolymerization process.

In another calorimetric study of isopolymolybdate formation (22), Jespersen reports four endpoints for the titration of molybdate (0.5mol L^{-1}) with a series of inorganic acids at mole ratios ($H^+:MoO_4^{2-}$) of 0.5, 1.14, 1.5, and 2.0. The reactions corresponding to each of these endpoints are reported as

$$\text{Endpoint 1:} \quad H^+ + 2MoO_4^{2-} = HMo_2O_8^{3-} \tag{6.7}$$

$$\text{Endpoint 2:} \quad 8H^+ + 7MoO_4^{2-} = Mo_2O_{24}^{6-} + 4H_2O \tag{6.1}$$

$$\text{Endpoint 3:} \quad 12H^+ + 8MoO_4^{2-} = Mo_8O_{26}^{4-} + 6H_2O \tag{6.8}$$

$$\text{Endpoint 4:} \quad 2H^+ + MoO_4^{2-} = MoO_3 + H_2O \tag{6.9}$$

The stoichiometry ($H^+:MoO_4^{2-}$) of endpoints 2,3, and 4 agree numerically with those observed by Kiba and Takeuchi (16) at various molybdate concentrations. A comparison of the assignment of reactions in each report is facilitated by rewriting Eqs. (6.1) and (6.7)–(6.9) in the form of successive reactions instead of relating each to the original molybdate ion. Therefore, we have

$$\text{Endpoint 1:} \quad H^+ + 2MoO_4^{2-} = HMo_2O_8^{3-} \tag{6.7}$$

$$\text{Endpoint 2:} \quad 7HMo_2O_8^{3-} + 9H^+ = 2Mo_7O_{24}^{6-} + 8H_2O \tag{6.10}$$

$$\text{Endpoint 3:} \quad 8Mo_7O_{24}^{6-} + 20H^+ = 7Mo_8O_{26}^{4-} + 10H_2O \tag{6.11}$$

$$\text{Endpoint 4:} \quad Mo_8O_{26}^{4-} + 4H^+ = 8MoO_3 + 2H_2O \tag{6.12}$$

The principal differences in the data are the endpoint at $H^+:MoO_4^{2-} = 0.5$ (not observed by Kiba and Takeuchi) and the endpoint at $H^+/MoO_4^{2-} = 1.42$ (not observed by Jespersen). Both can be attributed to intermediate protonation steps not observed in each case. For example, Eq. (6.7) and (6.10) can be summed to give Eq. (6.1), the equation representing the $H^+:MoO_4^{2-} = 1.14$ endpoint in both data sets. Similarly, if a protonation of heptamolybdate is interposed as an intermediate step before the formation of octamolybdate (between endpoints 2 and 3), Eq. (6.13) is obtained, that is,

$$8H^+ + 7MoO_4^{2-} = Mo_7O_{24}^{6-} + 4H_2O \quad (6.1)$$

$$Mo_7O_{24}^{6-} + 2H^+ = H_2Mo_7O_{24}^{4-} \quad (6.5)$$

$$10H^+ + 7MoO_4^{2-} = H_2Mo_7O_{24}^{4-} + 4H_2O \quad (6.13)$$

Therefore, an endpoint at $H^+:MoO_4^{2-} = 1.42$ would be predicted as observed by Kiba and Takeuchi.

Clearly, enthalpograms obtained for the acid-induced polymerization of molybdate should be interpreted with caution since the nature of the data will depend on the sensitivity of the instrument and the initial concentration of molybdate.

6.2.4. Nitrite—Oxidation with Sulfamic Acid

The oxidation of nitrite ion by sulfamic acid

$$NH_2SO_3H + NO_2^- = N_2 + HSO_4^- + H_2O \quad (6.14)$$

has a relatively large enthalpy change (-400.0 kJ mol^{-1}). Accordingly, a sensitive calorimetric determination of nitrite can be developed based on this reaction. Schifreen et al. (23) have reported a continuous flow method in which 120 μL samples of nitrite solution can be determined to a level of 1×10^{-5} mol L^{-1} at a maximum rate of 60 samples/h.

6.2.5. Sulfate

6.2.5a. Reaction with Barium (II)

The determination of sulfate by monitoring the enthalpy associated with the precipitation of barium sulfate has been documented since 1924. In a recent systematic study (24), the effect of foreign ions, titrant flow, and sample concentration have been examined in an attempt to develop a reliable calorimetric titration procedure and to more fully understand the precipitation mechanism. From a practical standpoint, one of the interesting recommendations of this report is that "neither the accuracy (stoichiometric ratio) nor the precision (reproducibility) of the thermometric titration is necessarily improved by the addition of alcohol." This is in contrast to the findings of Williams and Janata (29), who conclude that at least 50% alcoholic solutions are required to preclude negative errors caused by the adsorption of soluble sulfate on the precipitate. Dube and

Kimmerle (24) argue that, although the lower dielectric constant of the alcoholic solvent will reduce the degree of occlusion of sulfate salts, it will also increase the coprecipitation of barium salts resulting in increased interference from this source. Moreover, thermal noise from the mixing of an aqueous barium chloride titrant with alcoholic (>40%) sulfate solutions is also cited as having a deleterious effect on endpoint clarity.

Non-stoichiometric endpoints were obtained in this study. However, it was discovered that the concentration of sulfate in the sample ($[SO_4^{2-}]$) could be related to the titration time (t) by the equation

$$[SO_4^{2-}] = \mathcal{K}(t - \mathcal{T}) \cdot \text{titrant flow rate} \cdot [BaCl_2] \qquad (6.15)$$

where $[BaCl_2]$ is the concentration of barium chloride titrant, $\mathcal{K}$ is a constant, and $\mathcal{T}$ is a non-zero intercept associated with a time lag in the titrator and titration vessel. The authors conclude that the highest titrant flow rates consistent with a total titration time >30 s will yield near stoichiometric endpoints, $\mathcal{K} \approx 1$, and a relative standard deviation of ~1%. In practice, very high titration flow rates of 6 mL min^{-1} were used. In highly acidic media, interference from potential precipitants—fluoride (present as HF), bicarbonate (removed as CO_2), phosphate (present as H_3PO_4 or $H_2PO_4^-$), and chromate (soluble at pH<2)—can be neglected. The optimized procedure was ultimately applied to the determination of sulfite and thiosulfate in the spent alkaline liquors of an exhaust gas scrubbing system after permanganate oxidation to sulfate. The correlation of thermometric data with an indirect atomic absorption spectroscopy procedure is presented in Table 6.2.

TABLE 6.2. Comparison of Thermometric Titration and AAS Data for the Determination of Sulfate in Gas Scrubber Solutions[a]

Thermometric Titration (g L^{-1} Na_2SO_4)	Precision (% rsd)	AAS (g L^{-1} Na_2SO_4)	Precision (% rsd)	Difference (%)
6.4	1.5	6.5	2.2	−1.5
10.4	1.4	10.5	2.5	−1.9
15.8	1.1	15.5	2.8	+1.9
20.8	0.8	20.7	2.4	+0.5
25.4	1.1	25.5	2.7	−0.4
30.3	0.9	30.0	3.0	+1.0

Source: Reference 24, by permission of the American Chemical Society.

[a] 25-mL scrubber solution treated with 2 mL concentrated HCl and $KMnO_4$.

6.2.5b. "Precipitation Exchange" with Barium Chromate

The limit of detection for the enthalpimetric determination of sulfate is only about 10^{-3} mol L^{-1} because the enthalpy of precipitation of $BaSO_4$ (-19.25 kJ mol^{-1}) is relatively small (25). This detection limit can be improved by a factor of 10 by utilization of a "precipitation exchange" reaction between barium chromate and barium sulfate (25, 26). In this procedure, the analyte solution containing sulfate is added to a suspension of insoluble barum chromate. Chromate ions pass into solution in a molar amount equivalent to that of the sulfate in the analyte according to the equation

$$BaCrO_{4(s)} + SO_4^{2-} = BaSO_{4(s)} + CrO_4^{2-} \tag{6.16}$$

The solubilized chromate can then be determined by reduction to chromium (III) with iodide ion under acid conditions, that is,

$$2CrO_4^{2-} + 6I^- + 16H^+ = 3I_2 + 2Cr^{3+} + 8H_2O \tag{6.17}$$

The enthalpy change associated with the reaction represented by Eq. (6.17) is ca. -240 kJ mol^{-1}, resulting in a more than 10-fold increase in heat output per mole of sulfate in the sample versus precipitation of sulfate. An indirect method in which excess chromate ion is determined by use of the same redox reaction is also described in this chapter.

6.2.6. Sulfide

The calorimetric determination of sulfide in the presence of thiosulfate and polysulfides has been effected by a two-step procedure in which the sulfides are first protonated by addition of excess ammonium chloride at pH 8,

$$NH_4^+ + S^{2-} = NH_3 + HS^- \tag{6.18}$$

The analytical signal is then generated by monitoring the heat change associated with the formation of thiosulfate upon the injection of bisulfite solution (27),

$$2HS^- + 4HSO_3^- = 3S_2O_3^{2-} + 3H_2O \tag{6.19}$$

The enthalpy change associated with this reaction was determined to be -134 kJ mol^{-1}. The relatively slow rate of reaction of polysulfides prevents interference from this source.

6.2.7. Tungstate—Polymerization and Protonation

The potentiometric titration curves for the protonation of tungstate all have a single inflection point at a molar ratio (acid:tungstate) of 1.2 (16). However, the endpoint becomes indistinct as the concentration of tungstate decreases. At tungstate concentrations greater than 1×10^{-2} mol L^{-1}, a light yellow precipitate is formed during the course of the titration. The calorimetric titration curves, which have two breaks at acid-to-tungstate ratios of 1.17 and 1.5 at 5×10^{-3} mol L^{-1} tungstate, allow an elucidation of the polymerization mechanism (16). The principal processes are

$$6WO_4^{2-} + 7H^+ = HW_6O_{21}^{5-} + 3H_2O \quad \text{(mole ratio 1.17)} \qquad (6.20)$$

and

$$HW_6O_{21}^{5-} + 2H^+ = H_3W_6O_{21}^{3-} \quad \text{(mole ratio 1.50)} \qquad (6.21)$$

The latter product will precipitate at elevated concentration resulting in a third thermal response. In a single component system, the determination of tungstate according to the above reactions compares favorably in terms of accuracy and precision with a gravimetric procedure. The enthalpies of reaction for the processes described by Eq. (6.20) and (6.21) were determined to be -266.1 kJ mol^{-1} and -31.0 kJ mol, respectively (28).

6.2.8. Vanadate—Polymerization and Protonation

The perchloric acid titration enthalpograms for orthovanadate (Fig. 6.2a) and metavanadate (Fig. 6.2b) are complex, and their shape depends on the initial concentration of vanadate (16).

At an orthovanadate concentration of 2×10^{-2} mol L^{-1}, only one inflection point (acid:vanadate mole ratio = 1.0) has been observed, corresponding to the simple protonation reaction,

$$VO_4^{3-} + H^+ = HVO_4^{2-} \qquad (6.22)$$

At a vanadate concentration of 2×10^{-1} mol L^{-1} two additional break points can be observed at acid:vanadate mole ratios of 1.50 and 3.0. These changes in titration slope have been attributed to the reactions

$$2HVO_4^{2-} + H^+ = HV_2O_7^{3-} + H_2O \qquad (6.23)$$

and

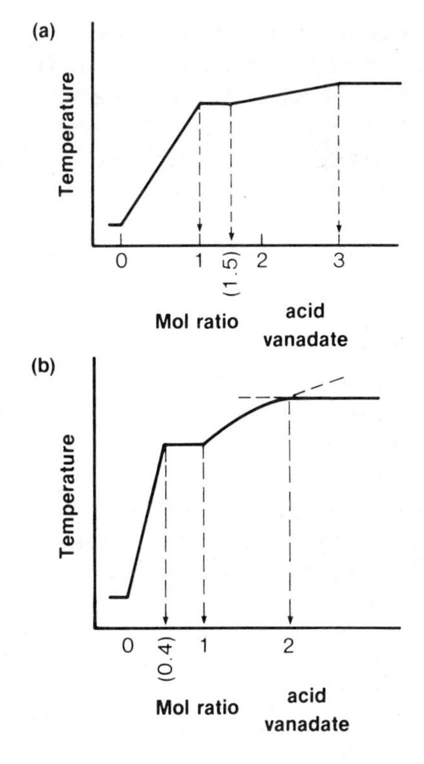

Figure 6.2. Enthalpogram for the titration of (a) orthovanadate (2.5×10^{-1} mol L^{-1}) and (b) metavanadate (1.0×10^{-1} mol L^{-1}) with perchloric acid (16). (By permission of Pergamon Press.)

$$5HV_2O_7^{3-} + 15H^+ = H_6V_{10}O_{28} \qquad (6.24)$$

The thermometric titration curve for metavanadate also contains three breaks at elevated concentrations of vanadate. The last endpoint is, however, vague. The three inflections are at acid:vanadate mole ratios of 0.4, 1.0, and 2.0. At all metavanadate concentrations the first endpoint is thought to correspond to the reaction

$$5V_4O_{12}^{4-} + 8H^+ = 2V_{10}O_{28}^{6-} + 4H_2O \qquad (6.25)$$

The suggested mechanism for the observation of the second and third endpoints is

$$\text{Second inflection:} \quad V_{10}O_{28}^{6-} + 6H^+ = H_6V_{10}O_{28} \qquad (6.26)$$

$$\text{Third inflection:} \quad H_6V_{10}O_{28} + 10H^+ = 10VO_2^+ + 8H_2O \qquad (6.27)$$

The rationale for the dependence of the second and third inflections on vanadate concentration is a kinetic argument based on the premise that the last step may proceed in two stages, namely, $H_6V_{10}O_{28} = 5V_2O_5 + 3H_2O$ followed by $V_2O_5 + 2H^+ = 2VO_2^+ + H_2O$. Since the formation of V_2O_5 is independent of pH but dependent on vanadate concentration, it can be argued that it will not be formed at any significant rate at low concentrations and, therefore, the second and third endpoints will not be seen. The invariance of the first endpoint at all vanadate concentrations allows its use as an analytical signal.

6.3. ANION MIXTURES

An often quoted criticism of calorimetric methods of analysis has been that the lack of selectivity in the detection system limits the application of the technique to simple systems. Historically, the determination of one or more components in a multicomponent system has only been attempted by the use of a sequence of selective reactions in which each component reacts with one in a series of carefully chosen reagents. Alternatively, thermodynamic selectivity has been used to successively titrate two or more components with the same reagent. In either case, the restrictions on reagent choice severely limit the application of calorimetry for the determination of complex mixtures.

Recently, however, the problem of multicomponent analysis has been addressed utilizing the concept of "partial molar enthalpies" (see Chapter 2) or "partial molar temperature pulses." This technique has been used extensively in the determination of multicomponent anionic solutions. A summary of the recently reported applications in this area is presented in Table 6.3.

The concept is quite simple; the total enthalpy change engendered by the reaction of a stoichiometric excess of reagent with one or more reactants is equal to the sum of the partial enthalpies of each reaction, assuming that the reactions are independent of one another. Therefore, for i components, i reagents (or sets of conditions producing i different thermal responses) are required to generate i simultaneous equations that can be solved for the concentration of each reacting specie. The appropriate equations are discussed in detail in Chapter 2.

6.3.1. Chloride/Bromide/Iodide, Bromide/Iodide

Bark and Nya (30) used the concept of a "partial molar temperature pulse" to achieve the simultaneous determination of chloride, bromide, and iodide. The reagents used in this study were silver ion and chloramine-T (sodium *p*-toluene sulfochloramide), the latter being used under two sets of conditions, namely, pH 2.5–3.5 and pH 8.0. Chloramine-T will react with both bromide and iodide at pH 2.5 but only with iodide at pH 8.0. An entirely empirical approach was taken in that a "molar temperature pulse" (in practice a recorder-pen deflection) was documented for each specie with *all* the reagents. This signal is in fact a

TABLE 6.3. Application of Calorimetry to the Determination of Anionic Mixtures

Mixture	Reagents	Detection Limit (DL), Range (R) of Any One Specie	Reference	Text Section Reference
Chloride, iodide, bromide	Ag^+ Chloramine T at pH 2.5/ pH 8.0	0.1 mmol (DL)	30	6.3.1
Iodide, bromide	MnO_4^- Ag^+	1×10^{-4}–1×10^{-3} mol L^{-1} (R)	31	6.3.1
Phosphate, arsenic (III), arsenic (V)	Ag^+ Ba^{2+} S^{2-}	0.01 mmol (DL)	32	6.3.2
Sulfate, sulfide, sulfite, thiosulfate	Ba^{2+} Ca^{2+} I_2 Ag^+	0.01 mmol (DL)	33	6.3.3
Sulfide, thiosulfate	I_2 Br_2	5×10^{-4}–1.5×10^{-3} mol L^{-1} (R)	34	6.3.4

characteristic of the particular equipment since it will be a function of the heat capacity of the system, the sensitivity of the Wheatstone bridge, the ambient temperature, and so on. It will also include any thermal response from the dilution/mixing of the reagent. The temperature pulse obtained for a series of standards of each component with each reagent was plotted versus the concentration of the standards, and the "molar temperature pulse" obtained from the slope of the graph. The relevant matrix for binary or tertiary mixtures of chloride, bromide, and iodide is shown in Table 6.4. The technique can be best illustrated by considering some of the data from Ref. 30 (Table 6.5.).

The total temperature change engendered by reaction of all species with Ag^+ is given by the sum of the "partial molar temperature pulses," that is,

$$100 = x(61) + y(70) + z(100) \tag{6.28}$$

where x, y, and z are the number of moles of Cl^-, Br^-, and I^-, respectively, in the sample cell.

TABLE 6.4. Calibration Matrix for the Determination of Binary and Tertiary Mixtures of Chloride, Bromide, and Iodide

	Molar Temperature Pulse[a]		
Reagent	Cl^-	Br^-	I^-
Ag^+	61	70	100
Chloramine-T			
at pH 2.5	0	35	125
at pH 8.5	0	0	40

Source: Ref. 30, by permission of Elsevier Science Publishers.
[a]Corrected for dilution response.

TABLE 6.5. Extract of Data from Ref. 30 for the Determination of a Tertiary Mixture of Chloride, Bromide, and Iodide

Amount Halide Added (mmol)			Molar Temperature Pulse[a]		
				Chloramine-T	
Cl^-	Br^-	I^-	Ag^+	pH 2.5	pH 8.5
0.20	0.40	0.60	100	89	24

Source: Ref. 30, by permission of Elsevier Science Publishers.
[a]Corrected for dilution response

At pH 2.5, the analogous equation is

$$89 = y(35) + z(125) \tag{6.29}$$

and at pH 8.5

$$24 = z(40) \tag{6.30}$$

Starting with Eq. (6.30) the values of x, y, and z can be calculated by successive substitution to be 0.2, 0.4, and 0.6, respectively. This process illustrates one disadvantage to this approach; the error in z is transmitted to the value of y and both these errors are incorporated into the value of x. The maximum error in x

over the range of concentrations studied is reported as ±2%. It is also clear from Eq. (6.28) that an approximate equality in the molar temperature pulses for each specie is a prerequisite in order to avoid inordinate propagation of error. Results were generated in binary and tertiary mixtures in which the components were present in molar ratios of 1:2:3.

In a similar vein, MnO_4^- and Ag^+ were used to generate simultaneous equations for the determination of a binary mixture of bromide and iodide by measuring the heat effect generated in a thermometric titration (31). The error in determining $10^{-4} - 10^{-3}$ mol L^{-1} of both anions was quoted as ca. 10%.

6.3.2. Phosphate/Arsenate (As^{5+})/Arsenite (As^{3+})

The principle of a "partial molar temperature pulse" has been extended to a so-called "equivalent system heat pulse" (SHP), which is defined as "the heat pulse which would be obtained in that particular titration system, using a sensitivity such that full scale deflection (FSD) 1 mV is obtained when 1 mmol of analyte is contained in the fixed volume (VmL) of the sample" (32). In other words, the partial temperature pulse described earlier (30) is normalized to a particular instrumental sensitivity (1 mV, FSD) and concentration of analyte (1 mmol/cell volume). This allows the calculation of analyte concentration from temperature pulses obtained with a variety of instrumental sensitivities and sample dilutions used in order to obtain an appropriate recorder-pen deflection. This procedure has been utilized to determine binary and ternary mixtures of phosphate, arsenic (III), and arsenic (V). The reagent matrix used is shown in Table 6.6. The kinetics of the reaction between sulfide and arsenic (III) are such that the reaction does not proceed to any significant extent during the course of the experiment.

TABLE 6.6. Reagent Matrix and "System Heat Pulse" (SHP) for the Determination of Phosphate, Arsenic (III), and Arsenic (V)

	SHP (1 mmol Analyte/20 mL at 1 mV FSD)		
Analyte	Ag^+	Ba^{2+}	S^{2-}
Phosphate	750	150	175
Arsenic (III)	300	0	0
Arsenic (V)	375	250	300

Source: Ref. 32.

An exothermic S^{2-}/As^{3+} reaction has been observed after a time delay of 20–30 s following reagent injection. However, this delay was sufficient to allow the delineation of this reaction from others and to allow the approximation that it does not contribute to the SHP. The interaction between phosphate and sulfide, although not predicted or explained, was observed to be reproducible and was included in the analytical calculation. The pH of the system was maintained around neutral to prevent the precipitation of Ag_2O. The method was extended to the determination of insoluble phosphates and arsenic (V) following conversion to soluble sodium salts by metathesis with sodium carbonate. The accuracy of the technique at the 1 mmol L^{-1} level was quoted as 1.5% and the total time for the determination of a mixture of all these anions is less than 10 min after dissolution of the sample.

6.3.3. Sulfate/Sulfide/Sulfite/Thiosulfate

Bark and Nya (33) have reported the determination of binary, ternary, and quaternary mixtures of sulfate, sulfide, sulfite, and thiosulfate using the reagent matrix shown in Table 6.7. Acceptable accuracy was obtained even in the case of a quaternary mixture that incurs the inevitable error propagation of solving four simultaneous equations.

6.3.4. Sulfide/Thiosulfate

Marik-Korda (34) has used the determination of sulfide and thiosulfate to assess the correlation between the expected and observed precision of the partial temperature pulse concept. The combination of reagents used in this case is iodine (iodine–potassium iodide solution) and bromine water (saturated). The relevant reactions are, therefore,

TABLE 6.7. Reagent Matrix and "System Heat Pulse" (SHP) for the Determination of SO_4^{2-}, S^{2-}, SO_3^{2-}, and $S_2O_3^{2-}$

	SHP			
Analyte	Ba^{2+}	Ca^{2+}	I_2	Ag^+
Sulfide	0	410	654	1000
Thiosulfate	0	0	100	500
Sulfite	50	−64	386	200
Sulfate	108	0	0	0

Source: Ref. 33, by permission of Elsevier Science Publishers.

I_2 (neutral medium)

$$2S_2O_3^{2-} + I_2 = S_4O_6^{2-} + 2I^- \tag{6.31}$$

$$S^{2-} + I_2 = S + 2I^- \tag{6.32}$$

Br_2

$$S_2O_3^{2-} + 4Br_2 + 5H_2O = 2SO_4^{2-} + 8Br^- + 10H^+ \tag{6.33}$$

$$S^{2-} + 4Br_2 + 4H_2O = SO_4^{2-} + 8Br^- + 8H^+ \tag{6.34}$$

The endothermic heat of dilution of the I_2/KI reagent was partly compensated for by the inclusion of methanol in the reagent.

Matrix theory (Chapter 2) predicts that the relative standard deviation for the determination of any two components can be represented by

$$\%\text{rsd}_{n_1} = \frac{\sigma_1}{n_1} = \frac{\sigma_q}{n_1}\left(\frac{\Sigma \Delta H_2^2}{D}\right)^{1/2} \tag{6.35}$$

$$\%\text{rsd}_{n_2} = \frac{\sigma_2}{n_2} = \frac{\sigma_q}{n_2}\left(\frac{\Sigma \Delta H_1^2}{D}\right)^{1/2} \tag{6.36}$$

when n_1, σ_1, and n_2, σ_2 are the determined amounts (mol) and standard deviations, respectively, for the determination of components x and y in a binary mixture and σ_q, a constant, is the standard deviation of the quantity of heat evolved, q. ΔH_1 and ΔH_2 are the *total* molar enthalpies of reaction of each component with *both* iodine (1) and bromine (2) as represented by Eq. (6.31)–(6.34). Therefore, $\Delta H_1 = H_{1\,(1)} + H_{1\,(2)}$ where subscript 1 denotes S^{2-}. D is the determinant

$$D = \begin{vmatrix} \Delta H_{1\,(1)} & \Delta H_{2\,(1)} \\ \Delta H_{1\,(2)} & \Delta H_{2\,(2)} \end{vmatrix} \tag{6.37}$$

Equations (6.35) and (6.36) predict that the uncertainty in the determination of one component in a binary mixture increases with increasing enthalpy of reaction of the other component and decreases with increasing values of D. The precision is, however, independent of the n_1/n_2 ratio. A generalized error diagram shown in Fig. 6.3 can be constructed from Eq. (6.35) and (6.36). For the $S_2O_3^{2-}/S^{2-}$

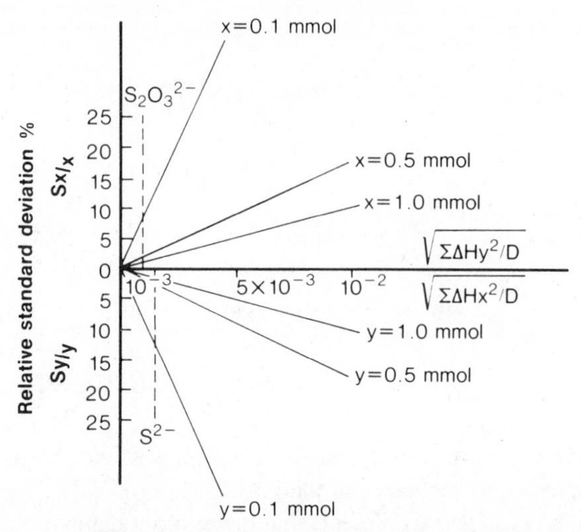

Figure 6.3. Relative error diagram for the simultaneous determination of thiosulfate and sulfide by direct-injection enthalpimetry; data constructed according to Eqs. (6.35) and (6.36). n_1 and n_2 are mol S^{2-} and mol $S_2O_3^{2-}$, respectively. The dashed line represents an extrapolation of the term $(\Sigma\Delta H^2/D)^{1/2}$ for both analyte reactions (34) (by permission of Pergamon Press).

sample, the diagram shows that a relative standard deviation of 0.74% per mmol $S_2O_3^{2-}$ and 0.79% per mmol S^{2-} can be expected. In practice, however, 45% of the determinations of $S_2O_3^{2-}/S^{2-}$ mixtures containing 0.01–0.3 mmol of each component had considerably poorer precision than predicted (34).

6.4. CATIONS

A summary of the application of calorimetric methods to the determination of cations classified according to the periodic table is shown in Table 6.8. Discussion has been limited to those reports in which the determination of a particular cation or series of cations is the primary objective. In order to avoid excessive duplication, the evaluation of individual thermochemical titrants for the determination of a variety of anions, cations, and functional groups has been dealt with separately (Section 6.13.3)

6.4.1. Group (II)A—Reaction with Linear Polycarboxylic Acids

Thermometric titrimetry has been used to determine the thermodynamic parameters, K_{eq}, ΔH_R, and ΔS_R for the complexation of a series of linear polycarboxylic

TABLE 6.8. Application of Calorimetry to the Determination of Cations

Periodic Classification	Analyte	Reagent	Mode[a]	Reference	Text Section Reference
Group IIA	Ca^{2+}, Mg^{2+}, Sr^{2+}, Ba^{2+}	Linear poly-carboxylic acids	T	35	6.4.1
	Ca^{2+}, Mg^{2+}	EDTA	T	36	—
Group VA	Sb^{3+}	BrO_3^-	T	37	6.4.2
Group VIA	Se^{4+}	$S_2O_3^{2-}$	T	38	6.4.3
Transition elements	Mixtures of Zn^{2+} and Cd^{2+}	EGTA	T	39	6.4.4a
	Mn^{2+} in the presence of Ca^{2+} and Ni^{2+}/mixtures of Mn^{2+} with Cd^{2+} and Zn^{2+}	EDTA	T	40	6.4.4b
	La^{3+}, Ti^{3+}	EDTA	T	41	6.4.4c
	Fe^{2+}	$S_2O_8^{2-}$	T	42	—
	Fe^{3+} in the presence of Cr^{3+}, Mn^{3+} and Al^{3+}	EDTA	T	43	6.4.4d, 6.4.4e
Actinides	Th^{4+}	EDTA	T	44	6.4.5a
	U^{4+}, U^{6+}	EDTA	T	41	6.4.5b
	U^{6+}	$B_4O_7^{2-}$	T	45	—

[a]T = titration; I = injection; F = flow.

acids with the alkaline earth cations (35). In particular the reactions of ethylenediamine-*N*, *N'*diacetic acid, trimethylenediamine-*N*, *N'*-diacetic acid, ethylenediamine-*N*-*N'*-di-α-propionic acid, and *N*, *N'*-dimethyl-ethylene-diamine-*N*, *N'*-diacetic acid were examined and the analytical potential of each complexing agent discussed. For each complexing agent the log K_{eq} of complexation decreases in the order $Mg^{2+} > Ca^{2+} > Sr^{2+} > Ba^{2+}$.

6.4.2. Group (V)A—Oxidation of Antimony (III) with Hypobromite

Antimony (III) in the form of Sb_2O_3 has been determined by thermometric titration with hyprobromite (37). The precision is comparable with a classical volumetric procedure for the determination of milligram amounts of Sb_2O_3.

6.4.3. Group (VI)A—Reduction of Selenium (IV) with Thiosulfate

The use of thiosulfate as a thermochemical titrant is often avoided because of its relatively high enthalpy of dilution (8.8 kJ mol^{-1}), which will diminish the sensitivity of a calorimetric procedure based on an exothermic reaction of thiosulfate with the analyte. However, the reaction of selenium (IV) with thiosulfate to form the selenopentathionate has been used successfully to determine submilligram amounts of selenium (IV) (38). An approximate enthalpy of -173 kJ mol^{-1} was assigned to the analytical reaction, that is,

$$H_2SeO_3 + 4S_2O_3^{2-} + 4H_3O^+ = Se(S_2O_3)_2^{2-} + S_4O_6^{2-} + 7H_2O \quad (6.38)$$

after subtraction of the endothermic dilution enthalpy for thiosulfate. Positive interferences were observed from Cu^{2+}, Pb^{2+}, and Hg^{2+}. A detection limit of 8.4×10^{-5} mol L^{-1} Se^{4+} is quoted for this procedure.

6.4.4. Transition Elements

6.4.4a. Simultaneous Determination of Zinc (II) and Cadmium (II) with EGTA

Predictably, the application of calorimetry to transition metal chemistry has been dominated by studies of EDTA and its analogues.

The feasibility of the successive titration of transition element mixtures with EDTA depends on there being a difference in the magnitude of *both* the equilibrium constant and the enthalpy of reaction for each reaction in the sequence.

The scope of this procedure can be widened by the incorporation of an auxiliary complexing agent into the system prior to titration. The overall titration reaction of each species then becomes the sum of two processes, namely,

$$MA_n \rightleftharpoons nA^- + M^{n+} \tag{6.39}$$

$$M^{n+} + Y^{n-} \rightleftharpoons MY \tag{6.40}$$

where A^- is the auxiliary complexing agent and Y^- is the complexation titrant. The thermodynamics of the auxiliary complex dissociation reaction can, in some instances, introduce the necessary thermodynamic differences required for a successive titration. This situation is best illustrated by considering the problem of a simultaneous determination of zinc (II) and cadmium (II) (39). Some relevant thermodynamic data are presented in Table 6.9. Inspection of these data shows that the successive titration of zinc (II) and cadmium (II) with EDTA is not feasible. Moreover, endpoint curvature could be predicted for a direct titration with EGTA as the difference in log K_{eq} for the two reactions is less than 3. Pretitration complexation of each specie with ammonia resolves this problem. Assuming that the tetraammine complex is formed in each case, the overall log K_{eq} for the concomitant dissociation of the zinc (II)–tetraamine and formation of zinc (II)–EGTA will be ca. 5. The analogous calculation for the cadmium titration reaction produces a log K_{eq} equal to 9.

Inclusion of the appropriate enthalpy data shows that the cadmium (II) reaction will occur first and will be exothermic. The succeeding zinc (II) substitution reaction will, in contrast, be endothermic.

TABLE 6.9. Thermodynamic Characteristics of the Reaction of Zinc (II) and Cadmium (II) with some Selected Complexing Agents

		Zn^{2+}		Ca^{2+}	
Complex	Ligand A	log β_{eq}[a]	ΔH_R[b] (kJ mol^{-1})	log β_{eq}[a]	ΔH_R[b] (kJ mol^{-1})
MA	EDTA	16.3	−23.4	16.3	−42.3
MA	EGTA	14.4	−20.9	16.6	−59.0
MA_4	Ammonia	9.46	−61.9	7.11	−58.6

Source: Ref. 39.

[a] $\beta_{eq} = [MA_n^{2+}]/[M^{2+}][A]^n$.

[b] ΔH_R for reaction $M^{2+} + nA = MA_n^{2+}$.

6.4.4b. Simultaneous Determination of Manganese (II) and Zinc (II) or Cadmium (II) with EDTA

The same auxiliary reagent principle has been extended to determine manganese (II) in the presence of copper (II) and nickel (II), and for the simultaneous determination of binary mixtures of manganese (II) with zinc (II) and cadmium (II) (40). In this instance, ethylenediamine, $(en)_2$, is used as the auxiliary reagent and EDTA is the titrant. The key to this determination is the relative instability (and hence small enthalpy of dissociation) of the manganese (II)–$(en)_2$ complex

TABLE 6.10. Equilibrium Constants and Reaction Enthalpies for the Reaction of Some Selected Transition Elements with EDTA and Ethylenediamine

		EDTA		$(en)_2$	
Cation	Complex (Ligand A)	$\log \beta_{eq}$ [a]	ΔH^b (kJ mo^{-1})	$\log \beta_{eq}$ [a]	ΔH^b (kJ mol^{-1})
H^+	HA	10.2	−23.8	10.1	−48.1
	H_2A	16.4	−41.8	17.6	−91.2
Cu^{2+}	MA	18.8	−34.3	10.7	−54.4
	MA_2	—	—	20.0	−106.3
	MA_3	—	—	—	—
Ni^{2+}	MA	18.7	−35.6	7.5	−37.7
	MA_2	—	—	13.8	−74.9
	MA_3	—	—	18.3	−114.6
Zn^{2+}	MA	16.5	−20.5	5.9	−21.8
	MA_2	—	—	7.8	−43.5
	MA_3	—	—	12.9	−65.3
Cd^{2+}	MA	16.5	−38.1	5.6	−25.9
	MA_2	—	—	10.2	−51.9
	MA_3	—	—	12.2	−77.8
Mn^{2+}	MA	13.8	−19.2	2.8	−11.7
	MA_2	—	—	4.9	−25.1
	MA_3	—	—	5.8	−46.4

Source: Ref. 40, by permission of Elsevier Science Publishers.

[a] $\beta_{eq} = [H_nA]/[H]^n[A]$; $[MA_n]/[M][A]^n$.

[b] ΔH for protonation $nH^+ + A \rightleftharpoons H_nA$; complexation $M + nA \rightleftharpoons MA_n$.

compared to the analogous zinc (II) and cadmium (II) complexes. The relevant thermodynamic data are presented in Table 6.10.

6.4.4c. Thermochemistry and Stoichiometry of Lanthanum (III)– and Titanium (III)–EDTA Complex Formation

The stoichiometry of the reactions of EDTA with lanthanum (III) and titanium (III) has been determined by thermometric titration (41) to be 1:1, that is, consistent with the reactions

$$La^{3+} + H_2Y^{2-} \rightleftharpoons LaY^- + 2H^+ \tag{6.41}$$

and

$$Ti^{3+} + H_2Y^{2-} \rightleftharpoons TiY^- + 2H^+ \tag{6.42}$$

Both reactions were observed to be endothermic.

6.4.4d. Thermodynamics of Iron (III)–EDTA Complex Formation

The thermodynamic parameters of the reaction

$$Fe^{3+} + Y^4 \rightleftharpoons (FeY)^- \tag{6.43}$$

where Y^- is the unprotonated EDTA anion, have been determined by direct thermometric titration (43). The thermochemistry of other reactions occurring during the course of the titration was established and taken into account in the calculation. The data are summarized in Table 6.11. The primary conclusion of this report is that the stability of the Fe (III)–EDTA chelate ($\log K_{eq} = 25$) is due primarily to the large entropy change occurring with its formation.

6.4.4e. Determination of Iron (III) by Titration with Na_4EDTA

Doi (43) has observed that iron (III) can be determined in the presence of chromium (III), manganese (II), aluminum (III), and phosphate ion by titration with Na_4EDTA in hydrochloric, sulfuric, or phosphoric acid media. Endpoint curvature observed with nitric or perchloric acid present in the reaction cell was attributed to the competitive effect of the complexing nitrate and perchlorate ions on the rate of Fe (III)–EDTA formation.

TABLE 6.11. Thermodynamics of Fe (III)–EDTA Complex Formation and Associated Reactions

Reaction	ΔH^a (kJ mol^{-1})	ΔG^a (kJ mol^{-1})	ΔS^a (J mol deg^{-1})
$Fe^{3+} + Y^{4-} \rightleftharpoons (FeY)^-$	-11.5 ± 0.5	-143	440
$Fe^{3+} + Y^{4-} + H^+ \rightleftharpoons FeHY$	-11.6 ± 0.5	—	—
$Fe^{3+} + H_2O \rightleftharpoons Fe(OH)^{2+} + H^+$	42.3 ± 6	—	—
Y^{4-} (titrant) $\rightarrow$ Y^{4-} (titrand)	3.8[b]	—	—

Source: Ref. 43.

[a]$\mu = 0.1$.

[b]Enthalpy of dilution of Na_4EDTA

6.4.5. Actinides

6.4.5a. Titration of Thorium (IV) with EDTA

The presence of large amounts of neutral alkali metal salts generally interferes in the classical EDTA titration of thorium (IV). In a recent report (44), Doi has shown that definitive thermometric endpoints can be obtained under these conditions. The enthalpy of reaction for the process

$$Th^{4+} + Y^{4-} = ThY \tag{6.44}$$

was shown to vary significantly with the ionic strength of the medium, ranging from -9 to -21 kJ mol^{-1} for $\mu = 0$ (extrapolation) to $\mu = 0.5$ ($NaClO_4$ ionic buffer), respectively. At high ionic strengths, the titrant was spiked with ionic salts to offset the effect of dilution enthalpies. By use of this procedure 4 mol L^{-1} of sodium chloride can be tolerated in the titration without any deleterious effect on endpoint precision. If 1.5 mol L^{-1} iodide ion is added to the solution as a masking agent, no interference is observed for lanthanum (III), manganese (II), bismuth (III), copper (II), cobalt (II), zinc (II), cadmium (II), lead (II), uranium (IV), nickel (II), and aluminum (III) at pH 1.5.

6.4.5b. Stoichiometry of Uranium (IV)/Uranium (VI)–EDTA Complexes

Thermometric titration studies of the uranium (IV)–EDTA and uranium (VI)–EDTA reaction show that the stoichiometry of either reaction is 1:1 in the presence of an excess of either reagent (41).

6.5. ACIDS AND BASES (NEUTRALIZATION CHEMISTRY)

The advantages of using calorimetric titration procedures for the determination of acidic or basic functionalities are well documented (1–4). The salient feature is that the feasibility of a successful titration does not depend *solely* on the magnitude of the free energy change ΔG_R (and hence the dissociation constant, pK) associated with the reaction. The difference in enthalpy change for the neutralization of a strong and weak acid (or base) is often much less significant than the difference in pK values because enthalpies of proton dissociation are typically small compared with neutralization enthalpies. In general, it is possible to calorimetrically titrate an acid or base with a p$K < 10$ with better than 1% precision (3). The restrictions imposed on a calorimetric titration by incomplete equilibria are discussed in detail in Chapter 2. A summary of the application of calorimetry to neutralization chemistry is presented in Table 6.12.

6.5.1. Inorganic Acids (Non-Aqueous Media)

Perchloric, hydroiodic, nitric, and hydrobromic acids have been titrated thermochemically with 1,3-diphenylguanidine in pyridine (46). Although linear enthalpograms were obtained, significant errors (4–40%) were obtained at titrant concentrations less than 1 mol L^{-1}. The choice of solvent is clearly critical; in an earlier report, Forman and Hume (57) titrated perchloric, hydroiodic, and hydrobromic acids in a less basic solvent, acetonitrile, with the same reagent and obtained satisfactory results.

6.5.2. Organic Acids (Aqueous Media)

6.5.2a. Diprotic Carboxylic Acids

The titration of diprotic carboxylic acids with a strong base in aqueous solution is a classic example of the ability of the calorimetric detection system to differentiate between reactions with very similar thermodynamic characteristics. For example, it has been shown that the first and second ionization reactions of oxalic, malonic, succinic, and phthalic acids can be delineated in a titration with sodium hydroxide (58). The most remarkable result observed in this study was that the first and second ionization stages of succinic acid, which differ in pK_a by only 1.45, could be distinguished. In the same report it was noted that intermediate inflection points could not be observed for the titration of glutaric, adipic, maleic, tartaric, and fumaric acids which have pK_a differences between first and second ionization stages of 1.1, 1.0, 4.4, 1.3, and 1.5 respectively. This work has recently been repeated by Burgot (47), who reports that intermediate endpoints can be seen for the titration of the above-mentioned acids. In

TABLE 6.12. Application of Calorimetry to the Study of Neutralization Reactions and the Determination of Acids and Bases

Classification	Analyte(s)	Reagent	Mode[a]	Solvent	Reference	Text Section Reference
Inorganic Acids (aqueous media)	H_3BO_3/H_2SO_4 mixtures	OH^-	T	Water	36	—
Inorganic acids (non-aqueous media)	HNO_3	$B_4O_7^{2-}$	T	Bu_3PO_4/ dodecane/CCl_4	45	—
	$HClO_4$, HI, HNO_3, HBr	1,3-Diphenylguanidine	T	Pyridine	46	6.5.1
Organic acids (aqueous media)	Tartaric, adipic, maleic, fumaric, glutaric	OH^-	T	Water	47	6.5.2a
	Glutaric, citric	OH^-	T	Water	48	6.5.2a
	Acetic/bromoacetic mixtures	OH^-	T	Water	49	6.5.2a
	Thioether-dicarboxylic acids	OH^-	T	Water	50	6.5.2b
	β-Aryl-α-mercaptopropenoic acids	OH^-	T	Water	51	6.5.2c
	Maleimide monoxime, dioxime	OH^-	T	Water	52	6.5.2d
	2-Mercaptopyridine, mercaptopyrimidine	OH^-	T	Water	53	—

Organic acids (non-aqueous media)	Mono- and polyprotic acids	OH^-	T	Water–methanol	54	6.5.3a
	N-*m*-Tolyl-*p*-methoxybenzo- and *N*-*m*-tolyl-*p*-chlorobenzohydroxamic acids	OH^-	T	Acetone	55	6.5.3b
	Picric, *o*-nitrobenzoic, acetic, benzoic acids, 2,4- and 2,5 dinitrophenol	1,3-Diphenylguanidine	T	Pyridine	46	6.5.3c
Organic bases	Thiocarbamides	OH^-	T	Acetone	56	—

[a]T = titration; I = injection.

TABLE 6.13. Thermochemical Data for the Neutralization of Diprotic Carboxylic Acids with Sodium Hydroxide

	Enthalpy of Neutralization[a] ΔH_N			Enthalpy of Dissociation[a] ΔH_d		
Acid	Total[c] ΔH_{NT}	First Stage ΔH_{N1}	Second Stage ΔH_{N2}	First Stage ΔH_{d1}	Second Stage ΔH_{d2}	$\Delta(\Delta H_d)^{a,b}$
Adipic	−118.3	−58.5	−59.8	−2.8	−4.0	−1.2
Tartaric	−110.4	−54.3	−56.0	−1.5	−0.3	−1.8
Glutaric	−120.4	−58.5	−61.9	−2.8	−6.1	−3.3
Fumaric	−120.0	−59.8	−63.1	−3.8	−7.4	−3.6
Maleic	−117.5	−56.4	−62.3	−0.6	−6.7	−6.1

Source: Ref. 47, by permission of Pergamon Press.

[a] kJ mol^{-1}.

[b] $\Delta H_{d2} - \Delta H_{d1}$.

[c] $\Delta H_{N1} + \Delta H_{N2}$.

addition, the enthalpies of ionization for each stage were determined by the usual extrapolation procedures. The relevant data are presented in Table 6.13. The enthalpy of dissociation ΔH_d was calculated from the equation

$$\Delta H_d = \Delta H_N - \Delta H_w \tag{6.45}$$

where H_w and H_N are the enthalpies associated with the reactions $H_3O^+ + OH^- = 2H_2O$ and $H_nA + OH^- = H_{n-1}A^- + H_2O$, respectively.

A value of -55.74 kJ mol^{-1} was used for ΔH_w. Significantly, only the pK difference for the dissociation of both maleic acid protons would permit successive titration by the classical "free-energy" methods. In a similar study (48), Bernard and Burgot have shown that three distinct slopes can be identified in the enthalpogram resulting from the titration of citric acid by sodium hydroxide. It had previously been stated that only one end-point could be observed in this titration (4). Burgot developed a theoretical framework to derive n equivalent equations which describe n different segments of the calorimetric titration curve of a polyprotic acid H_nA. For example, it was shown that in the titration of a diprotic acid, the heat generated by the neutralization of each proton $-q$ can be represented by

First Proton

$$-q = c_B Ft\Delta H_1 - [H_3O^+]_0 V_0(\Delta H_1 - \Delta H_w) \tag{6.46}$$

Second Proton

$$-q = c_B Ft\Delta H_2 + c_A V_0(\Delta H_1 - \Delta H_2) + [H_3O^+]_0 V_0 (\Delta H_w - \Delta H_1) \tag{6.47}$$

where c_A and c_B are the concentrations of diacid and titrant, respectively, $[H_3O^+]_0$ is the *initial* concentration of hydronium ions, F is the titrant flow rate, t is the titration time, ΔH_1 and ΔH_2 are the molar enthalpies associated with the reactions $H_2A + OH^- \rightleftharpoons HA^- + H_2O$ and $HA^- + OH^- \rightleftharpoons H_2O + A^{2-}$, respectively, and V_0 is the initial volume in the reaction cell. Insertion of the appropriate parameters for a typical diacid (glutaric) into Eq. (6.46) and (6.47) shows that the term $[H_3O^+]_0 V_0(\Delta H_1 - \Delta H_w)$ and its volume corrected version $[H_3O^+]_0(V_0 + V_t)(\Delta H_1 - \Delta H_w)$ are always negligible. Under the conditions of the experiment the value of the former was 0.030 J. In practice, therefore, the line representing the titration of the first proton can be fitted to the equation

$$-q = c_B Ft\Delta H_1 \tag{6.48}$$

Given the usual coordinates of $-q$ and Ft, the slope of this line is, therefore, $c_B\Delta H_1$. Similarly, the second titration line can be fitted to

$$-q = c_B Ft\Delta H_2 + c_A V_0(\Delta H_1 - \Delta H_2) \tag{6.49}$$

which has a slope of $c_B\Delta H_2$ for the same coordinates. The intersection of these lines corresponds to the half-neutralization point of the diacid. Moreover, insertion of the relevant parameters for glutaric acid into these equations showed that the difference in slope was within the sensitivity limits of a thermometric apparatus; this was confirmed by experiment. The authors have extended this simplified treatment to include corrective terms that compensate for competing equilibria, for example, the presence of significant amounts of A^{2-} after the neutralization of the first proton. When all such terms were taken into account, the shape of the enthalpogram, predicted by theory, correlated with experimental data.

The same theory governing successive neutralization equilibria of diprotic acids can be extended to determine the feasibility of the simultaneous determination of binary mixtures of monoprotic acids with similar pK_a values (49).

6.5.2b. Thioethercarboxylic Acids—Thermodynamics

The neutralization thermochemistry of three thioether acids(methylthio)acetic, thiodiacetic, and 3,3′-thiopropanoic—has been studied in order to establish the optimal conditions for their calorimetric titrations with sodium hydroxide (50). The enthalpies of neutralization and dissociation were calculated by two approaches: (1) the "overall heat capacity" approach, in which the integral heat change is determined and ΔH is calculated based on the assumption that the reaction goes to completion and (2) a point-by-point procedure in which the pK_a of the acid neutralized at any point is calculated. The latter results in a simultaneous determination of ΔH and pK_a as described in Chapter 5. The relevant data, which are presented in Table 6.14, follow the general trend for dicarboxylic acid deprotonation; the enthalpy of neutralization of the second proton is more exothermic, a result of the exothermic dissociation enthalpy.

6.5.2c. β-Aryl-α-mercaptopropenoic Acids—Thermochemistry

Substituted mercaptopropenoic acids have been titrated thermometrically with strong base after dissolution in 50% (*V/V*) ethanol/water solution (51). The

TABLE 6.14. Thermochemistry of Thioethercarboxylic Acid Neutralization[a]

(a) Neutralization Enthalpies

Acid	pK_a First Stage	pK_a Second Stage	Enthalpy of Neutralization, ΔH_N[b] "Heat Capacity" Method	"Point-by-Point" Method First Stage	"Point-by-Point" Method Second Stage
(Methylthio) Acetic	3.55	—	−60.31 ± 0.44	−60.53 ± 0.28	—
Thiodiacetic	3.14	4.01	−118.53 ± 0.66	−50.02 ± 1.54	−65.97 ± 1.81
3,3′-Thiopropanoic	3.87	4.68	−116.93 ± 0.65	−54.61 ± 0.98	−61.76 ± 0.76

(b) Dissociation Enthalpies

Acid	Enthalpy of Dissociation ΔH_d[b,c] First Stage	Second Stage
Methylthioacetic	−3.33 ± 0.35	—
Thiodiacetic	5.07 ± 1.54	−9.14 ± 0.65
3,3′-thiopropanoic	2.48 ± 0.98	−4.95 ± 0.70

Source: Ref. 50.

[a]At 25°C, $\mu = 0.1$

[b]$kJ\ mol^{-1}$.

[c]ΔH_d calculated by Eq. (6.45). ΔH_W in this case $= -56.44\ kJ\ mol^{-1}$.

successive neutralization of carboxylic (pK_a 3–3.5) and thiol (pK_a 8–8.5) groups was observed in the enthalpograms and the enthalpy of neutralization of both functionalities determined for a series of methoxyphenylmercaptopropenoic acids. The results are given in Table 6.15.

6.5.2d. Imidoximes

The determination of 0.02 mol L^{-1} maleimidemonoxine **1** by calorimetric titration of the oxime with sodium hydroxide in aqueous solution is possible

```
      N–OH             N–OH
HC—C//           HC—C//
||     \NH       ||     \NH
HC—C/            HC—C/
     \\O               \\N–OH

   1                 2
```

with an accuracy of ±1% (52). The authors report, however, that quantitative determination of the corresponding dioxime **2** by the same procedure is not

TABLE 6.15. Enthalpy of Neutralization for the Reaction of Methoxyphenylmercaptopropenoic Acids with Sodium Hydroxide

	Enthalpy of Neutralization $\Delta H_N{}^a$	
—Mercaptopropenoic Acid	-carboxyl	-thiol
3-(2-Methoxyphenyl)-2-	−46.9[b] −47.7[c]	−26.8[b] −27.2[c]
3-(4-Methoxyphenyl)-2-	−47.7[b] −47.7[c]	−26.8[b] −26.8[c]
3-(2,3-Dimethoxyphenyl)-2-	−48.5[b] −48.9[c]	−27.6[b] −28.4[c]
3-(3,4,5-Trimethoxyphenyl)-2-	−48.1[b] −48.9[c]	−25.9[b] −26.8[c]

Source: Ref. 51.

[a]kJ mol^{-1}.

[b]Chemical calibration.

[c]Electrical calibration.

possible. Following a distinct inflection point corresponding to the neutralization of 1 mol of hydronium ions, an attenuation in slope is observed accompanied by excessive enthalpogram curvature. In view of the symmetrical nature of the molecule, these data are surprising. The enthalpies associated with the neutralization of the monoxime and the first proton on the dioxime are approximately equal to -31.2 ± 0.3 kJ mol^{-1} and -30.1 ± 0.4 kJ mol^{-1}, respectively. An approximate calculation of the enthalpy associated with the neutralization of a second acid in the dioxime results in a value of -27.2 kJ mol^{-1}.

6.5.3. Organic Acids (Non-Aqueous Media)

6.5.3a. Thermochemistry of Mono- and Polyprotic Acids in 50% Water–Ethanol Medium

The use of non-aqueous solvents to enhance the acidity or basicity of an analyte relative to water is a well-established procedure irrespective of the endpoint detection method. For a calorimetric measurement, the use of a non-aqueous solvent system ion in some instances provides an added bonus of extra sensitivity (versus the sensitivity in water) because of the lower heat capacities involved. The significance of this effect will, of course, depend on the relative magnitudes of the reaction enthalpies in the two systems. In the case of a neutralization reaction where solvent equilibria play an important role in determining the overall enthalpy change, the reaction enthalpies may be significantly different in each system.

Cerda et al. (54) have documented the effect of a 50% methanol content on the aqueous neutralization enthalpies of several organic acids. The results are tabulated in Table 6.16. Evidently, the presence of 50% methanol by volume decreases the aqueous neutralization enthalpy by 65–80%. This is compared with a ca. 50% decrease in pure methanol (58). This reduction is rationalized by the increasing dominance of the methoxide ion as the titrant in methanolic solution. The neutralization enthalpy is therefore primarily governed by the reaction

$$H^+ + CH_3O^- = CH_3OH \tag{6.50}$$

A survey of these data show that little advantage is gained by using the compromise solvent of 50% methanol/water versus either pure solvent. For example, only one endpoint is observed for the diprotic acids—oxalic, malic, and tartaric—in 50% methanol, whereas both proton neutralizations are observed with either pure solvent. The only perceptible advantage is in the case of analytes that have limited solubility in water, for example, benzoic and salicylic acids. However, Harries (58) reports acceptable data in pure methanol for both these acids.

TABLE 6.16. Thermochemistry of Mono- and Polyprotic Acids in Aqueous And Methanol–Water Solvent Systems

	Neutralization Enthalpy, ΔH_N[a]					
	Water		Water–Methanol (50%)		Methanol[b]	
Acid	First	Second	First	Second	First	Second
Hydrochloric	−56.37	—	−45.23	—		
Boric	−42.68[b]	—	−55.31	—	−58.16	—
Benzoic	−58.24	—	−40.58	—	−29.45	—
Salicylic	−55.23	—	−42.09	—	−29.71	—
Sulfanilic	−37.24[c]	—	−35.16	—		
Anthranilic	−44.14[d]	—	−36.27	—		—
Nicotinic			−41.80	—		
Ascorbic		—	−46.61	—		
2-Thiophene-carboxylic	−61.25[e]	—	−46.98	—		
Oxalic	−56.48	−61.10	−45.23	Not observed	−35.15	−25.94
Malonic	−61.09	−63.18	−38.87	−43.72	−27.20	−30.12
Succinic	−55.23	−58.58	−36.27	−39.29	−26.36	−30.54
Maleic	−56.40	−62.30	−44.22	Not observed	−36.82	−19.25
Malic			−38.62	−40.92		
Phthalic	−59.50	−57.32	−38.45	−53.18	−27.20	−30.12
Hydrogen phthalate			−47.40	—		
Glutamic			−41.88	—		
Tartaric	−53.93	−55.98	−42.13	Not observed	−27.61	−25.52

Source: Ref. 54.

[a] kJ mol^{-1}

[b] Data from Ref. 58.

[c] Data from Ref. 59.

[d] Data from Ref. 60.

[e] Data from Ref. 61.

6.5.3b. Substituted Benzohydroxamic Acids

Data from the calorimetric, conductometric, and potentiometric titration of milligram quantities *N-m*-tolyl-*p*-methoxybenzohydroxamic acid and *N-m*-tolyl-*p*-chlorobenzohydroxamic acid with potassium hydroxide dissolved in isopropanol showed that potentiometric detection provides the most accurate (<1.0%) results (55).

TABLE 6.17. Calorimetric Determination of Organic Compounds

Classification	Analyte(s)	Mode[a]	Direct (D) or Indirect (ID)	Reagent	Reference	Text Section Reference
Alkenes/Alkynes[b]		I	D	Br_2water	62	—
Alcohols/glycols/ phenols	Ethylene glycol, glycerol	I	D	IO_4^-	63	—
	Ethylene glycol, glycerol	I	D	IO_4^-	64	6.6.1
	Ethylene glycol, glycerol	I	D	MnO_4^-	65	6.6.1
Amines	Sulfanilic acid, *p*-aminobenzoic acid,aniline, *o*-toluidine, α-napthylamine, benzidine, and diphenylamine	I	D	NO_2^-	66	6.6.2a
	Dihydrazinophthalazine sulfate	T	D		67	—
	Girard-T reagent [(carboxymethyl) trimethylammonium chloride hydrazide]	I	D	I_2 (aq)	68	—
	Hydrazine, aniline, hydroxylamine, methylhydrazine	T	D/ID	OBr^-, IO_3^-, $HClO_4$	69	6.6.2b
Carbohydrates	Arabinose, dulcitol, galactose, glucose, mannitol, mannose, rhamnose, ribose, sorbitol, xylose	T	ID	IO_4^-/hydrazine	70	6.6.3a
	Glucose, galactose, mannose, xylose, arabinose, ribose, rhamnose	T	ID	IO_4^-/mannitol, IO_4^-/hydrazine	71	6.6.3a
	Arabinose, galactose, glucose, mannose, xylose	T	ID	IO_4^-/mannitol	72	6.6.3a
	Arabinose, fructose, galactose, glucose, ribose, xylose	I	D	IO_4^-	73	6.6.3b
	Glucose, fructose, maltose	I	D	IO_4^-	74	6.6.3b

Carbonyls						
(i) aldehydes	Formaldehyde	I	D	MnO_4^-	65	—
	Aliphatic/aromatic aldehydes	T	ID	Unsymmetrical dimethylhydrazine	75	6.6.4a
	Acetaldehyde, formaldehyde, propionaldehyde, crotonaldehyde, salicylaldehyde, benzaldehyde	I	D	Hydroxylamine	76	6.6.4a
(ii) ketones	Acetone, 2-butanone, cyclopentanone, cyclohexanone and methylcyclohexanone	I	D	Hydroxylamine/ EtOH/KOH	77	6.6.4b
Carboxylic acids (non-neutralization reactions)	Tartaric, lactic and salicylic	I	D	MnO_4^-	64	—
Ethers	Macrocyclic, 18-crown-6 and 18-crown-7 ethers	T	D	Ba^{2+} Ag^+ K^+ Pb^{2+}	78	6.6.5a
Nitro/Nitroso compounds	*p*-Nitrophenol, picric acid, ammonium dinitro-*o*-cresolate (Nitrosan)	I	D	Ti^{3+}, Cr^{2+}	79	6.6.6
	Aromatic nitro compounds	T	ID	Ti^{3+}/Fe^{3+}	80	6.6.6
Polymers	Polyethylene glycols	T	D	$(Ph)_4B^-/Ba^{2+}$	81	6.6.7a
	Polyvinylamine, polyiminoethylene hydrochlorides	T	D	OH^-	82	6.6.7b
	Polyacrylic and polymethacrylic acids	T	D	OH^-, La^{3+}, Cu^{2+}, Ag^+	83	6.6.7c
Surfactants	Anionic/cationic and nonionic surfactants	T	D	H_2O	84	6.6.8a
	Anionic/cationic surfactants	T	D	Poly(vinylpyrrolidone)	85	6.6.8b

[a]T = titration; I = injection.
[b]All other reports of alkene/alkyne determinations are based on the technique of gas enthalpimetry and will be discussed in Section 6.12.

6.5.3c. Titration of Monovalent Acids with 1,3-Diphenylguanidine

The thermometric titration of picric, *o*-nitrobenzoic, acetic, and benzoic acids, and 2,4- or 2,5-dinitrophenol in pyridine with 1,3-diphenylguanidine produced generally unacceptable results, primarily due to enthalporgram curvature (46). As with the inorganic acids (Section 6.5.1) acetonitrile is the preferred solvent for this titrant.

6.6. ORGANIC COMPOUNDS

The relatively slow rates of molecular organic reactions are critical limiting parameters in the design of any "wet chemical" method for the determination of organic compounds. Kinetic considerations are particularly germane to the linear titration methods which are subject to error if the rate of the titration reaction approaches the rate of addition of titrant. In the special case of calorimetric analytical techniques there are essentially three methodological approaches to this problem:

1. An indirect method, in which a stoichiometric excess of reagent is titrated after the primary reaction has been allowed to reach equilibrium.
2. An incremental or discontinuous titration in which the titration is stopped at regular intervals in order to allow the titration reaction to reach equilibrium before further addition of titrant.
3. Direct-injection enthalpimetry, in which a stoichiometric excess of the reagent is injected into the analytical solution in a single pulse. The analytical signal in this case is not an endpoint but the magnitude of the resultant heat pulse as the reaction is allowed to reach thermal and chemical equilibrium. The acceptable time period for the reaction to come to completion will depend on the "adiabaticity" of the apparatus or the type of instrument (isoperibol versus isothermal). Such considerations are discussed in detail in Chapter 3.

Options 1 and 3 are the most often quoted methods for the calorimetric determination of organic compounds. From a practical standpoint, the efficient use of an incremental titration demands a programmable buret, which is still not a common accessory on most calorimeters.

In the following review, classification has been made according to the major classes of compounds studied rather than by a rigid adherence to functional group nomenclature. A summary of reported calorimetric determinations of organic compounds is presented in Table 6.17.

6.6.1 Alcohols/Glycols

The primary reaction used for the calorimetric determination of diols and polyhydroxyl compounds has been the oxidation of the hydroxyl functionality to the corresponding aldehyde by periodate under acid conditions (63, 64). Ethylene glycol and glycerol are presented as typical examples of this reaction in Eq. (6.51) and (6.52), respectively.

$$\underset{\text{OH}\ \ \text{OH}}{\text{H}_2\text{C—CH}_2} + IO_4^- = 2HCHO + IO_3^- + H_2O \qquad (6.51)$$

$$\underset{\text{OH}\ \ \text{OH}\ \ \text{OH}}{\text{H}_2\text{C—CH—CH}_2} + IO_4^- = H_2C{=}0 + \underset{\text{O}\quad \text{OH}}{\text{CH—CH}_2} + IO_3^- + H_2O \qquad (6.52)$$

The relatively slow rate of these reactions demands that injection or indirect titration procedures be employed. However, the sensitivity of the thermochemical determination based on these reactions is high because of the large enthalpy of reaction involved; for perspective, some typical ΔH_R values for the periodate oxidation of some selected compounds in this class are shown in Table 6.18.

Jeffries and Fresco (64) make the interesting observation that the reaction enthalpies for the compounds studied are approximately equal when expressed

TABLE 6.18. Enthalpy of Reaction—Periodate Oxidation of Diols, Triols, and Diones

Compound	ΔH_R (kJ mol^{-1})
Ethylene glycol	−207.1
2,3-Butanediol	−213.8
Trans-1,2-cyclohexanediol	−198.2
Cis/trans mixture cyclohexanediol	−193.3
Glycolaldehyde	−237.2
Glyceraldehyde	−501.6
2,3-Butanedione	−253.5
1,4-Dibromo-2,3-butanedione[a]	−252.7
Glycerol	−466.5
Mannitol	−1086.2

Source: Ref. 64.

[a]In 40% ethanol.

as kJ/equiv., suggesting that the enthalpy change can be used to determine the number of α-diol functions in a mixture. In the same report, this chemistry was applied to the determination of total glycols (ethylene and propylene) in windshield-washer fluid and antifreeze and glycerol in hand lotion. Since periodate oxidation is non-selective, α-diketones, α-ketoalcohols, etc., will interfere if present. Vulterin et al. (65) have shown that potassium permanganate can also be used as the oxidant if the reaction is carried out under alkaline conditions.

6.6.2. Amines

6.6.2a. Diazotization of Primary and Secondary Aromatic Amines

The feasibility of a direct thermochemical titration of aromatic amines by means of a diazotization reaction with sodium nitrite has long been established (3). Stastny et al. have recently described a "double injection" procedure utilizing the same reaction (66). This concept is in fact a reagent blank compensation method; the principle can be illustrated by using the diazotization of an aromatic amine as an example. In order to circumvent decomposition, the diazotization reagent, nitrous acid, is typically prepared *in situ* by injection of a stoichiometric excess of a solution of sodium nitrite into an acidified solution of the sample amine. The reactions contributing to the overall heat of reaction are, therefore,

$$NaNO_2 + HCl = HNO_2 + H_2O + NaCl \tag{6.53}$$

and

$$RNH_2 + HNO_2 + HCl = RN_2^+Cl^- + 2H_2O \tag{6.54}$$

Any heat effects associated with the dilution of sodium nitrite will also be included in the total enthalpy change. If a second aliquot of sodium nitrite is subsequently injected into the same reaction cell, *and a molar excess of hydrochloric acid (wrt $NaNO_2$) is still present,* the temperature change observed will be that associated with reaction (6.53) and any attendant dilution/mixing processes. This signal can therefore be subtracted from that obtained by the reaction of the analyte resulting in a response which is proportional only to the concentration of sample amine. The blank signal will, of course, be slightly attenuated by the increase in heat capacity caused by the increased volume due to the initial reagent injection. However, if the blank signal is minimized by judicious choice of experimental conditions and the reagent volumes are kept small, this factor can be neglected. By use of this procedure the determination of millimole amounts of *p*-aminobenzoic acid, sulfanilic acid, *o*-toluidine, aniline, 1-napthylamine, benzidine, and diphenylamine has been reported with a relative precision of 0.8% at the 95% confidence level.

6.6.2b. Binary Mixtures of Hydrazine and Selected Nitrogen Bases

Both the basic and reducing properties of hydrazine can be used for a thermometric determination, alone and in admixture with other nitrogen bases (69). Mixtures of hydrazine and aniline were determined by serial titration with potassium bromate in the presence of bromide ion. The redox potentials of these reactions are such that quantitative serial endpoints were obtained with hydrazine titrating first. The relevant stoichiometries are

$$2\ BrO_3^- \equiv 3N_2H_4 \quad \text{and} \quad 1BrO_3^- \equiv 1C_6H_5NH_2.$$

Poor hydroxylamine oxidation kinetics hindered the serial determination of hydrazine and hydroxylamine by direct bromate titration. However, quantitative results were obtained from a combination of the results of an indirect determination of total base content (by titration of excess bromate with standard hydrazine sulfate) and a direct bromate titration of hydrazine. Similarly, hydrazine/ammonia mixtures can be assayed by subtraction of the hydrazine content, obtained in a bromate titration, from the total base content obtained in a conventional neutralization titration with mineral acid. The determination of hydrazine and unsymmetrical dimethylhydrazine (UDMH) also required a two-titration sequence. A stoichiometric excess of an aldehyde was added to the mixture to form the corresponding hydrazone derivatives. In glacial acetic acid solution, the neutral UDMH derivative remained in solution, while the more acidic simple hydrazone precipitated. Oxidimetric bromate titration of this mixture therefore allowed a determination of UDMH alone. Subsequent acid–base titration to ascertain the total base content resulted in a calculation of hydrazine content.

6.6.3. Carbohydrates

6.6.3a. Indirect Titration of Excess Periodate

The oxidation of hydrazines by periodate forms the basis of an indirect thermometric titration method for the determination of carbohydrates (70, 71). Following equilibration of the carbohydrate with excess periodate in acidic solution (pH 1.5–2.0), the pH is adjusted to 7–7.5 and the residual periodate is then titrated with hydrazine sulfate solution. Typically, carbohydrate oxidation is complete within 30 min. As shown in Table 6.19, the results compare favorably with the classical "Malaprade Method" in which iodate, formed in the oxidation of the sugar, is determined by volumetric thiosulfate titration. The relative precision of the thermometric technique based on the determination of ten 25-mg aliquots of galactose is 0.8% (71). This procedure is an excellent example of the simplification of an analytical method by adaptation to calorimetric detection. In the Malaprade procedure, the analytical solution becomes quite complex because a sequence of reagents are employed to produce an analytical signal. Molybdate is added to mask excess periodate, and excess iodide ion is finally introduced to produce iodine which is subsequently titrated with thiosulfate. The probability of interference caused by interaction of reagents and reaction products obviously increases with the number of steps in the analysis. As indicated in Table 6.18, the enthalpy of reaction for the oxidation of mannitol by periodate is remarkably high at over -1000 kJ mol^{-1} (64) and offers the opportunity for a sensitive thermochemical determination of carbohydrates by back-titration of excess periodate with mannitol. The feasibility of this procedure has been evaluated by Bark and co-workers (70, 72). This study showed that rate of the mannitol–periodate reaction was fast enough to allow the use of mannitol as a direct titrant. Precision and accuracy data obtained for the determination of arabinose, galactose, fructose, mannose, and xylose are comparable with the hydrazine titration shown in Table 6.19. Slow periodate oxidation kinetics preclude the application of this technique to the determination of fructose or sorbose within an acceptable experiment time. For perspective, equilibration of sorbose with periodate for periods up to 18 h was required to approach complete reaction. The authors report the routine determination of 10^{-4} mol of carbohydrate by use of a simple "enthalpimeter." Nanomole detection limits could be predicted if this chemistry was monitored in an isoperibol or heat-conduction apparatus.

6.6.3b. Direct Determination by Periodate Oxidation

The direct calorimetric determination of carbohydrates based on the injection of concentrated periodate solution into carbohydrate solutions has been reported (73, 74).

De Oliveira and Rodella (73) observed that at pH 7.2 the oxidation reaction appears to proceed via two distinct stages; the first stage of the reaction is fast, the second significantly slower. Therefore, at the expense of sensitivity, the enthalpy change associated with only the first-stage reaction was monitored as the analytical signal for periods ranging between 2 and 5 min. The temperature–time plot returned to baseline slope after this time.

The relatively large exothermic enthalpy of dilution of sodium periodate at this pH proved to be a source of considerable imprecision, since at the levels of carbohydrate studied it accounted for 40% of the observed signal. This effect was effectively nullified by the use of a partially neutralized solution of periodic acid as the reagent. The endothermic dilution enthalpy of the acid is approximately equal and opposite in sign to that of periodate. This problem was overcome

TABLE 6.19. The Determination of Selected Carbohydrates by Indirect Thermometric Titration of Excess Periodate with Hydrazine Sulfate

	% Recovery[a]	
Compound	"Malaprade[b] Titration"	Thermometric[c] Titration
Arabinose	98.9	99.9
Dulcitol	99.2	100.05
Galactose	98.2	98.5
Glucose	99.6	101.0
Mannitol	99.4	99.4
Mannose	99.0	99.0
Rhamnose	99.7	100.2
Ribose	99.2	100.8
Sorbitol	98.5	98.9
Xylose	99.2	99.3

Source: Ref. 71.

[a]Each result is an average of four to five determinations.
[b]Based on a 100 mg sample.
[c]Based on a 25 mg sample.

in another report by use of the "double-injection" procedure (74). Both studies report precision and accuracy data comparable with the indirect thermometric titration procedure (see Section 6.6.3a).

6.6.4. Carbonyls

6.6.4a. Aldehydes

The scope of calorimetric titration methods for the determination of aldehydes has been limited by the scarcity of reagents that will react quickly with either the aldehyde itself or even with the comon aldehyde derivatization reagents, for example, hydrazines or hydroxylamines. Typically, the methods involve the use of a standard aldehyde solution to back-titrate the excess derivatization agent. Bark and Prachuabpaibul (75) have taken a different approach by utilizing the fact that most hydrazones are less basic than the corresponding hydrazine. Consequently, the excess reagent can be determined either alone or serially with the hydrazone by titration with strong acid. Specifically, the method describes the titration of unsymmetrical dimethylhydrazine (UDMH) with hydrochloric acid in methanolic solution.

Predictably, the time necessary for derivatization depends on the nature of the aldehyde. At room temperature, aliphatic aldehydes had to be refluxed with UDMH to achieve the same degree of reaction as that obtained by simply mixing the reagent with aromatic aldehydes for the same time. Three types of analyte were identified from the shape of the enthalpogram (see Figure 6.4). Type 1 enthalpograms were produced when the basicity of the hydrazone was so low that no titration was observed. In contrast, serial titration enthalpograms (type 2) were obtained when the basicity of the hydrazone was sufficient to allow quantitative reaction with hydrochloric acid but nonetheless significantly less than UDMH. Type 3 enthalpograms, requiring endpoint extrapolation, were obtained as the basicity of the hydrazone approached that of UDMH. The determination of butanal was not possible because its hydrazone derivative titrates simultaneously with UDMH. The list of compounds titrated, recovery data, and the type of enthalpogram observed with each compound are presented in Table 6.20.

The "double-injection" procedure has been used for the direct determination of acetaldehyde, formaldehyde, propionaldehyde, crotonaldehyde, salicylaldehyde, benzaldehyde, and furfural by oximation with hydroxylamine catalyzed by hydroxyl ion (76). Relative standard deviations of 0.7%, 1.4%, and 2.6% were reported for the determination of 10^{-2}, 10^{-1}, and 10^{-3} mol L^{-1} solutions, respectively. The method was extended to the determination of aldehydes in pharmaceutical products and essential oils.

6.6.4b. Ketones

The oximation procedure mentioned in Section 6.6.4a. has also been employed to determine 10^{-4}–10^{-2} mol L^{-1} of acetone, 2-butanone, cyclopentanone, cyclohexanone, and methylcyclohexanone (77). The operational enthalpies of reaction

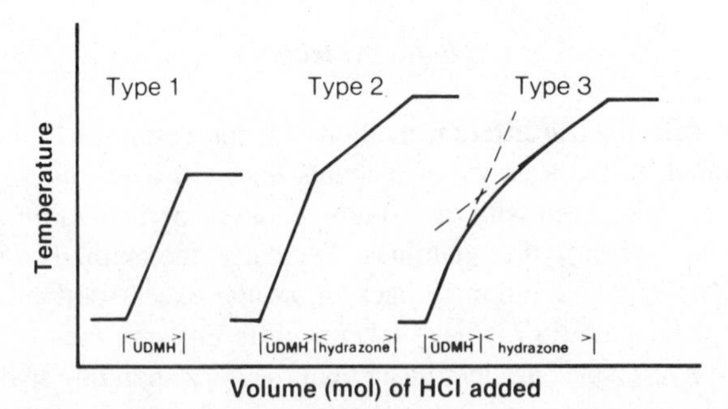

Figure 6.4. Types of enthalpogram obtained for the indirect determination of aldehydes by titration of excess unsymmetrical dimethylhydrazine with hydrochloric acid (75) (by permission of Elsevier Science Publishers).

TABLE 6.20. Recovery Data and Enthalpogram Type for the Indirect Thermometric Determination of Aldehydes

Compound	Type of Enthalpogram	Mass Taken (g)	Mass Found (g)	Percentage Recovery
Acetaldehyde	2	0.0165	0.0162	98.2
		0.0275	0.0268	97.45
3-Phenylpropenal (cinnamaldehyde)	3	0.1609	0.1575	97.9
		0.1249	0.1207	96.6
Butanal			Not Possible	
2-Butanal (crotonaldehyde)	2	0.0161	0.0143	88.8[a]
		0.0404	0.0354	87.6
Furfural	3	0.1507	0.1484	98.4
		0.1294	0.1282	99.0
Benzaldehyde	2	0.1414	0.1426	100.9
		0.1216	0.1223	100.6
2-Hydroxybenzaldehyde (salicylaldehyde)	1	0.1392	0.1376	98.9
		0.2884	0.2836	98.4
3-Hydroxybenzaldehyde	2	0.0608	0.0617	101.5
		0.1004	0.1021	101.7
4-Hydroxybenzaldehyde	3	0.1004	0.1001	99.7
		0.1410	0.1431	101.5
2-Methoxybenzaldehyde	2	0.0740	0.0741	100.2
		0.1130	0.1138	100.7
4-Methoxybenzaldehyde (anisaldehyde)	2	0.1384	0.1440	101.15
		0.1661	0.1680	101.1
3,4-Dimethoxybenzaldehyde veratraldehyde	2	0.1574	0.1597	101.4
		0.1614	0.1629	101.0
3,4-Methylenedioxybenzaldehyde (piperonaldehyde)	2	0.1272	0.1266	99.55
		0.1939	0.1914	98.71
3-Methoxy-4-hydroxybenzaldehyde (vanillin)	2	0.1099	0.1133	103.09
		0.1439	0.1459	101.38
3-Nitrobenzaldehyde	1	0.1136	0.1156	101.76
		0.2102	0.2130	101.33
4-Chlorobenzaldehyde	2	0.1274	0.1281	100.5
		0.1421	0.1447	101.8

Source: Ref. 75, by permission of Elsevier Science Publishers.

[a]The purity of this sample was found to be 88–90% by other methods.

for these compounds were reported as 55.6, 57.7, 59,0, 59.4, and 59.4 kJ mol^{-1}, respectively.

6.6.5. Ethers

6.6.5a. Titration of Macrocyclic Ethers With Metal Cations

A calorimetric titration procedure has been used to measure equilibrium constants and enthalpy changes associated with the reaction of a series of macrocyclic ether ligands with various cations (78). The class of compound examined in this study is typified by 18-crown-6-ether (**3**):

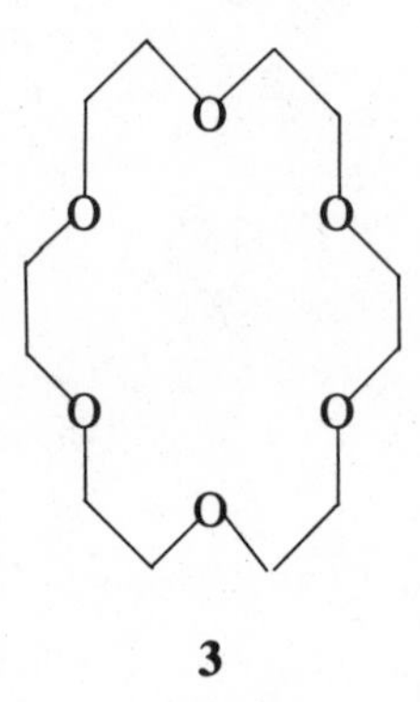

3

If the value of log K_{eq} ([complex]/[ligand][cation]) is greater than 3.5, the enthalpogram has enough endpoint definition to allow the titrimetric determination of either the ligand or the cation. Since these compounds have potential as metal ion extraction reagents, a thermometric titration was used to determine the partition coefficient of 18-crown-6 and dicyclohexano-18-crown-6 ethers between water and chloroform. In view of the scarcity of titrimetric reagents for alkali-metal determination, these compounds offer considerable promise in this context. Although not pursued in this report, the authors recommend 21-crown-7 ether as a titrant for the determination of rubidium or cesium. Some typical thermodynamic data are presented in Table 6.21.

6.6.6. Nitro and Nitroso Compounds

The chemistry of organic nitro and nitroso compounds can differ markedly since it is determined, at least in part, by the carbon framework to which these groups are attached. Nonetheless, these compounds are often classified together for analytical purposes because of the redox chemistry which is common to both

TABLE 6.21. Thermodynamic Data for the Interaction of Macrocyclic Ethers with Metal Cations

Macrocycle	Solvent	Cation	log K_{eq}	ΔH_r (kJ mol^{-1})
18-Crown-6	Water	Ba^{2+}	3.87	−31.71
	Water	Pb^{2+}	4.27	−21.59
	Methanol	K^+	6.06	−56.11
Thia-18-crown-6	Methanol	Ag^+	75.5	−51.46
1,10-Dithia-18-crown-6	Methanol	Ag^+	75.5	Not Reported
Pyridino-18-rown-6	Methanol	Ag^+	75.5	−34.85
2,6-Diketopyridino-18-crown-6	Methanol	K^+	4.66	−38.91
21-Crown-7	Methanol	K^+	4.22	−35.94

Source: Ref. 78.

functions. Two calorimetric studies on this subject have cited transition-metal reduction as the analytical reaction (79, 80). Volf et al. (79) have determined *p*-nitrophenol, picric acid, and ammonium dinitro-*o*-cresolate (Nitrosan) by direct injection of titanium (III) or chromium (II) reagents. An error of 1% was quoted for each of these determinations.

A more extensive range of nitro and nitroso compounds was determined indirectly by thermometric titration of excess titanium (III) with ammonium iron (III) sulfate (80). In order to prevent air oxidation of the unstable titanium (III) reagent, nitrogen was introduced into the reaction cell during the sample reduction stage. The gas flow was stopped during the titration to avoid exaggerating the thermal effects produced by the evaporation of solvent (ethanol). However, since the titration was performed in a sealed vessel, the nitrogen remaining over the surface of the titrate was sufficient to inhibit reagent oxidation. The reported accuracy of this procedure is ±1% for 0.4 mol of either functionality. The compounds titrated by means of this technique are presented in Table 6.22.

6.6.7. Polymers

6.6.7a. Polyethylene Glycols

It has been shown that thermometric titration methodology can be applied to the determination of polyethylene glycols (PEGs) of the general formula

TABLE 6.22. Nitro and Nitroso Compounds Determined by Indirect Thermometric Titration of Excess Titanium (III) With Iron (III)

Nitro Compounds	Nitroso Compounds
1,3-Dinitrobenzene	1-Nitroso-2-napthol
2,4-Dinitrobenzoic acid	2-Nitroso-1-napthol
4-Nitrobenzoic acid	4-Nitroso-*N, N'*-diethylaniline
4-Nitrophenol	4-Nitrosodiphenylamine
4-Nitroaniline	
8-Nitroquinoline	
4-Nitroacetanilide	

Source: Ref. 80.

$H(OCH_2CH_2)_nOH$ in aqueous media (81). Specifically, PEGs in the relative molecular mass range between 400 and 4000 g mol^{-1} were determined by titration of the PEG–Ba (II) complex with sodium tetraphenylborate (NaTPB) with a precision and accuracy better than 1% and 3%, respectively. The precipitation titration, if performed in an excess of barium (II), produces linear and well-defined enthalpograms that require little or no extrapolation. The stoichiometry of the Ba (II)–NaTPB reaction, 1:2, was observed to be independent of polymer chain length. The mole ratio of ethylene oxide to Ba (II) varied with chain length; values of 10.72, 10.29, and 8.68 were determined for PEG4000, 1000, and 400, respectively. The substantial decrease in this parameter for the low-molecular-mass polymer was attributed to the inability of the shorter-chain-length polymer to envelop the complexed barium ion. Tetraiodomercurate (HgI_4^{2-}) and molybdophosphoric acid were evaluated as alternative titrants to the relatively expensive NaTPB; neither gave satisfactory results. The NaTPB procedure is, however, recommended as being more sensitive, convenient, and less subject to interference than the analogous potentiometric procedure.

6.6.7b. Polyvinylamines and Polyiminoethylenes

Lewis et al. (82) have determined the thermodynamic parameters ΔG, ΔH, and ΔS, for the proton ionization of the hydrochloride salts of polyvinylamine at 298 K and of polyiminoethylene at 298 K and 323 K by calorimetric and potentiometric titration with sodium or potassium hydroxide. Potentiometric data were obtained both independently and, in some cases, simultaneously by incorporation of a combination of glass and Ag/AgCl microelectrodes in a 50 mL calorimeter cell. Changes, in ΔG, ΔH, and ΔS were observed as a function of charge state (fraction of charged amines) and ionic strength. As simple electrostatic theory failed to explain the variation in thermodynamic parameters, an alternative theory based on nearest-neighbor effects and on structural features of the polymers at different stages of protonation is presented. The dramatic temperature sensitivity of ΔH and $T\Delta S$ for polyiminoethylene protonation is explained on the basis of conformational changes which occur with variation of temperature and the effect that such structure changes have on the heat capacity of the system.

6.6.7c. Polyacrylic Acids

As a prelude to the thermometric titration of humic substances (see Section 6.8.1), Khalaf et al. (83) have reported the titration of polyacrylic and polymethylacrylic acids (PAA and PMA) with base and with a series of transition metals. The authors point out that these polyelectrolytes possess a formal similarity to humic acid, in that they both have a relatively high content of carboxyl groups and as such can be used as simple models. Thermometric titration of PAA and PMA with sodium hydroxide results in linear enthalpograms with well-defined endpoints. The enthalpies of neutralization for PAA and PMA were determined to be -54.8 and -54.0 kJ mol^{-1}, respectively. Calculation of the enthalpies of dissociation in the usual fashion results in values of 1.3 kJ mol^{-1} for PAA and 2.1 kJ mol^{-1} for PMA in accord with the relative acidities of these two polymers. The stoichiometries of the reactions of each polymer with La^{3+}, Cu^{2+}, and Ag^{+} determined from the enthalpogram were 3:1, 2:1, and 1:1 (COO^{-}:M^{n+}), respectively.

6.6.8. Surfactants

6.6.8a. Thermochemistry and Determination of Critical Micelle Concentration

The determination of critical micelle concentrations (CMC) in the context of surfactant–protein interactions is discussed in Chapter 7. Kresheck and Hargraves have studied the effect of head group, chain length, and temperature on the thermodynamics of micelle formation by thermometric titration of anionic, cationic, and nonionic surfactants into water (84).

The data extrapolation procedure used in this work can be explained by reference to Fig. 6.5, which represents a series of enthalpograms obtained for the dilution of some anionic, cationic, and nonionic surfactants. The standard enthalpy of micelle formation, ΔH_{MC}, for a particular system can be calculated from

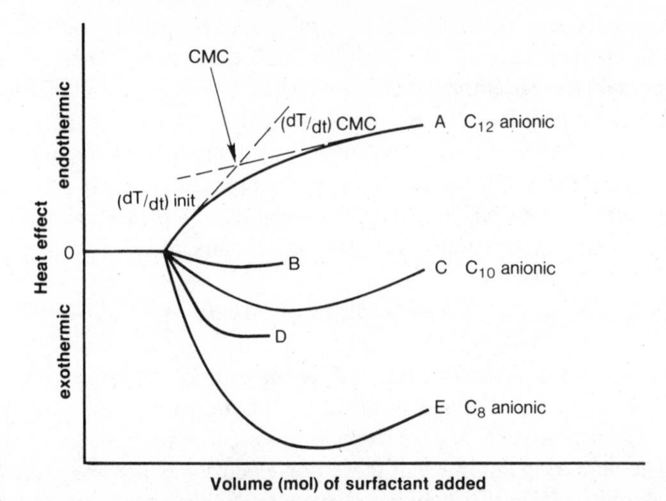

Figure 6.5. Enthalpograms obtained for the titration of several surfactants into water (84) (by permission of Academic Press): (A) 3.9×10^{-1} mol L^{-1} sodium dodecyl sulfate; (B) 1.0×10^{-1} mol L^{-1} dimethyldodecylphosphine oxide; (C) 6.7×10^{-1} mol L^{-1} sodium decyl sulfate; (D) 2.4×10^{-1} mol L^{-1} dimethyldecylphosphine oxide; (E) 1.07 mol L^{-1} sodium octyl sulfate.

$$\Delta H^{\ominus}_{\mathrm{MC}} = \frac{C_P}{n}\left[\left(\frac{dT}{dt}\right)_{\mathrm{init}} - \left(\frac{dT}{dt}\right)_{\mathrm{CMC}}\right] \tag{6.55}$$

where $(dT/dt)_{\mathrm{init}}$ and $(dT/dt)_{\mathrm{CMC}}$ are the experimentally observed slopes of the titration enthalpogram as the surfactant is added initially and just after the CMC, respectively. C_P is the heat capacity of the cell and its contents and n is the number of moles of surfactant added per unit time. The corresponding free energy and entropy terms, $\Delta G^{\ominus}_{\mathrm{MC}}$ *and* $\Delta S^{\ominus}_{\mathrm{MC}}$, can then be calculated from

$$\Delta G^{\ominus}_{\mathrm{MC}} = 2.303RT \log \mathrm{CMC} \tag{6.56}$$

$$\Delta S^{\ominus}_{\mathrm{MC}} = -\frac{\Delta G^{\ominus}_{\mathrm{MC}} + \Delta H^{\ominus}_{\mathrm{MC}}}{T} \tag{6.57}$$

Typical enthalpy and CMC data obtained in these experiments are presented in Table 6.23. Evidently, at a given temperature,the enthalpy of micellization for any surfactant can be positive or negative. However, a correlation does exist between the sign and magnitude of ΔH_{MC} and the chain length of the surfactant. Clearly, increasing chain length (at a given temperature) is accompanied by increasing exothermicity of ΔH_{MC}. This is illustrated in Fig. 6.5. Interestingly,

TABLE 6.23. Critical Micelle Concentration and Enthalpy of Micellization (ΔH_{MC}) Data for Selected Surfactants

Surfactant	Type[a]	Temperature (°C)	CMC[a] (m mol L^{-1})	ΔH_{MC} (kJ mol^{-1})
Sodium dodecylsulfate	A	25	3.8 (8.2)	−2.13
Sodium decylsulfate	A	25	24.5 (33.0)	3.14
Sodium octylsulfate	A	25	72.0 (130)	4.81
Sodium dodecoylsarcosinate	A	25	8.3	3.85
Dimethyldodecylphosphonium oxide	NI	30	0.5 (0.57)	1.67
Dimethyldecylphosphonium oxide	NI	30	4.0 (3.75)	8.28
Dodecylpyridinium bromide	C	25	16.4 (11.4)	−3.47
Dodecylpyridinium chloride	C	25	14.0 (14.7)	1.88

Source: Ref. 84.

[a]A = anionic; C = cationic; NI = nonionic.
[b]Literature values in parentheses.

the nature of the counter ion also affects the magnitude of ΔH_{MC}, which becomes more positive by substituting chloride for bromide in the cationic pyridinium halides. The authors observed that enthalpy changes did not parallel heat capacity changes, which appeared to reflect mainly solvent effects. CMC determinations produced acceptable data that were in reasonable agreement with data obtained from other methods. Discrepancies in anionic surfactant CMCs were attributed to the presence of impurities.

6.6.8b. Interaction with Polyvinylpyrrolidone

The interaction between a series of surfactants and polyvinylpyrrolidone (PVP) was investigated (85) in a similar fashion to the protein–surfactant studies discussed in Chapter 7. Changes in slope of the (dilution) titration enthalpogram performed in the presence and absence of the polymer and shifts in the apparent CMC were used as evidence of binding or non-binding. The alkyl sulfates (octyl, decyl, and dodecyl) showed indications of two types of binding. At low surfactant concentrations, monomeric alkyl surfactant binding to PVP was proposed. At higher concentrations, the data were consistent with direct micelle binding to the polymer. No evidence of binding was observed with sodium dodecylsarcosinate. Similarly, the dodecylpyridinium halides (chloride and bromide) and the alkyldimethylphosphine oxides exhibited only a general solvent effect in response to the presence of the polymer. Interaction between ionic portions of the polymer and the surfactant was invoked to explain differences in binding ability, which could be rationalized simply in terms of the hydrophobic character of a particular surfactant.

6.6.8c. The Effect of a Co-Surfactant on Calcium (II)/Anionic Surfactant Chemistry

The ease with which an anionic surfactant precipitates in the presence of calcium (II) clearly has important ramifications with respect to detergent design. The effect of a nonionic or cationic co-surfactant on this parameter can be monitored conveniently by isoperibol injection calorimetry as shown in Fig. 6.6A and 6.6B (86). Figure 6.6A shows the precipitation reaction that occurs when calcium (II) and a C_{14-15} alkyl sulfate surfactant interact.

The effect of a nonionic co-surfactant, in this case a Neodol^R alkylethoxylate, on the kinetics and thermodynamics of the precipitation is shown in Fig. 6.6B.

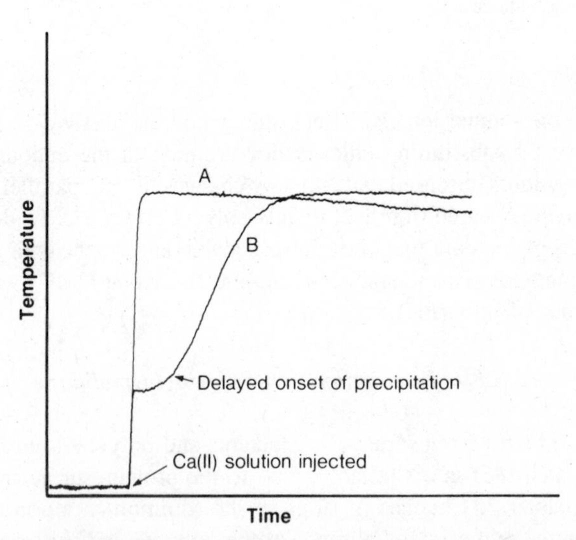

Figure 6.6. Direct-injection enthalpogram illustrating the effect of co-micelle formation on the precipitation of an anionic surfactant with Ca (II) (86). (A) C_{14-15} alkyl sulfate (300 ppm); (B) as in (A) + Neodol^R alkylethoxylate (80 ppm).

The delayed onset of precipitation can be rationalized by the formation of an anionic–non-ionic surfactant co-micelle hindering the formation of calcium (II)–carboxylate "bridges" necessary for calcium soap formation. The convenience of the calorimetric detection system for this type of study is that a series of co-surfactants of differing structure, charge, etc., can be screened for their ability to inhibit soap precipitation by use of the same technique.

6.7. PHARMACEUTICALS

6.7.1. Alkaloids

The formation of insoluble, large-molar-mass adducts between the bulky anions of alkaloids and appropriate high-mass cations that form the basis of the gravimetric determination of many nitrogen bases can also be used for a thermometric precipitation titration of such compounds. Silicotungstic acid is an anionic precipitant that meets the requirements of a thermochemical titrant, namely, high solubility and relatively fast precipitation kinetics. This reagent has been employed to determine a series of physiologically active alkaloids in pure and dosage forms (87). Following dissolution of crushed tablets, injection solutions or pure compounds in 0.1–0.5 mol L^{-1} mineral acid to produce the base anion, the alkaloids were titrated directly with aqueous silicotungstic acid. A minimum concentration of titrant (0.3 mol L^{-1}) is required to ensure that the analytical reaction proceeds to completion. Typical tablet excipients, magnesium stearate, lactose, and starch do not interfere with the accuracy of the method. A list of alkaloids titrated is presented in Table 6.24.

6.7.2. Barbiturates

Bark and Ladipo have applied two catalytic endpoint indicator reactions to the thermometric neutralization titration of several common barbiturate formulations with alcoholic potassium hydroxide (88). Both the base-catalyzed polymerization of acrylonitrile (in dimethylformamide) and the hydroxyl ion-induced condensation of acetone to produce diacetone alcohol produced satisfactory results as shown in Table 6.25. The theoretical and practical aspects of these reactions are discussed in detail in Sections 6.11.1 and 6.11.3, respectively. The authors note that chloroform (10%) should be added to the acetone system to avoid the precipitation of the potassium salts of the barbituric acid derivatives which cause the onset of the indicator reaction to be erratic and a loss of endpoint definition. Significant levels of lactose, magnesium stearate, starch, and talc can be tolerated without adversely affecting the accuracy of the data.

TABLE 6.24. The Application of Calorimetry to the Analysis of Pharmaceutical Chemicals and Preparations

Drug Classification	Analyte(s)	Reagent(s)	Mode[a]	Detection Limit (DL); Range (R)	Precision	Reference	Text Section Reference
Alkaloids	Atropine, brucine, cinchonine, codeine, morphine, pilocarpine, procaine, quinine, strychnine, thiamine	Silicotungstic acid	T	0.025–0.086 g (R)	—	87	6.7.1
Barbiturates	Barbitone, amylobarbitone, butobarbitone, hexobarbitone, pentobarbitone, phenobarbitone, quinalbarbitone	KOH in propan-2-ol	T (catalytic)	0.03–0.07 g (R)	2% maximum deviation from mean)	88	6.7.2
Benzodiazepines	Chlorodiazepoxide, oxazepam, diazepam, nitrazepam	HCl in 50% alcohol	T	—	—	89	—
Isoniazid	Isonicotinic acid hydrazide	$Fe(CN)_6^{3-}$	T	0.02–0.07 g (R)	0.1% rsd	90, 91	6.7.3
Panganin (asparaginates)	K^+ Mg^{2+}	$ZnSiF_4 \cdot 6H_2O$ K_2HPO_4	T, I	—	0.7–1.3% rsd 1.6–1.8% rsd	92	—

Phenothiazines	Promethazine, thioproperazine, aminopromazine, alimemazine	OH^-	T	0.2–3.4 mmol (R)	1.8–8.0% rsd	93	6.7.4
	As in Ref. 93, trifluoroperazine, prochlorpemazine, chlorpromazine, levomepromazine, diethazine, dimetotiazine, oxamemazine	Silicotungstic acid	T	0.3–0.9 mmol (R)	2.8–7.4% rsd	94	6.7.4
Sulfonamides	Sulfathiazole, sulfafurazole, sulfadiazine, sulfamethizole, sulfadimidine, sulfamerazine, succinylsulfathiazole, sulfapyridine	Ag^+	T	0.008–0.08 mg (R)	—	95	6.7.5a
	As in Ref. 95, phthalylsulfathiazole, sulfaquinoxaline, and sulfaurea	KOH in propan-2-ol	T (catalytic)	0.001 mmol (DL)	1–2% rsd	96	6.7.5b
	Various sulfanilamides	$NaNO_2$ or NaOH	T (differential)	—	—	97	—

TABLE 6.24. (*Continued*)

Drug Classification	Analyte(s)	Reagent(s)	Mode[a]	Detection Limit (DL); Range (R)	Precision	Reference	Text Section Reference
Vitamins	Ascorbic acid	ICl_2^-	T	0.009–0.065 mg (R)	0.7% rsd	98	6.7.6
	Ascorbic acid	$Fe(CN)_6^{3-}$	I	0.003–0.02 g (R)	2% rsd	99	6.7.6
	Thiamine	Silicotungstic acid	T	0.03–0.05 g (R)	—	87	—
Drug–protein interactions	Benzodiazepines	Human serum albumin (HSA)	F	—	—	100	6.7.7
	Salicylate, sulfaethidole, fenoprofen, indomethacin, flufenamic acid	HSA	F	—	—	101	6.7.7

[a]T = titration; I = injection; F = flow.

TABLE 6.25. Catalymetric–Thermometric Determination of Selected Barbiturates

	Acrylonitrile Indicator			Acetone Indicator		
Compound	Amount Taken (mg)	Recovery (%)	Precision[a] (% deviation from mean)	Amount Taken (mg)	Recovery (%)	Precision[a] (% deviation from mean)
Barbitone	36.8	100.0	1.35	46.1	100.4	0.2
	27.6	100.0	0.35			
Amylobarbitone	57.6	100.9	1.4	34.0	99.9	0.8
	67.9	99.9	0.3	45.3	100.2	1.1
Butobarbitone	63.7	98.9	1.2	31.8	100.3	0.95
				42.5	100.1	0.48
Hexobarbitone	70.9	99.9	0.7	47.3	100.0	0.0
Pentobarbitone	45.3	100.4	1.1	56.6	100.0	0.0
	67.9	100.0	0.0	67.9	100.2	0.6
Phenolbarbitone	58.1	99.9	0.7			
	69.7	100.6	2.0			
Quinalbarbitone	59.6	100.7	2.0	35.8	100.1	0.55
				47.7	100.0	0.0

Source: Ref. 88, by permission of the Royal Society of Chemistry.

[a]5–10 determinations.

6.7.3. Isoniazid

The calorimetric determination of isonicotinic acid hydrazide in isoniazid tablets has been reported (90, 91), based on a DIE procedure in which hexacyanoferrate (III) solution is injected into an alkaline solution of isoniazid. Crushed tablets were dissolved in 20% *w/v* potassium hydroxide and 1 mL of aqueous hexacyanoferrate solution (0.6 mol L^{-1}) was injected. Quantitation was achieved by means of a calibration graph. Although the procedure was applied only to typical dosage concentrations of isoniazid (mmol), the potential exists for a low detection limit since the enthalpy of reaction was determined to be -695 kJ mol^{-1}.

6.7.4. Phenothiazines—Neutralization Titration

The basic nitrogen functionality in the alkyl side chain (R) of phenothiazines **4** can be titrated calorimetrically with sodium hydroxide in aqueous solution.

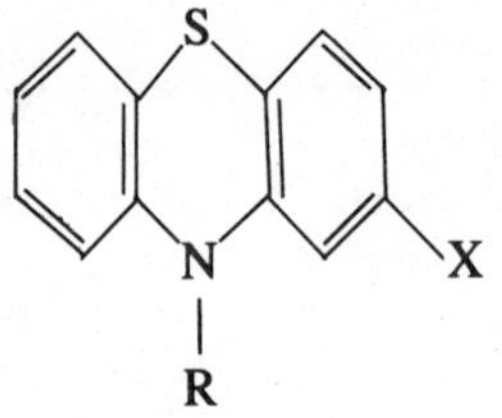

4

Titration of aminopromazine fumarate, thioproperazine bis-methane sulfonate, alimemazine tartrate, and promethazine chlorohydrate with sodium hydroxide produces enthalpograms with a single inflection point (93). The break corresponds to the neutralization of the tertiary amine(s) in the alkyl side chain (R) as indicated in Table 6.26. For example, both nitrogen functions in aminopromazine fumarate titrate simultaneously. The degree of enthalpogram curvature predicted from the magnitude of pK_b for promethazine is not observed. Evidently, the precipitation of the insoluble promethazine base shifts the neutralization equilibrium nearer to completion, resulting in a definitive endpoint. The enthalpies of neutralization and dissociation for promethazine chlorohydrate were determined to be -17.2 and $+38.5$ kJ mol^{-1}, respectively.

As mentioned earlier (Section 6.7.1), silicotungstic acid is a general precipitant for high-molar-mass nitrogen bases. Burgot (94) has used this reagent to titrate 11 phenothiazines in pure and dosage forms (tablets and ampoule solutions) (see Table 6.27). As with the neutralization described above, the technique does not discriminate between the nitrogen functionalities in the phenothiazine alkyl side chain. One inflection point is obtained, with each mole of silicotungstic acid consumed being equivalent to a mole of basic nitrogen in the side chain. The sulfonamide functions in dimetotiazine or thioproperazine ($X = -SO_2N\begin{smallmatrix}CH_3\\CH_3\end{smallmatrix}$) are not basic enough to titrate. Enthalpogram curvature is minimized by use of titrant concentrations not less than 1 mol L^{-1} and maintainance of the analytical solution pH between 0 and 1.

6.7.5. Sulfonamides

6.7.5a. Precipitation with Silver (I)

Under controlled pH conditions, sulfonamides can be made to precipitate in the presence of silver (I) allowing the thermometric titration of a variety of sulfonamides with silver nitrate (95). A pH of 9.2 (borate buffer) was shown to provide enough alkalinity for the formation of the silver sulfonamide without inducing

TABLE 6.26. Thermometric Neutralization Titration Data for Selected Phenothiazines

Compound	R	X	Amount Taken (mmol)	Recovery[a] (%)	Precision (% rsd)
Promethazine chlorohydrate	$-CH_2-CH(CH_3)-N(CH_3)_2$	$-H$	0.46	100	8.0
			1.02	97	6.8
			1.60	99	4.0
			2.17	101	1.8
			3.41	101	3.2
Thioproperazine bis-methane sulfonate	$-CH_2-(CH_2)_2-N(C_4H_8)N-CH_3$ (piperazine ring)	$-SO_2N(CH_3)_2$	0.276	97	3.7
Aminopromazine fumarate	$-CH_2-CH(N-CH_3)-CH_2-N(CH_3)_2$	$-H$	0.303	103	3.6
Alimemazine tartrate	$-CH_2-CH(CH_3)-CH_2-N(CH_3)_2$	$-H$	0.236	104	7.4

Source: Ref. 93.

[a] Four to six replicates.

TABLE 6.27. Thermometric Titration of Selected Phenothiazines with Silicotungstic Acid

Compound	Amount Taken (mol $\times$ 10^{-4})	Recovery (%)	Precision (% rsd)
Trifluoroperazine chlorhydrate	6.16	104	4.80
Prochlorpemazine bis-maleate	4.99	102	3.17
Aminopromazine fumarate	3.84	101	3.20
Thioproperazine bis-methanesulfonate	4.69	101	4.30
Chlorpromazine chlorhydrate	8.47	103	2.60
Promethazine chlorhydrate	9.35	97	3.40
Levomepromazine chlorhydrate	8.22	100	4.89
Diethazine chlorhydrate	8.96	101	5.50
Dimetotiazine methane sulfonate	6.09	96	7.40
Alimemazine tartrate	3.96	104	5.00
Oxamemazine chlorhydrate	8.09	99	2.77

Source: Ref. 94.

the concurrent precipitation of silver oxide. Typical tablet excipients caused no significant interference. The list of sulfonamides successfully titrated is shown in Table 6.24. Consistently high results were obtained for sulfapyridine; non-stoichiometric complexation of silver (I) with the pyridine functionality of this compound was cited as the probable cause of this discrepancy. The optimum range of determination, restricted at high concentrations by the ability of the instrument to effectively stir the precipitate, is quoted as 2.5×10^{-3}–2.5×10^{-2} mol L^{-1}.

6.7.5b. Catalytic Neutralization Titration

Greenhow and Spencer (96) have evaluated a series of bases as non-aqueous thermochemical titrants of the acidic hydrogen in the sulfonamide group. An endpoint indication reaction—the base-catalyzed polymerization of acrylonitrile—was incorporated into the procedure to increase the sensitivity of the method (see Section 6.11.1). No significant analytical advantage was found for any of the titrants studied including tetra-*n*-butylammonium hydroxide, sodium methoxide, potassium methoxide, and alcoholic potassium hydroxide. The reagent of choice, alcoholic potassium hydroxide, was selected on the grounds of convenience. Dimethylformamide was recommended as the solvent for the determination of sulfadiazine, sulfadimidine, sulfaisonazole, and sulfamethiozole tablets.

Dimethylsulfoxide is the preferred solvent system for sulfamethoxypyridiazine and sulfathiazole preparations. The inclusion of magnesium stearate, talc, and potato starch which can give positive errors of up to 0.4% in a non-aqueous neutralization titration with a visual endpoint, caused no significant interference with the calorimetric procedure. Analytical data and reagent specifications for the analysis of a series of common sulfonamides are shown in Table 6.28. The detection limit for the catalimetric technique, quoted as 0.1 μ mol, is some two orders of magnitude lower than any of the conventional thermometric titration methods reported in the literature.

6.7.6. Vitamins—Ascorbic Acid—Redox Reactions

Under acidic conditions, ascorbic acid will reduce iodine monochloride to iodide ion, that is,

$$ICl_2^- + 2e = I^- + 2Cl^- \qquad (6.58)$$

In the presence of excess iodine monochloride, iodide ion is then further oxidized to elemental iodine. Accordingly, the enthalpogram obtained for the titration of ascorbic acid with iodine monochloride has two inflection points, as shown in Fig. 6.7. This sequence of reactions has been utilized in the thermometric titration of ascorbic acid in pure and tablet form (98). In order to produce a sharper endpoint, the addition of mercury (II) to the solution is recommended. The iodide formed in the initial reaction is complexed as HgI_4^{2-}, inhibiting the formation of iodine. The effect on the enthalpogram is shown in Fig. 6.7.

A "subtractive" direct-injection enthalpimetric system has been discussed for the determination of ascorbic acid by reaction of the analyte with hexacyanoferrate (III) solution (99). In order to minimize the effect of the large enthalpy of dilution of this reagent, a double syringe system simultaneously injects 2 mL

TABLE 6.28. Precision and Accuracy Data for the Calorimetric Titration of Selected Sulfonamides using a Catalyzed Endpoint Indicator Procedure

Compound/ Preparation	Amount (mg)	Solvent,[a] Volume (mL)	Volume of Acrylonitrile (mL)	Titrant,[b] Concentration (mol L^{-1})	Precision (% rsd)	Recovery (%), Thermometric Titration	Recovery (%)[c], B.P. Procedure
Pure Compound							
Sulfadiazine	25.07	A, 1	2	C, 0.1	0.56	—	—
Sulfadimidine	27.94	A, 1	2	C, 0.1	0.62	—	—
Sulfamethizole	2.71	A, 1	2	C, 0.01	0.30	—	—
Sulfadiazine	0.25	A, 1	1	C, 0.001	2.00	—	—
Sulfaguanidine	10.09	B, 1	4	D, 0.1	0.23	—	—
Sulfaurea	100.20	A, 1	1	D, 1.0	0.50	—	—
Tablets							
Sulfadiazine	25.07	A, 1	3.5	C, 0.1	0.17	73.3	72.8
Sulfaisoxazole	24.61	A, 1	3.5	C, 0.1	0.27	82.8	82.4
Sulfaguanidine	16.13	B, 2	3	D, 0.1	0.37	79.1	78.3
Sulfathiazole	25.22	B, 1	4	D, 0.1	0.57	87.0	86.2
Sulfamethoxypyridazine	201.6	B, 1	5	D, 1.0	0.65	83.6	83.7
Sulfamethizole	25.95	A, 1	3.5	C, 0.1	0.54	80.0	80.3
Sulfadimidine	28.09	A, 1	4	C, 0.1	0.41	87.9	88.4

Source: Ref. 96, by permission of the American Chemical Society.

[a] A = dimethylformamide; B = dimethylsulfoxide.
[b] C = tetra-*n*-butylammonium hydroxide, D = potassium hydroxide.
[c] Standard British Pharmacopeia procedure.

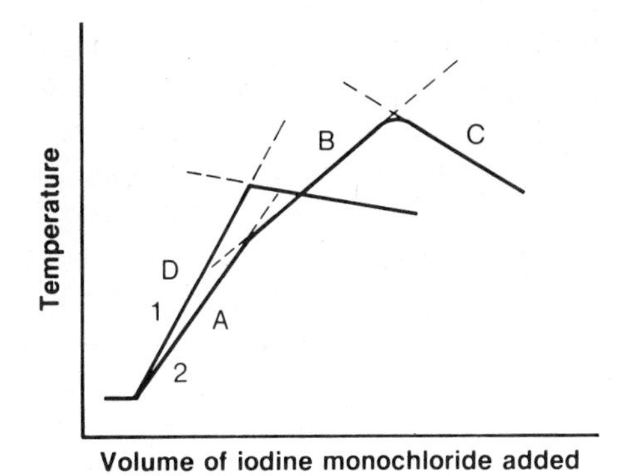

Figure 6.7. Enthalpogram obtained for the titration of ascorbic acid with iodine monochloride (98): (1) Hg^{2+} present; (2) no Hg^{2+} present. (A) ICl_2 + ascorbic acid = I^- + dehydroascorbic acid + $2Cl^-$; (B) I^- + ICl_2 = I_2 + $2Cl^-$; (C) reagent dilution/iodine evolution; (D) as in (A) plus Hg^{2+} + $4I^-$ = HgI_4^{2-}.

of potassium hexacyanoferrate (III) solution (1.0 mol L^{-1}) into the analyte solution and a blank solution. The "blank" signal is subtracted from the analytical signal by incorporation of matched thermistors in opposite arms of the Wheatstone bridge. Linear calibration graphs passing through the origin were obtained for pure ascorbic acid solutions and tablet slurries, indicating that no spurious heat effects from interaction of the reagent with the tablet matrix are obtained within the sensitivity of the instrument. It is possible to perform one assay every 3–4 min by this procedure.

6.7.7. Drug–Protein Interactions

Human serum albumin (HSA) is known to possess several different sets of binding sites for different classes of drugs. The equilibrium constants of complexation and the stoichiometries of the drug–HSA complexes have a fundamental role in determining the free drug concentration in plasma, which, in turn, influences the pharmacodynamic activities of these drugs (100). The thermodynamics of protein interaction with benzodiazepine derivatives (100) and salicylate, sulfaethiodole, fenoprofen, indomethacin, and flufenamic acid (101) have been studied by flow calorimetry. The relative contributions of entropy effects have been positioned as functions of ionic strength and support electrolyte concentrations. Both the iterative least squares technique (100) and the two-point method (101) were used to calculate K_{eq} and ΔH_R from flow calorimetric data. The principles of both these methods of data reduction have been discussed in detail in Chapter 5.

6.8. CHEMISTRY AT SOLID–LIQUID INTERFACES

The use of calorimetric methods for the determination of analytes in solutions containing large amounts of insolubles is well documented (1–4). A logical extension of this type of application is the study of chemistry occurring at the interface between a solid and a liquid. A solution calorimeter possesses an almost unique property ideally suited to this type of measurement, that is, the response is not significantly affected by the presence of an insoluble matrix as long as the efficient transfer of heat through the solution is not seriously attenuated. Consequently, it is possible to monitor directly the progress or extent of a reaction occurring at a surface without recourse to batch sampling or to indirect methods often used to avoid the deleterious effect of a solid on many analytical detectors. Moreover, thermodynamic information can often be obtained concurrently with analytical data, increasing our understanding of sometimes complex reaction chemistry. Importantly, the technique will quantitate and, in certain cases, delineate many types of surface interactions including adsorption/desorption, neutralization, chelation, etc. As will become evident, interpretation of curvature obtained in titration enthalpograms plays a key role in the qualitative assignment of reaction mechanisms in this application area.

Specific applications are listed in Table 6.29. Evidently, interest has focused on three types of matrix, namely, soils and clays, zeolites, and modified silica surfaces.

Steinberg (110) has recently evaluated the potential of flow microcalorimetry as a routine analytical tool for the study of many surface reactions including polymer adsorption, solvent competition, preferential adsorption, surfactant/clay interactions in tertiary oil recovery and many others.

6.8.1. Soils and Clays

Several calorimetric studies have been reported in which the primary objective was the characterization of the proton ionization chemistry of humic substances derived from peat (102), soil (103), and river waters (104). In each case, slurries of humic substances, isolated from natural sources by standard extraction procedures, were titrated with sodium hydroxide producing enthalpograms that were curved in the endpoint region. Khalef et al. (102) interpret this curvature as evidence of many different classes of ionizable groups by analogy with linear enthalpograms obtained from an identical titration of polyacrylates, which have a generic structural resemblance to humic acids (Section 6.6.7c) but contain a single class of carboxylic acids. No attempt was made to quantitate the endpoint. Perdue (103) has taken a different approach by comparing the humic acid enthalpogram with titration data for benzoic acid, *o*- and *p*-hydroxybenzoic acids, and phenol. The enthalpy change associated with the initial linear period in the humic acid enthalpogram is consistent with carboxylic acid neutralization (ΔH_N = ca. -55 kJ mol^{-1}). The non-linear region approximates the curvature obtained with the titration of *p*-hydroxybenzoic acid and is therefore attributed to phenolic acidity. By extrapolation of these two regions of the enthalpogram, Perdue

TABLE 6.29. The Application of Calorimetry to the Study of Chemistry at Solid–Solution Interfaces

Matrix	Reactant(s)	Reagent(s)	Mode[a]	Type of Interaction	Reference	Text Section Reference
		Soils and Clays				
Soil	Humic acid	OH^-, La^{3+}, Cu^{2+}, Ag^+	T	Neutralization complexation, precipitation	102	6.8.1
	Humic acid	OH^-	T	Neutralization	103, 104	6.8.1
Soil/Clay	Ca^{2+}	K^+	I	Ion exchange	105	6.8.1
Zeolites Na_2O/Al_2O_3 SiO_2/Al_2O_3	Bronsted acids	*n*-Butylamine, 2,6-dimethylpyridine	T	Neutralization	106	6.8.2
Linde molecular sieve (zeolite)	Lewis acids	Toluene, nitrobenzene, and aniline	T	Adsorption	107	6.8.2
		Modified Silica Gel/Glass				
Controlled-pore glass	Surface-bound ethylenediamine	Cu^{2+}	T	Complexation	108	6.8.3
Silica gel	Surface-bound silanols	*n*-Butylamine	T	Adsorption	109	6.8.3

[a]T = titration; I = injection; F = flow.

concludes that humic acid contains approximately 4.1 mmol g^{-1} and 0.7 mmol g^{-1} of carboxylic and phenolic functionality, respectively. In an extension of this study, a calorimetric titration procedure was used to evaluate the data obtained by the calcium acetate exchange reaction for the determination of the carboxylic acid level of humic substances. This procedure depends on the tendency of humic acids to precipitate an insoluble calcium salt in the presence of calcium acetate. After filtration, the titrated acetic acid is back-titrated with sodium hydroxide. The calorimetric titration data are shown in Fig. 6.8.

In this experiment, curvature in the humic acid titration enthalpogram, relative to an identical titration of acetic acid, is interpreted as evidence that acids other than acetic acid are titrated in the calcium acetate procedure. Quantitative evaluation of the data shows that the apparent carboxyl content is increased by approximately 11% by calcium binding. Titration of sodium humate with various metal-ion solutions shows that both lanthanum (III) and copper (II) initially react rapidly, although these data conflict with potentiometric measurements which indicate slow complexation kinetics (102). No thermochemical evidence was obtained to indicate a reaction between these ions and humic acid.

Talibudeen et al. (105) have described a microinjection system designed to incrementally deliver microliter volumes of reagent simultaneously into twin cells of a heat-conduction microcalorimeter. The incremental addition procedure was developed to monitor kinetically slow potassium–calcium exchange reactions in various soil slurries. This technology, based on the mechanical activation of microswitches, has effectively been replaced by programmable (stepper motor) buret systems, which provide a much wider range of reagent delivery sequences.

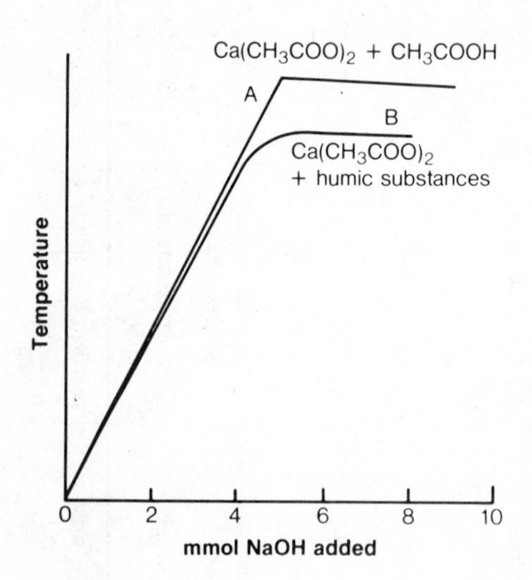

Figure 6.8. Enthalpograms obtained for the titration of (A) acetic acid (5×10^{-3} mol L^{-1}) and (B) humic substances (7.5×10^{-1} g L^{-1}) in 1.0×10^{-1} mol L^{-1} calcium acetate with sodium hydroxide (5×10^{-3} mol L^{-1}) (104) (by permission of Pergamon Press).

6.8.2. Zeolites

The titration of porous ion exchangers with organic bases to a visual endpoint in order to obtain theoretical capacity data suffers from the disadvantage that commonly used indicators often have molecular dimensions larger than the pores of the exchanger, resulting in "indicator error." A discontinuous thermometric titration, in which aliquots of *n*-butylamine were added incrementally at 15–20-s intervals to slurried decationized, acidic zeolites, has been used to assess the effect of pore size on maximum Bronsted acidity, that is, on available ion-exchange capacity (106). Predictably, a strong correlation between pore size/volume and the fraction of the theoretical Bronsted acidity titrated was observed. Phenomena such as "pore mouth blockage" by strongly adsorbed bases were observed and used to interpret low titration efficiencies of some zeolites.

In a similar vein, Eatough et al. (107) were able to thermodynamically characterize the adsorption of toluene, aniline, and nitrobenzene on to Linde molecular sieve 13X by incremental calorimetric titration. The least squares procedure described in Chapter 5 was used to calculate the number of moles of adsorption sites as well as K_{eq} and ΔH_R (Table 6.30).

6.8.3. Modified Silica Gel/Controlled-Pore Glass

In recent years there has been considerable interest in the chemistry of modified silica surfaces motivated by the use of such materials as chromatographic supports, chelation sites for metal preconcentration modules, or immobilized-enzyme matrices. In the light of an earlier discussion (Section 6.8) it is not surprising, therefore, that calorimetric procedures have made a contribution in this field of study.

Kvitek et al. (108) reported the first thermometric titration of a chemically modified porous glass in their study of the chelation chemistry of glass-bound

TABLE 6.30. Adsorption Data Obtained from the Discontinuous Thermometric Titration of Linde Molecular Sieve 13X

Titrant	log K_{eq}	ΔH (kJ mol^{-1})	Adsorption Capacity (n') (mmol g^{-1})
Toluene	1.17	−13.8	1.64
Aniline	3.08	−25.98	1.67
Nitrobenzene	3.21	−14.6	1.37

Source: Ref. 107.

ethylenediamine with copper (II): 0.12 g of modified glass in 35 mL of 0.1 mol L^{-1} sodium nitrate was titrated directly at constant rate with 0.4 mol L^{-1} copper sulfate solution. The resulting enthalpograms, although curved, could be extrapolated to give an accurate endpoint. Interestingly, the copper-binding capacity determined by thermometric titration, 1.64 μmol m^{-2}, differed by a factor of 2 from the capacity determined from the corresponding potentiometric titration, 3.12 μmol m^{-2}. This anomolous datum was attributed to the binding of a second mole of ethylenediamine with an enthalpy of reaction below the detection limit of the instrument. Variation of titration rates excluded kinetic dependence of the data.

In general, the assumption that a functional group attached to a silica (or any other) surface will react in exactly the same manner as it would in homogeneous solution is questionable. Moreover, it does not follow that the total number of bound functional groups determined by, for example, elemental nitrogen or carbon analysis is equal to the number available for reaction. These anomolies have been examined in a recent model study in which the protonation chemistry of *n*-butylamine in the presence and absence of silica gel was compared by isoperibol titration calorimetry (109). Two significant features of the interaction between *n*-butylamine and silica gel become apparent from the titration enthalpograms shown in Fig. 6.9. First, the mere presence of silica gel (no attempt was made to chemically bind the amine) induces curvature in the enthalpogram, which cannot be reduced by slower titration rates. The reversible hydrogen binding between surface silanol groups and the amine clearly has a significant effect on the basicity of the amine. This was confirmed by the fact that the identical titration of silica gel, treated with the silanol "capping" agent trimethylchlorosilane (TMCS), resulted in enthalpograms with considerably less curvature. Second, only 80% of the amine added is titrated in the presence of

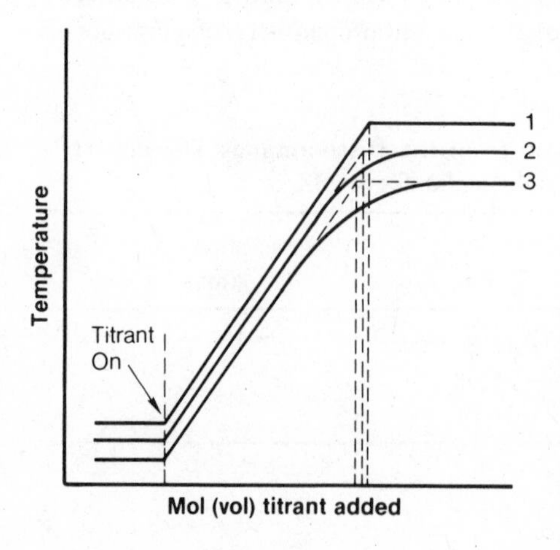

Figure 6.9. Enthalpograms recorded for the titration of *n*-butylamine (5×10^{-4} mol): (1) in aqueous solution, (2) in the presence of 0.5 g TMCS derivatized silica gel, (3) in the presence of 0.5 g of untreated silica gel (109) (by permission of the American Chemical Society).

untreated gel, whereas 93% is titrated in the presence of TMCS derivatized gel, indicating that a significant proportion of the amine is irreversibly "bound" or trapped on the silica gel. These data were confirmed by carbon analysis which showed that 20% of the amine remains "attached" to the silica gel after titration. The data do not allow any definitive conclusions as to whether the untitrated amine is chemically bound or physically entrapped. However, comparison of adsorption/desorption enthalpy data, obtained by direct titration of *n*-butylamine into a silica gel slurry with the determined titration reaction enthalpy, indicates that the enthalpy of the irreversible reaction is either below the detection limit of the instrument or zero. It was concluded that a chemical reaction accounting for 20% of the total reaction was unlikely in view of these data. This question clearly needs further study with pore size and the basicity and/or molecular dimensions of the amine as experimental variables. In the same report it was shown that the curvature in the enthalpogram, associated with reversible hydrogen bonding, can be fitted to the equation

$$\frac{1}{[n\text{-BuNH}^+_{3(\text{soln})}]_t} = \frac{1}{K_{eq}}\frac{1}{[\text{H}_3\text{O}^+]_t}\frac{1}{[n\text{-BuNH}^+_{3(\text{soln})}]_F} + \frac{1}{[n\text{-BuNH}^+_{3(\text{soln})}]_F} \tag{6.59}$$

which is a solution analog of the Langmuir equation. $[n\text{-BuNH}^+_{(soln)}]_F$ is the postendpoint concentration of protonated *n*-butylamine in solution, and $[\text{H}_3\text{O}^+]_t$ and $[n\text{-BuNH}^+_{3(soln)}]_t$ are the concentrations of hydronium ion and protonated amine in solution at any point during the course of the titration. K_{eq} is the equilibrium constant for the reaction

$$n\text{-BuNH}_2\text{ (ads)} + \text{H}_3\text{O}^+ \rightleftharpoons n\text{-BuNH}^+_{3(\text{ads})} + \text{H}_2\text{O} \tag{6.60}$$

where (ads) signifies that the amine is still adsorbed on the gel. The variables

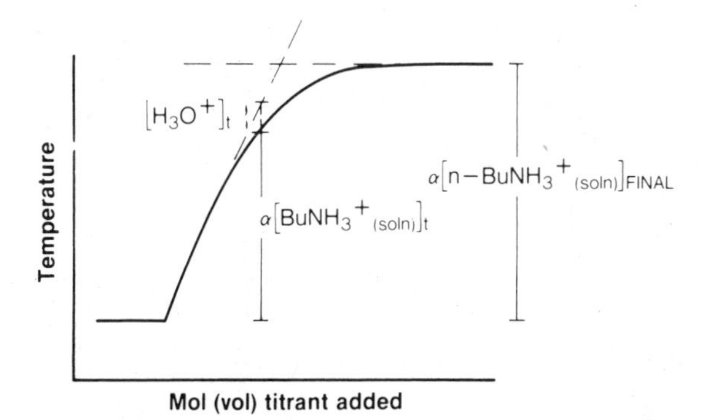

Figure 6.10. Extrapolation procedure for the determination of K_{eq} according to Eq. (6.59) (109) (by permission of the American Chemical Society).

TABLE 6.31. Application of Calorimetry to the Analysis of Raw Materials and Industrial Process Chemicals

Raw Material Process	Analyte(s)	Reagent(s)	Mode[a]	Reference	Notes
6.9.1. Alloys					
Ferrovanadium alloys/ vanadium carbides	V^{4+}	H_2O_2	I	111	Detection limit 10 μg V mL^{-1}
Brass	Al	HF, $SrCl_2$	I	112	Analytical reaction: $2(AlF_6)^{3-} + 3Sr^{2+} = Sr_3(AlF_6)_2$
Cast iron, steel	Si	HF	I	113	Analytical reaction: $H_4SiO_4 + 6F^- + 4H^+ + 2K^+ = K_2(SIF_6) + 4H_2O$ $\Delta H_R = -240$ kJ mol^{-1}
Barium–aluminum alloys	Ba	SO_4^{2-}	I	114	—
6.9.2. Bayer Process/ Aluminum Processing					
Bayer process solutions	$OH^-(Na_2O)$ Al_2O_3	HCl/F^-	T	115	1. Tartrate added to OH^-, that is, $Al(OH)_4^- + n(C_4H_4O_6)^{2-} = Al(OH)_3(C_4H_4O_6)_2^{2-} + OH^-$ Total OH^- titrated with HCl 2. F^{-1} added after titration: $Al(OH)_3(C_4H_4O_6)_n^{2-} + 6F^- = AlF_6^{3-} + n(C_4H_4O_6)^{2-} + 30H^-$ OH^- equivalent to Al_2O_3 then titrated with HCl

Bayer liquors	Na_2O Al_2O_3	HCO_3^- tartrate	F F	116 117	Evaluation of a thermometric flow analyzer for on-line analysis of Bayer liquors. Coefficient of variation for Al_2O_3/Na_2O ratio = ± 3%
Aluminum anodizing solution	Al, H_2SO_4	KOH in methanol	T	118	Titration solution 1:10 in DMF
6.9.3. Cements/ Concretes					
Cement	SiO_2	F^-, K^+, H_3O^+	I	119, 120	Silicate precipitated as K_2SiF_6: $SiO_2 + 6HF + 2K^+ = K_2SiF_6 + 2H_2O + 2H^+$; $\Delta H_R = -220$ kJ mol$^-$
Cement	SiO_2	HF	I	121	Three sample preparation procedures compared. Dissolution in HCl, treatment of filtered insoluble with Na_2Co_3, and refiltration preferred. HF containing urea used as reagent
Cement	Al_2O_3		I	122	—
Cement	SO_3^{2-} TiO_2	$BaSO_4$ seed + $BaCl_2$ H_2O_2	I I	123 123	SO_3^{2-} determination recommended for routine analysis. TiO_2 determination presents little advantage versus other (photometric) methods
Road concrete	SiO_2	F^-, K^+, H_3O^+	I	124	—
Portland cement	Al_2O_3	HF, Ba^{2+}	I	125	Sample dissolved in HCl, HNO_3, and NH_4Cl. NH_3 precipitates SiO_2, TiO_2. After dissolution of precipitate in HCl, HF is added. Ba^{2+} solution is then added in excess

TABLE 6.31. *(Continued)*

Raw Material Process	Analyte(s)	Reagent(s)	Mode[a]	Reference	Notes
6.9.4. Ceramics/ Glasses					
Glass	Ba, Ca, K, Zn, Pb, Th, La, W	Various	I	126	Review. Application of DIE to the analysis of silicate and non-silicate glasses
Glass	W^{6+}	H_2O_2	I	127	Glass treated with orthophosphoric acid Analytical reaction: $WO_4^{2-} + H_2O_2 = (WO_5)^{2-} + H_2O$; $\Delta H_R = -34.3$ kJ mol^{-1}. Method suitable for silicate glasses
Glass	Al^{3+} Ti^{4+}	K^+, F^- H_2O_2	I	128	Al^{3+} and Ti^{4+} determined by precipitation of K_3AlF_6 and K_2TiF_6 simultaneously. Ti^{4+} alone determined by addition of H_2O_2: $Al^{3+} + 6F^- + 3K^+ = K_3AlF_6$, $\Delta H_R = -113.0$ kJ mol^{-1} $Ti^{4+} + 6F^- + 2K^+ = K_2TiF_6$, $\Delta H_R = -92.0$ kJ mol^{-1} $TiO^{2+} + H_2O_2 = TiO_2^{2} + H_2O$, $\Delta H_R = -44.35$ kJ mol$^-$
Glass	Th^{4+}, La^{3+}	IO_3^-, Oxalate	I	129	Method applied to non-silicate glasses. Th^{4+} and La^{3+} determined by precipitation of $Th(C_2O_4)_2$ and $La_2(C_2O_4)_3$ simultaneously. Th^{4+} determined by precipitation of $Th(IO_3)_4$

Glass	PO_4^{3-}	Ammonium molybdate, H_2O_2	I	130	Samples dissolved by fusion with KOH. Excess ammonium molybdate added to precipitate phosphomolybdate. Excess molybdate determined by formation of peroxomolybdate. ΔH_R for analytical reaction, -49.0 kJ mol^{-1}
Glass	Na_2O	$KAlF_5$	I	131	Differential procedure
Water glass	Na_2O, SiO_2	HCl, HF	I, I	132, 132	Total alkalinity determined by adding 1 mol L^{-1} HCl, then SiO_2 determined by addition of HF
Alkali calcium silicate glass	CaO	EDTA	I	133	—
Lead silicate glasses	PbO	$Cr_2O_7^{2-}$	I	134	—
6.9.5. Detergents	SiO_2	HF, K^+	I	135	Determination based on precipitation of K_2SiF_6.
6.9.6. Fertilizers	K^+ (K_2O)	$B(Ph)_4^-/Br_2$	I	136	Excess tetraphenylborate oxidized by Br_2. NH_4^+ will interfere unless removed by boiling under alkaline conditions
	$K^+(K_2O)$, N (ammonia/urea), PO_4^{3-}	$(Ph)_4B^-$, BrO^- Mg^{2+}, NH_4^+	I	137	K^+ precipitated as tetraborate, NH_4^+ removed by boiling. N (in the form of ammonia or urea) is determined by oxidation to elemental nitrogen with hypobromite. PO_4^{3-} is precipitated as magnesium ammonium phosphate. K^+ and N reactions are non-stoichiometric.
	K_2O MgO SO_4^{2-}	ClO_4^- $(NH_4)_2SO_4$ Ba^{2+}	F	138–141	Technicon "Thermometric Analyzer" evaluation procedures
	P (as P_2O_5)	$ZrOCl_2$	I	142	Differential procedure to offset heat of dilution of $ZrOCl_2$. The presence of 2 mol L^{-1} H_2SO_4 prevents interference from calcium and magnesium

TABLE 6.31. (*Continued*)

Raw Material Process	Analyte(s)	Reagent(s)	Mode[a]	Reference	Notes
	K^+, NH_4^+	$B(Ph_4)_4^-$	T	143	NH_4^+ masked with NaOH and HCHO, K^+ titrated with $B(Ph_4)_4^-$. NH_4^+ and K^+ determined by titration of unmasked solution.
6.9.7. Foodstuffs/ Beverages					
Preserves, chocolate	Sorbitol	IO_4^-	T	144	Foodstuff dissolved in water, diluted with 0–1 mol L^{-1} H_2SO_4 and titrated with $NaIO_4$. The reaction is non-stoichiometric, and quantitation is achieved by calibration graph versus standards
Fruit juice, vinegar	Total acidity/ acetic acid	OH^-	F	145, 146	Evaluation of Technicon Flow Analyzer for foodstuffs and beverage analysis. Ethanol content of beverages determined by measuring the enthalpy of dilution of an alcoholic perchlorate solution on mixing with the analyte sample.
6.9.8. Nuclear Fuel Processing					
Simulated nuclear fuel solutions	Free acids	OH^-	T	147	Hydrolyzable ions are precipitated or complexed with an excess of potassium fluoride (5 mol L^-). No interference from Al^{3+}, Fe^{3+}, Ni^{2+}, Cr^{3+}, Ce^{4+}, Ba^{2+}, Zr^{4+}, Th^{4+}, and U^{6+}
Spent nuclear fuel solutions	Free H_3O^+, UO_2^{2+}	Pyridine (aq)	T	148	After free HNO_3 is determined, an excess of H_2O_2 is added. The H^+ liberated in the reaction, $UO_2^{2+} + H_2O_2 + 2H_2O = UO_4 \cdot 2H_2O + 2H^+$ is then titrated to determine UO_2^{2+}. The method is applicable to analysis of organic extracts of spent nuclear fuels containing Pu and other

6.9.9. Minerals/Ores					
Bauxites (acidic and calciferous)	SiO_2, Al_2O_3, Fe_2O_3, TiO_2, MnO_2, CaO, MgO	HF, $K^+/Na^+/$ HF, $S_2O_8^{2-}$, H_2O_2, MnO_4^-, oxalate, and Na_2HPO_4	I	149, 150	Evaluation of "Directhermom" enthalpimetric analyzer for analysis of bauxites. Error for each analyte estimated to be $\pm 0.4\%$
Dolomite	Mn^{2+}, Ca^{2+}, Mg^{2+}, Al^{3+}, Fe^{3+}, SiO_2	MnO_4^-, oxalate, $(NH_4)_3PO_4$, KF, S_2O_8, and HF	I	151, 152	Powdered sample dissolved in HCl and hydrated SiO_2 separated. Reagents added serially to filtrate
Hydraulic limes	CaO	EDTA	T	153	—
Potassium–magnesia granules	K^+	ClO_4^-	F	154	Method used to determine raw salts containing 6–10% K_2O, K–magnesia granules containing 30% K_2O, K_2SO_4, with 51% K_2O, and KCl with 60% K_2O.
6.9.10. Spinning Baths/Rayon Processing	H_2SO_4	NH_3	F	155	Continuous flow analysis
	$ZnSO_4$	aq. MeOH	F	156	Determination based on the heat of resolvation on transferring a strong electrolyte from water into aqueous methanol

[a]T = titration; I = injection; F = flow.

in Eq. (6.59) can be obtained directly from the titration enthalpogram as shown in Fig. 6.10. The value of K_{eq} can be obtained from a plot of $1/[n\text{-BuNH}^+_{3(soln)}]_t$ versus $1/[H_3O^+]$ or by the procedures described in Chapter 5; correlation of K_{eq} values obtained by three different extrapolation procedures was favorable.

The information obtained in this report underlines the utility of calorimetry as a complementary tool for the analysis of heterogeneous systems. The matrix need not, of course, be restricted to silica gel nor need the bound group be an amine. The universal nature of calorimetry allows the study of any functional group chemistry; all that is required is a suitable reagent.

6.9. RAW MATERIALS/INDUSTRIAL PROCESS ANALYSIS

In the industrial sector, the primary motivation for the choice of an enthalpimetric procedure is the ability of the detection system to tolerate solutions that would seriously inpair the accuracy and precision of other analytical transducers, for example, solutions with a high solids or electrolyte content, corrosive liquids, or solutions with an inherent color. The applications calorimetry to the analysis of raw materials and industrial process solutions are summarized in Table 6.31. Many of the same reactions have been applied to different processes, for example, the determination of silicate by precipitation of the silicofluoride appears in the analysis of cements, detergents, ceramics, and ores. Accordingly, in the interests of brevity and clarity, the review is presented entirely in tabular form. Significant features of each application appear as Notes in Table 6.31.

6.10. ENVIRONMENTAL ANALYSIS

Traditionally, calorimetric methods of analysis have not made a significant impact in the area of environmental analysis. The principal reason for this is that, in the past, calorimetry has not been able to provide the sensitivity which is a prerequisite of most analyses of this type. However, with the advent of small-cell technology and the incorporation of computer-based data correction facilities, the calorimetric determination of nanomole levels of analyte has now become a reality, and several reports of the qualitative and quantitative characterization of environmental samples from calorimetric data have appeared in the literature (Table 6.32).

6.10.1. Chlorinity, Alkalinity, and Sulfate in Sea Water

Millero et al. (157) have suggested the use of thermometric titrimetry for ship-board analysis of electrolyte content in marine or estuarine water samples. They

TABLE 6.32. The Application of Calorimetry to the Analysis of Environmental Samples

Analyte	Reagent(s)	Sample/Matrix	Mode[a]	Reference	Text Section Reference
Cl^-, alkalinity SO_4^{2-}	Ag^+, H_3O^+, Ba^{2+}	Sea water	T	157	6.10.1
S^{4+}, SO_4^{2-}	$Cr_2O_7^{2-}$, Ba^{2+}	Airborne particulates	T,I	158	6.10.2
S^{4+}, SO_4^{2-}	$Cr_2O_7^{2-}$, Ba^{2+}	Aerosols from flue lines, from stacks of copper and lead smelters, from coal- and oil-fired power plants, and from New York City	T,I	159	6.10.2
S^{4+}, SO_4^{2-}, acidic species	$Cr_2O_7^{2-}$, Ba^{2+}, OH^-	Copper and lead smelter flue dusts	T,I	160	6.10.2, 6.10.3
Nitrite	Sulfamic acid	Airborne particulates samples from Provo, St. Louis, and New York City; fly ash and plume from an oil-fired power plant	I	161	6.10.2
Acidic and basic species	OH^-, H_3O^+	Copper and lead smelter flue dusts and New York City airborne particulates samples	T	162	6.10.3

[a]T = titration; I = injection; F = flow.

argue that the apparatus is suitable for a variety of analyses and that it avoids the use of procedures such as precipitate digestion which are impracticable for a mobile laboratory. This application has been demonstrated by the determination of total sulfate, total alkalinity, and chlorinity of sea water by titration with barium chloride, perchloric acid, and silver nitrate, respectively, in a ship-board laboratory.

6.10.2. Sulfur (IV), Sulfate, and Nitrite in Airborne Particulates

A relatively fast and inexpensive thermochemical procedure has been developed for the determination of sulfur (IV) and sulfate in airborne particulate samples (158). An aliquot of the solution used to extract the particulates from the sampling filter is injected into a 3-mL calorimeter vessel and titrated with potassium dichromate in 0.1 mol L^{-1} hydrochloric acid and 5 mmol L^{-1} iron (III) chloride solution. After completion of the redox titration, 0.25 mL of 0.1 mol L^{-1} barium chloride, in the same solvent system, is injected over a period of 1 min. Rapid precipitation kinetics were promoted by coating the inside walls of the Dewar cell with a suspension of barium sulfate. A typical enthalpogram is illustrated in Fig. 6.11. The level of sulfur (IV) in the original sample is calculated directly from the endpoint (X) of the dichromate titration. The amount of sulfate in the solution can be quantitated by calibration of ΔT from a standard curve constructed from data obtained for sodium sulfate solutions under identical conditions. In order to determine the level of sulfate in the particulate, the DIE result must be corrected by subtraction of the amount of sulfur (VI) formed in the redox titration. If reducing interferents are present, for example, iron(II) or arsenic (III), the titration slope of sulfur (IV) should be identified by the enthalpy change associated with the reaction

$$3H_2SO_3 + Cr_2O_7^{2-} + 5H^+ = 3HSO_4^- + 2Cr^{3+} + 4H_2O, \quad \Delta H_R = -124.7 \text{ kJ mol}^{-1} \tag{6.61}$$

Specific experiments to examine the effect of iron (II), arsenic (III), arsenic (V), hydrogen sulfide, colloidal sulfur, and organic aldehydes showed that under the conditions of the titration no interference from these species can be observed. The reduction potentials and enthalpy changes associated with each of these potential reactions are sufficiently different that delineation of sulfur (IV) oxidation is easily achieved.

Systematic errors in the barium sulfate precipitation process can be as large as 5% in the presence of excess dichromate, iron (III) chloride, or potassium chloride. The precision of the technique is estimated at $\pm 5\%$ of the total sulfur (IV) + 3 nmol and $\pm 10\%$ of the total sulfate $+30$ nmol. This procedure has

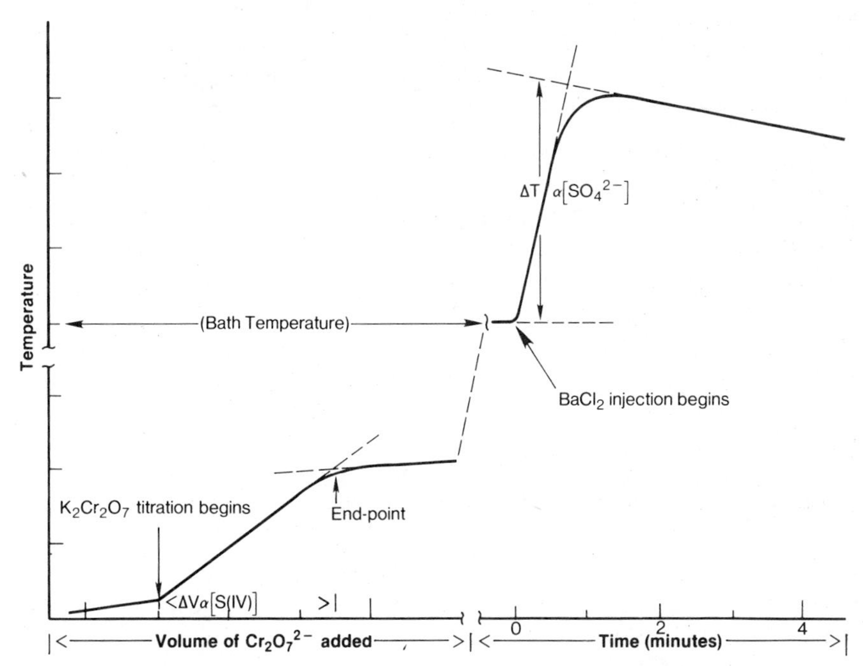

Figure 6.11. Typical enthalpograms for the serial thermometric titration of sulfur (IV) with $K_2Cr_2O_7$ and the direct-injection enthalpimetric determination of sulfate (158) (by permission of the American Chemical Society).

been applied successfully to the analysis of airborne particulate samples obtained from smelter flues, oil- and coal-fired power plants, and ambient air samples from New York City (159, 160).

A direct-injection enthalpimetric method has been described for the analysis of environmental samples containing nitrite or compounds that will produce nitrous acid in 0.1 mol L^{-1} hydrochloric acid (161). The method consists of injecting 0.15 mL of 0.05 mol L^{-1} sulfamic acid in 0.1 mol L^{-1} hydrochloric acid into 2 mL of analyte solution extracted from the sampling filter. Nitrite quantitation can be achieved by calibration of the resultant temperature pulse from standard sample data or by Joule heating calibration and subsequent calculation from known reaction enthalpies. In the presence of excess hydrochloric acid, the reactions occurring in the calorimeter are

$$NH_2SO_3H + HNO_2 = N_2 + SO_4^{2-} + H^+ + H_2O, \quad \Delta H_R = -421.9 \text{ kJ mol}^{-1} \qquad (6.62)$$

and

$$H^+ + SO_4^{2-} = HSO_4^-, \quad \Delta H_R = +21.92 \text{ kJ mol}^{-1} \tag{6.63}$$

In 0.1 mol L^{-1} hydrochloric acid, $[SO_4^{2-}][HSO_4^-] = 0.12$, therefore, the operational enthalpy of reaction is -402.5 kJ mol^{-1}.

If required, the nitrite determination can be included as a third step in the sulfur (IV)/sulfate procedure described earlier. In this case, the operational enthalpy must be adjusted to include the precipitation of barium sulfate which results from the reaction of the sulfate produced in Eq. (6.62) with excess barium (II); under the conditions of the experiment this adjusted enthalpy is -448.94 kJ mol^{-1}. A precision of $\pm 5\%$ rsd and a detection limit of 3 nmol was achieved for the determination of nitrite in more than 50 air filter and flue-dust samples.

6.10.3. Characterization of Acidic and Basic Species in Airborne Particulates

Eatough et al. have used parallel pH and thermometric titration data to characterize the acidic and basic species present in airborne particulates collected from copper and lead smelter flues and from New York City air (160, 162). As can be seen from Fig. 6.12, the data can be complex owing to the diversity of species in the sample. However, groups of acid–base titratable species can be determined from the inflections in the pH titration curve and corresponding changes in slope in the titration enthalpogram. Qualitative analysis can be achieved

TABLE 6.33. Probable Species Assignments for Copper Smelter Flue-Dust Sample based on Correlation of Potentiometric and Calorimetric Titration Data

Extraction pH	$pH_{1/2}$	$-\Delta H_B$ (kJ mol^{-1})	Probable Species	Found (μeq/mg)	Predicted[a] (μeq/mg)
4.2	4.3	57.3	Fe^{3+}	0.18 ± 0.01	0.18
	5.1	7.5		0.13	
	6.0	46–29	Fe^{2+}, Cu^{2+}, Zn^{2+}	2.76 ± 0.20	3.33
	7.9	19.7		0.34 ± 0.07	
	10.8	13.0	Ca^{2+}	0.41 ± 0.08	0.40

Source: Ref. 160.

[a]Based on proton-induced X-ray emission, thermometric, ESCA, and X-ray techniques.

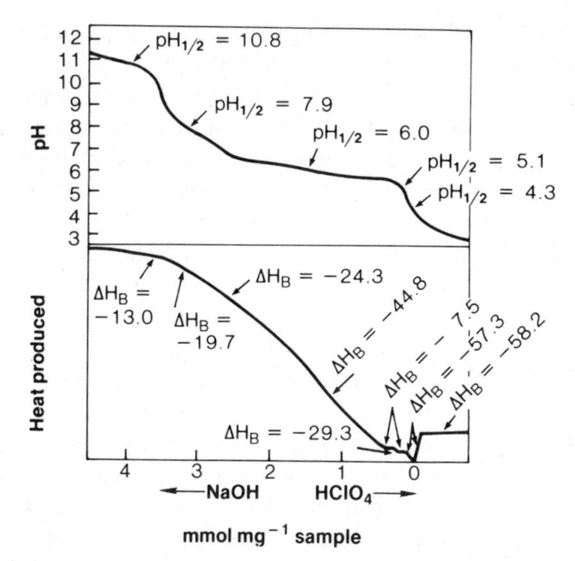

Figure 6.12. Calorimetric and potentiometric neutralization titration data for an aqueous extract of a copper smelter flue dust (ΔH_B in kJ mol^{-1}) (160) (by permission of Pergamon Press).

by correlation of the pH at the mid-point of each region, $pH_{1/2}$, and the corresponding enthalpy change, ΔH_B, with literature values. Divalent transition- and posttransition-metal ions frequently cannot be distinguished from each other by this procedure and species with a pK_a value <3 will titrate together as strong acids. Nonetheless, probable assignments can be made from such data as shown in Table 6.33, which represents the characterization of the sample titrated in Fig. 6.12. Clearly, agreement with levels predicted by a combination of techniques is favorable.

6.11. CATALYZED REACTIONS

As discussed in Chapter 2, the primary reason for the incorporation of a catalyzed reaction into a thermochemical determination is to increase the sensitivity of the technique. The analytical experiment can be designed to take advantage of the fact that certain classes of chemical reactions can be initiated by very small amounts of catalyst to produce relatively large heat changes. There are essentially two designs for such experiments.

1. The analyte is titrated (directly or indirectly) by the catalyst which initiates the catalytic (indicator) reaction as soon as it appears in excess in the titration solution. The onset of the indicator reaction signals the endpoint of the titration. Consequently, the temperature change accompanying the catalyzed reaction merely serves as an endpoint indicator; neither the extent of the reaction nor its rate are used directly to determine the analyte. An increase in sensitivity, versus a conventional titration, results from the fact that the concentration of the indicator reagents can be made large so that the temperature change associated with their reaction is of the order of several degrees celsius. On the other hand, the titrant (catalyst) and analyte concentrations can be relatively small since the temperature change associated with the titration reaction itself is unimportant, and only a small amount of an efficient catalyst is required to initiate the indicator reaction. Most titrations of this nature have been performed on simple "thermometric" instrumentation which has adequate environmental temperature control for the measurement of these large temperature changes. Accordingly, this procedure is usually referred to as "catalytic thermometric titrimetry."

2. The analyte itself is the catalyst and is added (usually injected) into a solution containing the indicator reaction reagents. In this type of experiment, the *rate* of the resulting reaction, as manifested by the slope of the temperature–time output, is usually the analytical signal because over a certain range of catalyst concentrations the initial rate of the indicator reaction is proportional to the concentration of the catalyst (analyte), that is,

$$\text{initial rate of indicator reaction} = k[\text{catalyst}]f([\text{A}]_0, [\text{B}]_0) \qquad (6.64)$$

where k is a proportionality (rate) constant, $[A]_0$ and $[B]_0$ are initial indicator reagent concentrations, and f is an arbitrary function. Quantitation is achieved by means of a calibration graph. Greenhow (14) has written a comprehensive review on the application of catalyzed reactions to calorimetric analyses reported up to 1977.

Applications are most conveniently classified according to the nature of the indicator reactions. The most commonly used reactions are:

1. Vinyl polymerization (acid/base or I_2 catalyzed).
2. Ketone or aldehyde condensation/dimerization/addition.
3. Acetylation.
4. Iodide–azide reaction.

5. Cerium (IV)–arsenic (III) reaction.
6. Manganese (III)–arsenic (III) reaction.
7. Transition-metal-catalyzed oxidation and decomposition reactions involving hydrogen peroxide.

6.11.1. Ionic (Vinyl) polyermization—Acid or Base Catalyzed

The use of vinyl monomers that undergo rapid ionic polymerization initiated by strong acids or bases, as thermochemical indicators for the determination of bases and acids, has increased considerably since the introduction of the concept in 1972 (163, 164). A summary of applications and pertinent reagent data is presented in Table 6.34.

These systems function efficiently as indicator reactions because basic titrants either neutralize the acid catalyst directly or inhibit cationic polyermization initiated by the acid and conversely for acidic titrants (14). Therefore, the conditions for an ideal catalytic thermometric titration are in place, that is, the indicator reaction does not begin until the determinative reaction is complete. Acrylonitrile, alkyl acrylates,and dialkyl itaconates have been shown to be suitable thermometric indicators for the titration of organic acids, while α-methylstyrene and alkyl vinyl ethers have been used for the titration of bases. Generalized reaction schemes for the titration of a base (B) and an acid (HA) are shown in Eq. (6.65)–(6.70) (14):

Acid-catalyzed polymerization

Initiation: $$CH_2{=}CHR \xrightarrow{H^+} CH_3CH^+R \qquad (6.65)$$

Polymerization:

$$CH_3CH^+R \xrightarrow{nCH_2=CHR} CH_3CHR(CH_2CHR)_{n-1}CH_2CH^+R \qquad (6.66)$$

Inhibition by basic analyte:

$$CH_3CH^+B \rightarrow CH_2{=}CHR + BH^+ \qquad (6.67)$$

Base-catalyzed polymerization

Initiation: $$CH_2{=}CHR \xrightarrow{OH^-} HOCH_2CH^-R \qquad (6.68)$$

TABLE 6.34. The Application of Acid/Base-Catalyzed Ionic Polymerization Reactions as Endpoint Indicators for Titration Calorimetry

Analyte(s)	Titrant, Concentration (mol L^{-1})	Indicator	Solvent	Detection Limit (DL), Range (R)	Reference	Text Section Reference
Acids (Non-aqueous Media)						
Mono- and polycarboxylic acids; mono- and polyhydric phenols; imides; hydroxycarboxylic acids; cyanuric and thiocyanuric acids	Bu_4NOH in methanol/toluene, 0.001–0.1 or KOH in propan-2-ol, 0.001–1.0	Acrylonitrile	Dimethylformamide or toluene	0.0001 mmol (DL)	165	6.11.1b
Thiols	Bu_4NOH in toluene/propan-2-ol, 0.001–0.1; KOH in propan-2-ol, 0.1–1.0	Acrylonitrile	Dimethylformamide	0.0001 mmol (DL)	166	6.11.1b
Phenol–, resorcinol–, and phenol/resorcinol–formaldehyde resins (Novalacs and Resoles)	KOH in propan-2-ol, 0.1–0.5	Acrylonitrile	Dimethylsulfoxide	—	167,168	6.11.1b
Phenols and other weak acids in mineral insulating oils	Bu_4NOH in toluene/methanol, 0.001	Acrylonitrile	Dimethylformamide	0.00003 mmol (DL)	169	—
Polyfunctional carboxylic acids, phenols, and vegetable tannins	KOH in propan-2-ol, 0.1	Acrylonitrile	Dimethylsulfoxide	—	170	—
Carboxylic acids and phenols in petroleum bitumens	KOH in propan-2-ol, 0.1 or Bu_4NOH in	Acrylonitrile	Toluene/propan-2-ol, dimethylformamide	—	171	6.11.1b

	propan-2-ol, 0.03 or $(Me)_4NOH$ in pyridine/propan-2-ol, 0.1/0.03.					
Catecholamines (acidic function)	Bu_4NOH in toluene/methanol, 0.001–0.1	Acrylonitrile	Dimethylformamide	10–20 μg (DL)	172	6.11.1b
Acids (aqueous media) mono- and polycarboxylic acids, phenols, amino acids, phosphorous, phosphonic and boric acids	KOH in propan-2-ol or water, 0.5	Acrylonitrile	Dimethylformamide	0.1–0.2 mmol (R)	173	6.11.1b
Bases (non-aqueous media)						
Aliphatic amines, pyridine and aniline derivatives, difunctional aromatic amines, heterocyclic nitrogen compounds, amides, sulfur and phosphorous derivatives	$HClO_4$ in glacial acetic acid or BF_3 diethyletherate in dioxan, 0.001–0.1	α-Methylstyrene or isobutyl vinyl ether	Toluene, dicholoroethane, nitroethane, acetone, acetic acid	0.00001 mmol (DL)	174	6.11.1c
Alkaloids and alkaloidal salts	$HClO_4$ in acetic acid, 0.001–0.1	α-Methylstyrene	Acetic acid or 1,4-dioxan or 1,2-dichloroethane or chloroform	0.0001 mmol (DL)	175	6.11.1c
Catecholamines (basic function)	$HClO_4$ in acetic acid, 0.001–0.1	α-Methylstyrene	Formic acid/acetic acid/1,2-dichloroethane/propylene carbonate	0.0001 mmol (DL)	172	6.11.1c

Polymerization:

$$HOCH_2CH^-R \xrightarrow{nCH_2=CHR} HO(CH_2CHR)_nCH_2CH^-R \quad (6.69)$$

Inhibition by acidic analyte:

$$HOCH_2CH^-R + HA \rightarrow HOCH_2CH_2R + A^- \quad (6.70)$$

6.11.1a. Instrumentation

The relatively large temperature changes associated with the indicator reactions allow considerable flexibility in the type of apparatus required. Acceptable data have been obtained from a rudimentary apparatus as shown in Fig. 6.13 in which the titrant was added manually from a conventional buret and a −5 to +50°C thermometer was used to record the temperature changes point-by-point (175). In the interests of convenience and efficiency, the thermometer can be replaced by a thermistor and the manual buret by a motor-driven syringe pump. In either case, the "calorimeter" need not be more sophisticated than a conventional Dewar vessel surrounded by polystyrene insulation. As temperature changes of 5–10°C during the course of the indicator reaction are not unusual, the effects of heat transfer during the determination reaction and the fluctuations in the ambient temperature are insignificant by comparison. This is a luxury not associated with

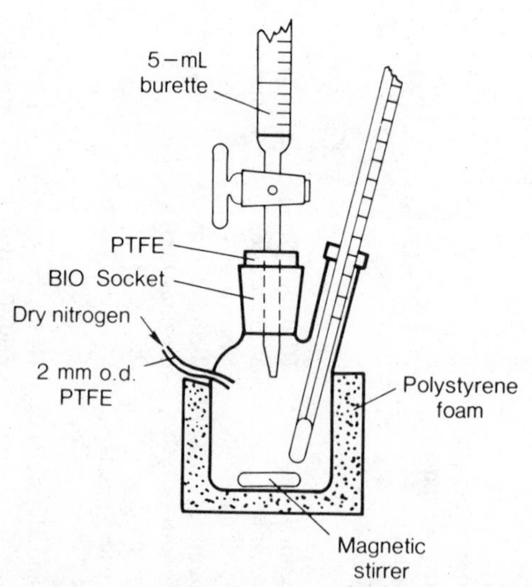

Figure 6.13. Manual catalytic thermometric titration apparatus (cell volume, 15 mL) (165) (by permission of the Royal Society of Chemistry).

any other form of solution calorimetry! The only extra precaution necessary to ensure accuracy is the exclusion of carbon dioxide from the cell by use of a closed system purged with dry nitrogen.

6.11.1b. Acids

In a series of reports on solvent effects (176–180) it has been shown that the choice of solvent and basic titrant is absolutely crucial in order to achieve the ideal endpoint inflection for the titration of acids by this technique. To quote a statement from one of the papers, ". . . . the aim is to minimize the extent of reactions other than the determinative reaction before the endpoint, and to maximize the rate of the indicator reaction immediately the titrant is present in excess" (180).

The concentration of titrant is varied commensurate with the concentration of the analyte. For maximum sensitivity, 0.001 mol L^{-1} titrants are used, increasing up to 0.1 mol L^{-1} for larger amounts of analyte. As the concentration of titrant is decreased, it becomes necessary to reduce the overall volume in the cell (sample + solvent + monomer) in order that the concentration of the catalyst remains at a level sufficient to initiate polymerization. Typically, 1 mL of monomer and 0.1 mL of sample are used for the most dilute titrants.

Two solvent-related issues have been identified as being key to the sharpness of the endpoint inflection. First, the solvating effects of alcohols and other cosolvents have a significant effect on the rate of the polymerization reaction since free ions are much more effective than ion pairs in inducing ionic polymerization. Therefore, tetrabutylammonium hydroxide, a bulky cation, which will tend to ionize completely to produce hydroxyl ions, produces sharp inflection points for base-catalyzed acrylonitrile indicator reactions regardless of the solvent used (176). However, when potassium hydroxide is used as the titrant, a considerable increase in endpoint clarity is obtained by the incorporation of a diprotic apolar solvent with a high dielectric constant such as dimethylformamide or dimethysulfoxide. The rationale is that both these solvents will solvate the potassium ion and liberate unsolvated anions more efficiently than propan-2-ol, the usual titration solvent. This effect is shown in Fig. 6.14, which represents the titration of potassium hydroxide in propan-2-ol (curves A–D) and tetrabutylammonium hydroxide (curves E and F) into blank acrylonitrile indicator reagent in different solvents. Since pyridine has a lower dielectric constant than acrylonitrile, its beneficial effect on endpoint definition (cf. curves A and C) is attributed to its nucleophilicity, which aids in the solvation of potassium ions and causes the release of unsolvated hydroxyl and propan-2-oxide ions (177). Another interesting solvent effect can be seen by comparing the blank titration volumes for 0.1 mol L^{-1} and 0.5 mol L^{-1} alcoholic potassium hydroxide. The effect of

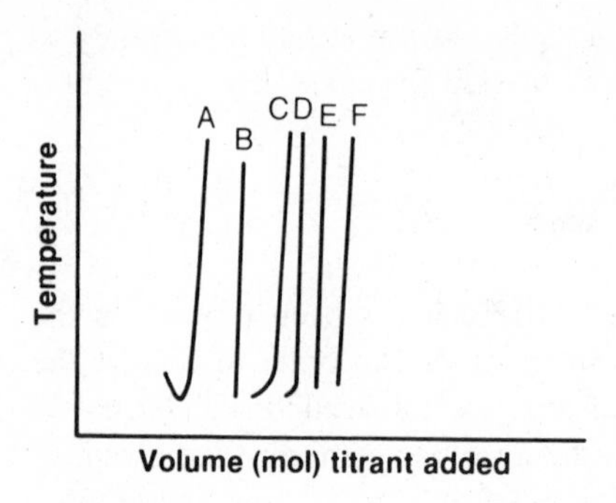

Figure 6.14. Effect of solvent compositions on the shape of catalytic thermometric titration enthalpograms (176): (A) Solvent: 4 mL acrylonitrile; titrant: 0.1 mol L^{-1} potassium hydroxide, (B) Solvent: 4 mL acrylonitrile, 1 mL dimethylformamide; titrant as in A. (C) Solvent: 3 mL acrylonitrile, 1 mL pyridine; titrant as in A. (D) Solvent: as in C; titrant: 0.5 mol L^{-1} potassium hydroxide. (E) Solvent: as in A; titrant: 0.1 mol L^{-1} tetrabutylammonium hydroxide. (F) Solvent: As in C; titrant as in E.

changing to the more concentrated titrant is to halve the blank volume, not reduce it by a factor of 5 as might be predicted (177). Evidently, the volume of propan-2-ol introduced into the titration mixture has an effect on the magnitude of the blank titration value. Additionally, it has been demonstrated that the presence of propan-2-ol as a solvent in the titration mixture decreases the slope of the indicator reaction signal and increases the temperature change ("heat of mixing") in the determinative reaction (cf. a pure dipolar aprotic solvent system). The latter effect, which has been attributed to the thermal contribution of hydrogen-bonding reactions involving the alcoholic hydroxyl function, can be minimized by use of appropriate mixtures of propan-2-ol and a dipolar solvent.

The composition of the solvent also has a significant effect on the titration data of polyfunctional compounds. In general, increasing the fraction of dipolar aprotic solvent in the titration mixture reduces the neutralization reaction stoichiometry. It is thought that this effect is related to the competition between the neutralization and polymerization reactions, the presence of aprotic solvents favoring the latter.

The second major reaction affecting endpoint definition is cyanoethylation caused by the presence of primary and secondary alcohols (178, 180). Cyanoethylation is the termination of the anionic polymerization process after only one monomer unit has been consumed, that is,

$$K^+OR^- + CH_2{=}CHCN = ROCH_2CH^-CNK^+ \tag{6.71}$$

$$ROCH_2CH^-CN + ROH = ROCH_2CH_2CN + OR^- \tag{6.72}$$

K^+OR^- is the alkoxide formed in the following equilibrium which exists in an alkaline alcoholic medium:

$$K^+OH^- + ROH \rightleftharpoons K^+OR^- + H_2O \quad (6.73)$$

Initially, it was postulated that the exothermic indicator reaction was caused by both cyanoethylation *and* polymerization, and that, kinetically, cyanoethylation preceded anionic polymerization when primary or secondary alcohols were present (163). Subsequently it has been shown that this hypothesis is likely to be valid (180). Indeed, there is evidence that when highly reactive alcohols are present some cyanoethylation may take place prior to the endpoint, adversely affecting the accuracy of the data. Ideally, therefore, the alcohol should be of moderate reactivity so that the heat effect accompanying the rapid cyanoethylation reaction will be incorporated into the indicator reaction without introducing titration error. Satisfactory inflection points have been obtained under conditions where cyanoethylation could not have taken place, for example, by use of a tertiary alcohol solvent. Cyanoethylation is therefore not a prerequisite for precise titration data. Greenhow and Dajer de Torrijos (180) have summarized the specifications of the ideal solvent/basic titrant system in the following manner: "Suitable reagent—solvent combinations will have, as the titrant, either a higher primary alkoxide in a higher primary alkanol or a tertiary alkoxide in a tertiary alkanol and, as the sample solvent, a mixture of acrylonitrile, a dipolar aprotic solvent and, when the tertiary alkoxide reagent is used, a higher primary alkanol to provide a rapid cyanoethylation reaction." Illustrations of these principles are shown in Fig. 6.15.

In a systematic study, Greenhow and Spencer (165) have evaluated the technique for the determination of a variety of mono- and polyfunctional acids. In addition to acrylonitrile, other monomers were examined for their efficacy as indicator systems: methyl acrylate and dimethyl itaconate gave satisfactory results: styrene, isoprene, and methyl methacrylate did not. Tetra-*n*-butylammonium hydroxide was found to be a satisactory titrant for most monofunctional acids having the advantage that it produced soluble salts in the solvent systems used, thereby reducing the possibility of interference from co-precipitation phenomena. Interference from this source must be considered when using potassium hydroxide in propan-2-ol (which produces insoluble salts) as the titrant. The number of acidic functionalities titrated in a polyfunctional compound also depends on the titrant used. The list of compounds titrated, reaction stoichiometries, and experimental conditions used in this study are presented in Tables 6.35a and 6.35b.

Alkyl and aryl thiols have been determined in the presence of carboxylic acids and phenols by a two-titration procedure utilizing acrylonitrile polymerization and acetone dimerization as the indicator reactions (166). The acrylonitrile procedure titrates only the carboxylic acids and phenols, the acetone procedure (see 6.11.3) will determine the total acidity of the mixture allowing the determination of thiol content by subtraction. Some of the heterocyclic thiols and

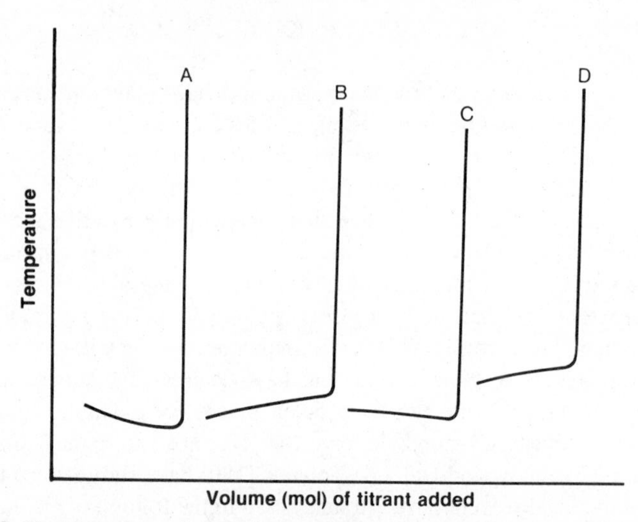

Figure 6.15. Ideal (experimental) catalytic thermometric titration enthalpograms. Sample: benzoic acid, A–C, 1×10^{-4} mol; D, 5×10^{-4} mol (180) (by permission of the Royal Society of Chemistry). (A) Solvent: 4 mL acrylonitrile, 2 mL dimethylformamide; titrant: 0.1 mol L^{-1} potassium *n*-butoxide in *n*-butanol. (B) Solvent: 3 mL acrylonitrile, 2 mL dimethylformamide, 1 mL *n*-butanol; titrant: 0.1 mol L^{-1} potassium tert-butoxide in tert-butanol. (C) Solvent: 4 mL acrylonitrile, 2 mL dimethylsulfoxide; titrant: as in A. (D) Solvent: 4 mL acrylonitrile, 1 mL dimethylformamide, 1 mL *n*-butanol; titrant: 0.5 mol L^{-1} potassium tertbutoxide in tert-butanol.

thioamide derivatives will titrate with both indicator systems. However, attempts to titrate thiols by use of the iodine-catalyzed polymerization of ethyl vinyl ether produced substoichiometric data. Relevant data from this report are summarized in Table 6.36.

In a similar two-titration procedure the phenolic hydroxyl groups in phenol and resorcinol resins and phenol–resorcinol–formaldehyde co-condensates have been determined (167, 168). The principle of the method is that the resorcinol in the mixture can be titrated as either a mono- or difunctional acid depending on the concentrations of the titrant, monomer, and solvent. The phenol/resorcinol ratio can be determined by extrapolation of a calibration graph relating the composition of standard mixtures to the ratio of the two-titration results designed to titrate resorcinol as a mono- and diprotic acid, respectively.

With pyridine as a co-solvent, the acid contents in commercial petroleum bitumens determined by catalytic thermometric titrimetry agree with those determined by potentiometric methods (171). The potassium hydroxide/acrylonitrile/pyridine system gave superior results to other titrant/solvent/co-solvent combinations investigated. The authors recommend the calorimetric procedure (versus potentiometry) on the grounds of convenience and speed, and the better tolerance

TABLE 6.35a. Monofunctional Organic Acids Titrated Thermometrically with 0.1 mol L^{-1} Tetra-*n*-Butylammonium Hydroxide Solution with Acrylonitrile as the Endpoint Indicator[a]

Conditions: 0.1 mmol acid, 1 mL toulene or dimethylformamide, 10 mL acrylonitrile
Phenylacetic acid, benzilic acid, hippuric acid, and cinnamic acid
Benzoic acid, *p*-toluic acid, *o*-nitrobenzoic acid, 3,5-dinitrobenzoic acid, *p*-aminobenzoic acid, *m*-aminobenzoic acid, *p*-methoxybenzoic acid, and acetylsalicylic acid
3,5-Xylenol, 2,6-xylenol, 2,6-di-t-butyl-4-methylphenol (0.5), *o*-nitrophenol, *o*-aminophenol, salicylaldehyde, methyl salicylate, and 3-hydroxypyridine
1-Naphthol, 1-amino-7-naphtol, 8-hydroxyquinoline, and 2-hydroxyquinoline (O)
Dimedone
Succinimide (0.17) and phthalimide (0.35)

Source: Ref. 165, by permission of the Royal Society of Chemistry.

[a]Figures in parentheses following the names of compounds indicate non-stoichiometric titration, the figure being the fraction of the acid function titrated at the endpoint.

TABLE 6.35b. Thermometric Titration of Polyfunctional Organic Acids with Acrylonitrile and Acetone as Endpoint Indicators

Compounds	Acidic Groups	Groups Titrated by[a]		
		Method 1	Method 2	Method 3
Succinic acid	2	2	2	(1)
Pyrocatechol	2	1	1	(1)
Resorcinol	2	1	2	(2)
Hydroquinone	2	1	1	(2)
Phloroglucinol	3	2	2	2
Pyrogallol	3	1	2	(2)
Salicylic acid	2	1	1	(2)
p-Hydroxybenzoic acid	2	2	2	2
3,4-Dihydroxybenzoic acid	3	2	2	2
3,4,5-Trihydroxybenzoic acid	4	2	2	2
Tannic acid	–	10	12	20
2,6-Pyridinedicarboxylic acid	2	2	2	1.74
Cyanuric acid	3	1	1	2
Trithiocyanuric acid	1	2	3	3
Dichloroisocyanuric acid	1 (3)	2	3	2

Source: Ref. 165, by permission of the Royal Society of Chemistry.

[a]Method 1: acrylonitrile endpoint indicator with tetra-*n*-butylammonium hydroxide titrant (0.1 mol L^{-1}). Method 2: acrylonitrile endpoint indicator with potassium hydroxide titrant (0.1 mol L^{-1}). Method 3: acetone method with potassium hydroxide titrant (1.0 mol L^{-1}); values in parentheses are from Ref. 185.

TABLE 6.36. Thiols and Thioamides Titrated with 1.0 mmol L^{-1} Potassium Hydroxide and 0.1 mmol L^{-1} Tetra-*n*-Butylammonium Hydroxide Solutions with Acetone and Acrylonitrile, Respectively, as Endpoint Indicators

Conditions: 1 mmol of thiol in 3 mL of acetone was titrated with 1.0 mol L^{-1} potassium hydroxide solution by using the acetone-indicator method, and 0.1 mmol of thiol in a mixture of 1 mL of dimethylformamide and 2 mL of acrylonitrile with 0.1 mol L^{-1} tetra-*n*-butylammonium hydroxide solution by using the acrylonitrile-indicator method.

Aliphatic thiols: Heptane-1-thiol (1:0.9:0)[a]; dodecane-1-thiol (1:0.9:0); 2,3-dimercaptopropan-1-ol (2:1.7:0); and mercaptosuccinic acid (3:2.5:1.9)

Aromatic thiols: Toluene-1′-thiol (1:0.8:0); toluene-4-thiol (1:0.8:0); 4-aminobenzenethiol (1:0.65:0); 2-mercaptobenzoic acid (2:2:1); salicylideneaminobenzene-2-thiol (2:1.9:1); and pyridine-2-thiol (1:1:0.18)

Heterocylic thiols: 2-Mercaptothiazoline (1:1:0); 2-thiohydantoin (1:1:1); 4-hydroxypyrimidine-2-thiol (2:1:1); 4,6-dihydroxypyrimidine-2-thiol (3:2:1); purine-6-thiol (1:1.7:0.8); 2-mercaptobenzimidazole (1:1:0.3); 2-mercaptobenzoxazole (1:1:1); and 2-mercaptobenzothiazole (1:1:1)

Thioamides: Thioacetamide (1:1:0.9[b]); thiourea (1(2):0.1:1[b]); thiocarbanilide (1(2):1:0.36 or 0.53[b]); dithiooxamide (rubeanic acid) (2:2:1); thiosemicarbazide (1(2):1.1:0.9); and diphenylthiocarbazone (dithizone) (1(2):1:0.8 or 1[b])

Source: Ref. 166, by permission of the Royal Society of Chemistry.

[a]Figures in parentheses denote the theoretical number of acidic functional groups in the molecule, the number of groups titrated by using the acetone-indicator method, and the number of groups titrated by using the acrylonitrile-indicator method, respectively.

[b]Values obtained in the titration of 1 mequiv. of sample by using the acrylonitrile-indicator method and 1.0 mol L^{-1} potassium hydroxide titrant.

of the thermistor probe to the bituminous solution. The precisions of both procedures (1–5% rsd) are comparable.

The phenolic functions of catecholamines have been determined by titration with tetra-*n*-butylammonium hydroxide to an endpoint indicated by acrylonitrile polymerization (172). Dimethylformamide was a suitable solvent for dopamine hydrochloride, adrenaline hydrogen tartrate, and α-methyldopa. It was necessary to add an accurately known amount of *p*-toluenesulfonic acid to dimethylformamide in order to dissolve adrenaline, *L*-noradrenaline, *L*- and *DL*-dopa and (+)-Corbasil. The (known) value of titrant consumed by *p*-toluenesulfonic acid was subtracted from the overall titration volume. The precision obtained for the determination of catecholamine formulations, for example, *L*-dopa tablets, is similar to that achieved for pure catecholamines (<1.6% rsd). The effect of magnesium stearate (free stearic acid) on the accuracy of the data is small. At the 2% level the titer is increased by only 0.1%.

All the catalyzed indicator reactions discussed so far have been non-aqueous systems. However, Greenhow and Shafi (173) have shown that the acrylonitrile polymerization reaction can be used effectively as a thermochemical indicator for the determination of organic and inorganic acids in aqueous solution; a dipolar aprotic solvent, dimethylsulfoxide, and aqueous potassium hydroxide were used as the co-solvent and titrant, respectively.

Although the inflection at the endpoint was less well-defined than the corresponding non-aqueous systems, acceptable enthalpograms were obtained for the titration of glycine, cysteine, valine, benzoic acid, phenol, resorcinol, oxalic acid, phosphoric acid, and boric acid at concentrations of 1–10% *w*/*v*. The lower temperature changes observed in the indicator reaction and the much lower sensitivity (versus non-aqueous solution) are presumed to be due to polymer-chain-termination processes induced by the presence of water. Nonetheless, the technique does offer advantages over a corresponding visual endpoint titration in that weak acids, such as boric acid and amino acids, can be titrated directly without any addition of reagents to enhance acidity.

6.11.1c. Bases

Cationic, proton-induced polymerization has formed the basis of the catalytic thermometric titration approach to the determination of bases. The titrant/monomer system cited almost exclusively in reports to date is perchloric acid (in glacial acetic acid)/α-methylstyrene as shown in Table 6.34. In an early study, two titrant/monomer combinations, perchloric acid/α-methylstyrene and boron trifluoride diethyletherate (in dioxan)/isobutyl vinyl ether, were evaluated for the determination of a large number of organic bases (174). In general, both systems proved satisfactory; however, the boron trifluoride system did result in more substoichiometric endpoints. For example, although both titrants can be used to determine primary, secondary, and tertiary aliphatic amines and simple pyridne and quinoline derivatives, only perchloric acid reacted stoichiometrically with the monofunctional aniline derivations studied. The bases titrated and the stoichiometries observed with each titrant are shown in Table 6.37. Dilute titrants were prepared by dilution of stock solution with dioxan or glacial acetic acid; however, endpoint clarity diminished at titrant concentrations <0.01 mol L^{-1}. This problem was rectified by the use of either nitroethane or 1,2-dichloroethane as diluents: the mechanism of this improvement in endpoint definition is unexplained.

Strychnine, nicotine, atropine, quinine, papaverine, caffeine, and theophylline have been determined by use of the perchloric acid/α-methylstyrene indicator reaction with a precision of about 1.5% rsd at the microgram level (175). Positive titration errors resulting from the occlusion of titrant in the precipitated papaverine perchlorate restrict the determination of papaverine to the 0.05–0.01 mmol range where the effect of this phenomenon was minimized. This problem was not

TABLE 6.37. Organic Bases Titrated with 0.1 mol L^{-1} Perchloric Acid in Acetic Acid and Boron Trifluoride Diethyl Etherate in Dioxan[a]

Condition: 0.1 mmol of base in 1 mL of solvent (toluene, dichloroethane, nitroethane, acetone, or acetic acid), 10 mL of α-methylstyrene or isobutyl vinyl ether

Aliphatic amines: *n*-Butylamine (1:1:1); morpholine (1:1:1); triethylamine (1:1:1); tris (hydroxymethyl) methylamine (1:1:0); and 6-aminocaproic acid (1:1:1)

Pyridine derivatives: Pyridine (1:1:1); α-picoline (1:1:1); 4-vinylpyridine (1:1:1); quinoline (1:1:1); 8-hydroxyquinoline (1:1.4:1); and 2,6-lutidine (1:1:0.9)

Aniline derivatives: *p*-Toluidine (1:1:0.7); *p*-nitroaniline (1:1:0); and *p*-hydroxyaniline (1:1.2:0)

Difunctional aromatic amines: *o*-Phenylenediamine (2:2:1); *m*-phenylenediamine (2:2:1.3); *p*-phenylenediamine (2:2:1.5); 2-amino-4-methylpyridine (2:1:1.3); 8-aminoquinoline (2:2:1.3); and hydrazobenzene (2:2:0)

Heterocyclic nitrogen compounds: Imidazole (2:1:1.2); benzimidazole (2:1:1); benzotriazole (3:1:1); quinoxaline (2:1.4:0.5); and 2,3-dichloroquinoxaline (2:0.8:0)

Amides and sulfur and phosphorus derivatives: Acetamide (1:1:0);′ dimethylformamide (1:1:0); diethylformamide (1:1:0.6); *N,N*-dimethylacetamide (1:1:0.4); dimethyl sulfoxide (1:1:0.7); and hexamethylphosphoramide (1:1.5:1)[b]

Source: Ref. 174, by permission of the Royal Society of Chemistry.

[a]Figures in parentheses following the name of the base denote the theoretical number of basic functional groups in the molecule, the number of groups titrated with the 0.1 mol L^{-1} perchloric acid, and the number of groups titrated with the 0.1 mol L^{-1} boron trifluoride, respectively.

[b]On the basis of the amido groups the functionality will be 3.

encountered with the other insoluble alkaloid perchlorates. The *L*-dopa content of normal tablets, sustained release tablets, and capsules has been determined by the same technique and the results compared with a standard BP assay procedure and an ultraviolet spectrophotometric method (172). Some selected data are presented in Table 6.38.

6.11.2. Ionic (Vinyl) Polymerization—Iodine Catalyzed

The cationic polymerization of some vinyl monomers, in particular alkyl vinyl ethers, can be catalyzed by the presence of iodine in the form of I^+ or I_3^+. The mechanism is analogous to the acid- or base-catalyzed reactions shown in Eq. (6.65)–(6.70) and can be represented as follows (14):

$$2I_2 \rightleftharpoons I^+I_3^-; \quad 3I_2 \rightleftharpoons I_3^+I_3^- \tag{6.74}$$

$$\text{Initiation: } CH_2{=}CH(OR) \xrightarrow{I^+ \text{ or } I_3} ICH_2CH^+(OR) \tag{6.75}$$

Polymerization:

$$ICH_2CH^+(OR) \xrightarrow{nCH_2=CH(OR)} I[CH_2CH(OR)]_nCH_2CH^+(OR) \tag{6.76}$$

Inhibition by analyte:

$$ICH_2CH^+(OR) + B \rightarrow CH_2{=}CH(OR) + BI^+ \tag{6.77}$$

The thermochemical titration of an iodine oxidizable moeity by means of a catalyzed indicator reaction endpoint is therefore feasible. Indeed, iodine in dimethylformamide has been used to titrate a variety of compounds as shown in Table 6.39. Dimethylformamide and acrylonitrile have been shown to be the solvents of choice for the iodine-catalyzed polymerization of alkyl vinyl ethers producing sharper endpoint inflections than propylene carbonate, toluene, 1,4-dioxan, tetrahydrofuran, diethylformamide, dimethylacetamide, and dimethylsulfoxide (184).

The mechanism of the reaction occurring when iodine (in dimethylformamide) and ethyl vinyl ether are mixed has been studied by termination of the reaction with sodium thiosulfate and examination of the non-aqueous layer by GC/MS (184). The principal components of the mixture were identified as

TABLE 6.38. Correlation Data for the Determination of the *L*-Dopa Content of Pharmaceutical Preparations[a]

Sample	BP Method of Assay[b]	UV Spectrophotometry	Catalytic Thermometric Titration
Levodopa tablets (normal)	67.9	68.7	67.2
Levodopa tablets (sustained release)	74.2	74.3	74.9
Levodopa capsules	92.3	93.2	91.3
L-Dopa	98.3	100.0[c]	99.5

Source: Ref. 172, by permission of the Royal Society of Chemistry.

[a]All values are percent *w*/*w* and are the average of three determinations.

[b]Non-aqueous perchloric acid titration to visual endpoint.

[c]Calibration standard.

TABLE 6.39. The Application of Iodine-Catalyzed Ionic Polymerization Reactions as Endpoint Indicators for Titration Calorimetry

Analyte(s)	Titrant, Concentration (mol L^{-1})	Indicator	Solvent	Detection Limit (DL); Range (R)	Reference	Text Section Reference
Pyridine, THAM, hydrazine derivatives, and water	I_2 in dimethylformamide, 0.01–0.05	Ethyl vinyl ether	Dimethylformamide	0.0005 mmol (DL)	181	6.11.2a
Dithiocarbamates and phosphorodithioates	I_2 in dimethylformamide, 0.05	Ethyl vinyl ether	Dimethylformamide or acrylonitrile	0.002 mmol (DL)	182	6.11.2b
0-Alkyldiethiocarbonates (xanthates) and metal iodides	I_2 in dimethylformamide, 0.05	Ethyl vinyl ether	Dimethylformamide, acrylonitrile, or chloroform	0.002 mmol (DL)	183	6.11.2c
Metal iodides, thiocyanates, thiols, and sodium azide	I_2 in dimethylformamide	Ethyl vinyl ether	Dimethylformamide/1,3-dioxolan	—	184	6.11.2d

solvent, unreacted monomer, $ICH_2CH(OC_2H_5)CH_2CH(OH)(OC_2H_5)$, and $I[CH_2CH(OC_2H_5)]_2CH_2CH(OH)(OC_2H_5)$. Comparison of these structures with the proposed products of the iodinium ion-catalyzed polymerization of an alkyl vinyl ether in Eq. (6.74)–(6.77) shows that the mechanism postulated in these equations is probably correct.

6.11.2a. Organic Bases, Hydrazines, and Water

In a systematic evaluation of the potential of the iodine-catalyzed polymerization reaction as a thermochemical indicator, Greenhow and Spencer established that this technique can be used to titrate not only oxidizable compounds, for example, hydrazines, but also compounds which form simple iodine addition products, for example, pyridines (181) and water. The list of compounds studied and some typical enthalpograms are shown in Table 6.40 and Fig. 6.16, respectively.

TABLE 6.40. Organic Bases and Quaternary Ammonium Halides Titrated with 0.05 mol L^{-1} Iodine in Dimethylformamide[a]

Conditions: 0.025 mmol of base or halide in 1 mL of dimethylformamide and 2 mL of ethyl vinyl ether

Aliphatic and alicyclic amines: *n*-Butylamine (1.8); benzylamine (1.8); 1,2-diaminoethane (3.37); morpholine (1.8); piperidine (1.8); triethylamine (2.1); tris(hydroxymethyl)methylamine (2.1); 2-*NN*-diethylaminopropionitrile (2.2); *N*-methylmorpholine (2.3); and *N*-ethylpiperidine (2.2)

Pyridine derivatives: Pyridine (2.0); -picoline (1.9); 2,6-lutidine (2.0); 2,6-pyridinedicarboxylic acid (0.04); pyridine *N*-oxide (1.1); 3-picoline *N*-oxide (1.1); quinoline (2.0); 4-methylquinoline (2.1); 8-hydroxyquinoline (1.0); and 2-hydroxyquinoline (0)

Aniline derivatives: Aniline (0.6); *o*-toluidine (0.7); *p*-toluidine (0.9); *p*-nitroaniline (0); diphenylamine (0); and triphenylamine (0)

Heterocyclic nitrogen compounds: Hexamethylenetetriamine (2.5); phthalazine (1.7); quinoxaline (0); and benzimidazole (2.0)

Quaternary ammonium halides: Tetra-*n*-butylammonium bromide (0.9); tetra-*n*-butylammonium iodide (0.7); cetyltrimethylammonium bromide (0.7); cetylpyridinium bromide (0.9); and benzyldimethylmyristylammonium chloride (1.2)

Phosphorus compounds: Triphenylphosphine (0.5); and triphenylphosphine oxide (0)

Source: Ref. 181, by permission of the Royal Society of Chemistry.

[a]Figures in parentheses following the name of the compound denote the number of moles of iodine combining with one mole of the compound on the basis of the iodimetric titration.

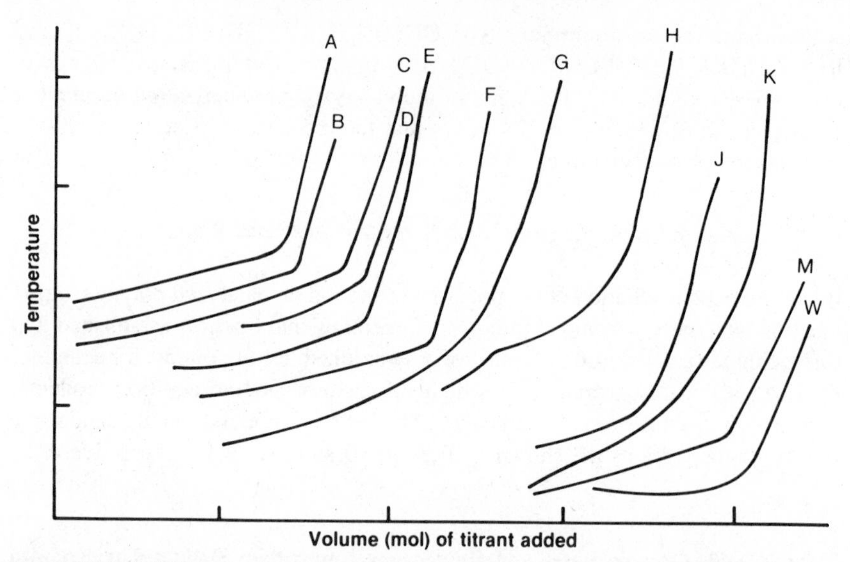

Figure 6.16. Enthalpograms for the catalytic thermometric titration of organic bases, water, and quaternary ammonium halides with 0.05 mol L^{-1} iodine (181). Analytes/mass (mg): A, *n*-butylamine, 0.84; B, tris (hydroxymethyl)methylamine, 1.2; C, pyridine, 1.3; D, hexamethylenetetramine, 1.6; E, morpholine, 1.89; F, 8-hydroxyquinoline, 2.9; G, pyridine *N*-oxide, 2.3; H, triphenylphosphine, 7.4, J. cetyltrimethylammonium bromide, 4.0; K, tetra-*n*-butylammonium iodide, 9.3; M, benzyldimethylmyristylammonium chloride, 5.0; and *W*, water, 10.0 (by permission of the Royal Society of Chemistry).

Most of the aliphatic amines and pyridine derivatives consumed 1.8–2.3 moles of iodine to the endpoint. Pyridine itself combined with 2 moles of iodine, consistent with the formation of the adduct $C_5H_5N{\cdot}2I_2$.

Aniline and the alkylamines show only a low reactivity toward iodine, which is reduced even further by the presence of electron-withdrawing groups in the aniline and pyridine ring.

The determination of water by this method is not sensitive and the relationship between the amount of water, and the volume of iodine required to produce an endpoint is dependent on the titration rate. A titration rate of 0.06 mL min^{-1} was determined empirically to produce a linear relationship between the titer and the amount of analyte. A blank titration is recommended in order to compensate for the water content of the solvents.

The reaction stoichiometries observed for the titration of oxidizable hydrazines are shown in Table 6.41. The precision of the quantitative data was of the order of 2% rsd when a 0.01 mol L^{-1} titrant was used. Other monomer systems examined in this report, including *n*-butyl vinyl ether, isobutyl vinyl ether, 2-chloroethyl vinyl ether, and divinyl ether, proved inferior to ethyl vinyl ether

TABLE 6.41. Hydrazine Derivatives Titrated with 0.05 mol L^{-1} Iodine in Dimethylformamide[a]

Conditions: 1 mL of dimethylformamide, 2 mL of ethyl vinyl ether
NN-Dimethylhydrazine (1.7); 2-hydroxylethylhydrazine (1.1); phenylhydrazine (0.5); 4-nitrophenylhydrazine (0.5); 2,4-dinitrophenylhydrazine (0.09); *NN*-diphenylhydrazine (0.43); benzoylhydrazine (0.7); isonicotinoylhydrazine (2.1); benzenesulfonohydrazide (0.16); 4,4′-oxybis (benzenesulfonohydrazide) (0.31); semicarbazide hydrochloride (0.9); and thiosemicarbazide (1.5)

Source: Ref. 181, by permission of the Royal Society of Chemistry.

[a]Figures in parentheses following the name of the compound denote the number of moles of iodine combining with one mole of the compound on the basis of the iodiometric titration.

as generally applicable indicator systems. Similarly, the interhalogen compounds—iodine bromide, iodine chloride, and iodine trichloride—gave less-well-defined enthalpograms when used as titrants in lieu of iodine.

6.11.2b. Dithiocarbamates and Phosphorodithioates

Ammonium, cadmium, bismuth, iron, lead, manganese, nickel, piperidinium, and zinc dithiocarbamates (**5**) and nickel and zinc phosphorodithionates (**6**) have been titrated directly with iodine to a catalyzed endpoint. The product of the titration reaction is the corresponding disulfide with ethyl vinyl ether as the indicator (182):

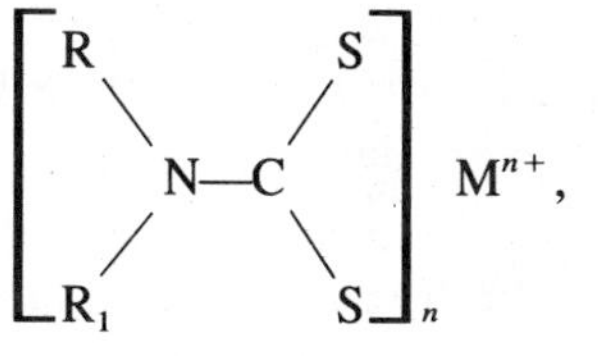

$$\left[(R)(R_1)N\text{—}C(S)(S) \right]_n M^{n+}, \qquad \left[(RO)(R'O)P(S)(S) \right]_n M^{n+}$$

5 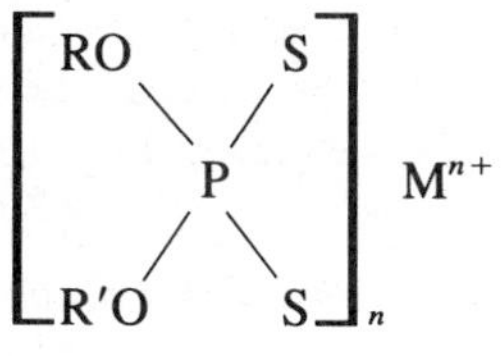 **6**

The number of endpoints observed depends on the amount of sample taken for both types of compounds. As the sample concentration is increased, a second inflection point is observed which has been assigned to a further oxidation reaction of the disulfide. The first endpoint, which is more well-defined than the second, should be used for quantitative data.

The stoichiometry of the reaction is also complex and depends on the structure of the analyte as well as the nature of the central ion. As a result of the inexact stoichiometry, a calibration procedure is required to obtain quantitative data.

These limitations notwithstanding, the accuracy and precision of the determination of technical and laboratory reagent grade compounds are good when compared to an aqueous iodometric back-titration procedure with a visual endpoint. In general, the results obtained by the two methods agree to within $\pm 1.2\%$; the precision (rsd) is in the range of 0.4–2.0%. The calorimetric method was extended successfully to the determination of some lead and zinc dithiocarbamates and phosphorodithioates in hydrocarbon oils.

6.11.2c. O-Alkyldithiocarbonates and Metal Iodides

Polyiodides are formed when sodium, potassium, lead, nickel, and zinc alkyldithiocarbonates are titrated with iodine in dimethylformamide (183). For the sodium and potassium derivatives, the reaction stoichiometry at the endpoint is consistent with the formation of the triiodide, that is,

$$3I_2 + 2MSC(:S)OR = 2MI_3 + [RO(:S)S]_2 \tag{6.78}$$

where M = sodium or potassium ion. The stoichiometries of the lead, nickel, and zinc reactions are less easily interpreted and are dependent on reaction conditions. Typical impurities found in technical grade dithiocarbonates, such as sulfides, sulfites, thiosulfates, sulfates, carbonates, and trithiocarbonates, are insoluble in dimethylformamide and can be removed by centrifugation. The titration of sodium, potassium, and nickel iodides was used to confirm the mechanism of the dithiocarbonate reactions by inference from the reaction stoichiometry. The formation of triiodides indicates that Eq. (6.78) is correct. The iodides can also be determined by a calibrative approach although the calibration graph is not linear. In a subsequent report (184), rubidium (I) and cesium (I) iodides were also determined by the same procedure.

6.11.2d. The Determination of Thiocyanates, Thiols, and Sodium Azide

The metal thiocyanates [sodium (I), potassium (I), mercury (II), zinc (II), barium (II), lead (II), cobalt (II), and iron (III)] and ammonium thiocyanate are all soluble in dimethylformamide and titrate with iodine to well-defined endpoints if the ethyl vinyl ether polmerization indicator reaction is used (184). With the exception of mercury (II) and the mixed ion mercury (II)–potassium (I) salts, the stoichiometries are consistent with the formation of polyiodides and iodine thiocyanate. Sodium azide can be dissolved in a mixed dimethylformamide/dimethylsulfoxide solvent and titrated in similar fashion. The endpoint stoichiometry is consistent with the equation,

$$2NaN_3 + 3I_2 = 2NaI_3 + 3N_2 \quad (6.79)$$

6.11.3. Ketone/Aldehyde Dimerization, Condensation, and Addition

The condensation of acetone in the presence of hydroxyl ion to form diacetone alcohol was the first reaction to be used as a thermochemical endpoint indicator (185). The mechanism of this reaction is usually represented as follows:

$$CH_3COCH_3 + OH^- \rightleftharpoons CH_3COCH_2^- + H_2O \quad (6.80)$$

$$CH_3COCH_2^- + CH_3COCH_3 \rightleftharpoons CH_3COCH_2CO^-(CH_3)_2 \quad (6.81)$$

$$CH_3COCH_2CO^-(CH_3)_2 + H_2O \rightleftharpoons CH_3COCH_2C(OH)(CH_3)_2 + OH^- \quad (6.82)$$

This original concept has been extended to include other ketone and aldehyde reactions and applied to the determination of a variety of acidic compounds as shown in Table 6.42.

Greenhow et al. have used the acetone system, with potassium hydroxide in propan-2-ol as the titrant, for comparison purposes in their reports on the determination of acids (165), thiols (166), and phenols in petroleum bitumens (171) by means of the base-catalyzed vinyl polymerization indicator reaction. Each indicator system has advantages and disadvantages associated with it and neither can be recommended for all determinations. Indeed, acrylonitrile and acetone indicator reactions should be considered as complementary rather than as alternatives. It is possible to employ both reactions in concert to introduce some degree of selectivity into a catalytic thermometric titration. This can be illustrated by reference to Table 6.35b in which data from both indicator systems are compared for the titration of polyfunctional acids (165). For example, the acetone procedure will titrate both acidic groups of hydroquinone and salicylic acid, while the acrylonitrile technique will only titrate one. Similar distinctions can be made with tannic acid. Another example of indicator reaction selectivity occurs with the determination of monofunctional alkyl and aryl thiols (166). When these compounds are introduced into a solution containing acrylonitrile, cyanoethylation occurs prior to the determinative reaction, consuming the thiol. Consequently, no endpoints are observed. The acetone procedure must be used in this instance. However, for a polyfunctional thiol, for example, 2-mercaptobenzoic acid or salicylideneaminobenzene-2-thiol, manipulation of the titration data obtained with both indicator systems allows a quantitative differentiation between carboxylic/phenolic acidity and thiol acidity.

Greenhow has shown that cyclopentanone and propiophenone can also be used as base-catalyzed thermochemical indicators (186), expanding the range of selective indicator systems, with the discovery that phenolic protons will not

TABLE 6.42. The Application of Base-Catalyzed Aldehyde/Ketone Dimerization, Condensation, and Addition Reactions as Endpoint Indicators for Titration Calorimetry

	Titrant, Concentration (mol L^{-1})	Indicator	Solvent	Detection Limit (DL); Range (R)	Reference	Text Section Reference
Hydroxybenzoic, tannic, cyanuric, thio-cyanuric isocyanuric, and pyridine carboxylic acids	KOH in propan-2-ol, 0.1	Acetone	Toluene	—	165	6.11.3
Thiols	KOH in propan-2-ol, 1.0	Acetone	Acetone	—	166	6.11.3
Phenols and carboxylic acids in petroleum bitumens	KOH in propan-2-ol, 0.1	Acetone	Acetone	—	171	6.11.3
Benzoic acid, succinimide, phenols	KOH in propan-2-ol 1.0	Cyclohexanone or propiophenone	Cyclohexanone or propiophenone	—	186	6.11.3
Dithiocarbamates	KOH in propan-2-ol, 0.1–1.0	Acetone	Propan-2-ol	0.5–16 mmol (R)	187	6.11.3
Amino acids, phenols, and boric and phosphoric acids	KOH in propan-2-ol, 0.5	Acetaldehyde	Water	—	188	6.11.3
Carboxylic, phosphoric and boric acid, and amino acids	KOH in propan-2-ol, 0.5	Acetaldehyde	Acetaldehyde	—	188	6.11.3
Long-chain aliphatic carboxylic, fumaric benzoic, and monochloracetic acids	Bu_4NOH in propan-2-ol, 0.07	Formaldehyde	Ethanol or propan-2-ol or methanol	8.25 μg (R)	189	6.11.3

titrate in solution with propiophenone. A two-titration technique, with propiophenone and acetone (or cyclohexanone) as endpoint indicators, can therefore be used to quantitate phenolic acidity in the presence of carboxylic acid groups.

Zinc, copper, and nickel dithiocarbamates have been determined as pure compounds and as residues on fruit by catalytic thermometric titration with alcoholic potassium hydroxide using the acetone indicator reaction (187). The dithiocarbamates were first decomposed with acid to liberate carbon disulfide, which was then "trapped" by a solution of ethylenediamine in propan-1-ol forming the titrand, dithiocarbamic acid. The flow rate of the nitrogen purge gas used to sweep the carbon disulfide into the trap proved to be a critical parameter since it determines the percentage of gas converted to the titrand during its residence time in the trap. Recoveries of 96–100% and precisions ranging from 2 to 5% rsd were obtained for the determination of 0.5–16 μmol of dithiocarbamate.

In their original publication (185) on acetone indicator reactions, Vaughan and Swithenbank reported that acetaldehyde could not be used as a thermochemical indicator because an immediate heat effect was observed when base was introduced into the solution, even in the presence of acid. It has since been shown that if the solution contains a significant amount of water, satisfactory endpoints can be obtained for the titration of acids in acetaldehyde when either aqueous or alcoholic potassium hydroxide is used as the titrant (188). The effect of water content on the shape of the enthalpogram is shown in Fig. 6.17. In every instance, the temperature change observed for the acetaldehyde indicator reaction (curves A–C) exceeds that of the corresponding acetone reaction. As the amount of water in the system is increased the initial temperature change (observed by Vaughan and Swithenbank) decreases. Greenhow concludes that this temperature change is related to the hydration of acetaldehyde or hemiacetal formation rather than aldol condensation.

Several other base-catalyzed aldehyde reactions have been evaluated as potential endpoint indicators (189). Formaldehyde polymerization in alcoholic medium was used for the determination of organic acids alone or free acid impurities (formic acid) in aqueous formaldehyde solutions. Compounds determined by this procedure included palmitic, lauric, benzoic, fumaric, myristic, stearic, and monochloracetic acids. Another new indicator reaction, the base-catalyzed aldolic addition of benzaldehyde to acetophenone in ethanol, was used to titrate milligram amounts of maleinic acid anhydride with tetrabutylammonium hydroxide. The fast autoxidation of benzaldehyde is cited as a limiting factor in the applicability of this reaction.

6.11.4. Acetylation

The acid-catalyzed acetylation of alcohols and phenols by acetic anhydride has been used to indicate the endpoint in the thermometric titration of tertiary aliphatic

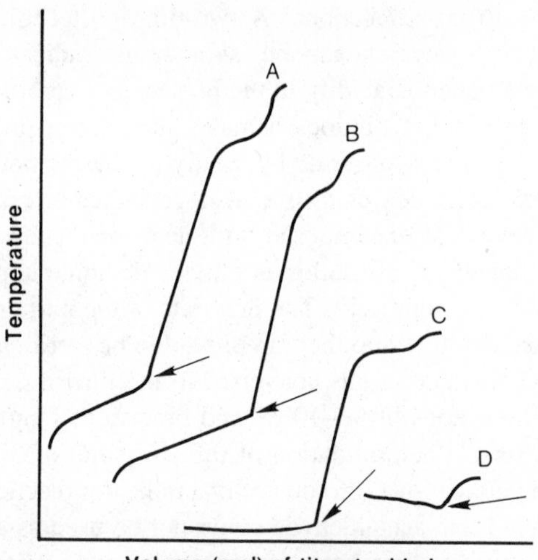

Figure 6.17. Enthalpograms for the catalytic thermometric titration of benzoic acid in aqueous and non-aqueous acetaldehyde with 0.5 mol L^{-1} potassium hydroxide in propan-2-ol (188). Arrow denotes endpoint. (A) 34.1 mg benzoic acid, 0% water. (B) 63.5 mg benzoic acid, 10% water. (C) 64.6 mg benzoic acid, 50% water. (D) 44.1 mg benzoic acid, 0% water (acetone solvent).

amines, alkaloids, heterocyclic bases, and metal carboxylates with 0.001–0.1 mol L^{-1} perchloric acid dissolved in acetic acid (190). The most satisfactory endpoints were obtained for the determination of weak bases when quinhydrone was used as the indicator reagent in combination with acetic anhydride. In addition, the sharpness of the endpoint was increased by inclusion of dichloromethane, nitromethane, or propylene carbonate to the analytical solution. With an optimized indicator system, coefficients of variation for the determination of microgram to milligram amounts of quinoline, antipyrine, caffeine, and potassium acetate ranged from 0.5%, with 0.01 mol L^{-1} titrant, to <1% for 0.01 and 0.001 mol L^{-1} titrants. The use of the acetylation indicators for the determination of primary and secondary amines is precluded by the relative ease of acetylation of these functional groups.

6.11.5. Iodine–Azide Reaction

The mechanism of the reaction between iodine and azide ion, which is catalyzed by the presence of sulfur in the -2 oxidation state, is thought to be as follows (191):

$$\text{R-S-Na} + I_2 = \text{R-S-I} + \text{NaI} \tag{6.83}$$

$$\text{R-S-I} + 2\text{NaN}_3 = \text{R-S-Na} + \text{NaI} + 3\text{N}_2 \qquad (6.84)$$

Therefore, in effect, the reaction is

$$2\text{NaN}_3 + \text{I}_2 = 2\text{NaI} + 3\text{N}_2 \qquad (6.85)$$

Although the details are vague, under certain conditions the concentration of sulfur specie is rate limiting with respect to the production of nitrogen, an exothermic reaction. This phenomenon has been utilized for the determination of several sulfur species, as summarized in Table 6.43. All the cited applications based on this reaction have utilized a direct-injection enthalpimetric approach in which the final component required to initiate the reaction is injected. The rate or extent of the catalyzed reaction, as manifested by its temperature profile, is then used as the analytical signal. Calibration standards provide the analytical data.

6.11.5a. Thioureas

Kiba et al. have reported the determination of thioureas alone (192) and in mixtures (193) by measurement of the extent of the iodine–azide reaction. In a

TABLE 6.43. Calorimetric Applications of the Iodide–Azide Reactions

Analyte(s)	Mode[a]	Detection Limit (DL); Range (R)	Reference	Text Section Reference
Thiourea, *N*-methyl-, 1,3-dimethyl-, 1,3-diethyl-, 1,3-diisopropyl-, 1,3-dibutyl-, 1,1,3-trimethyl-, and 1,1,3,3-tetramethylthiourea	I	5×10^{-8}–10^{-6} mol (R)	192	—
Thiourea, *N*-substituted thioureas and mixtures	I	0.05–0.5 μmol (R)	193	6.11.5a
Sulfide	I	0.02–0.5 μmol (R)	194	6.11.5b
Thiosulfate, hydrogen sulfide	I	112–1120 μg L^{-1}, 5–100 ppm	195	6.11.5b

[a]I = injection.

combined chromatographic/calorimetric procedure, the thiourea mixtures were first separated on silica gel-coated thin-layer chromatography plates. The individual spots were then transferred to the calorimeter cell and suspended in a pH-adjusted sodium azide solution. As a final step, iodine solution was injected into the stirred slurry. An effectively instantaneous temperature pulse was recorded and calibrated versus a similar pulse from thiourea standards. Typically (50–500) $\times 10^{-6}$ mol of thiourea was determined with a relative error and precision <6%. The concentration of azide and iodide and pH are critical parameters with respect to analytical precision.

6.11.5b. Thiosulfate and Sulfide

The use of the iodide–azide reaction for the direct calorimetric determination of sulfide ion in the presence of a precipitant, for example, Cu^{2+}, Cd^{2+}, or Ni^{2+}, is obviously not possible. However, this problem can be solved by decomposition of the metallic sulfide in strong acid to produce hydrogen sulfide which can then be swept into a sodium hydroxide trap and determined. This type of procedure has been used to determine the sulfide ion content in copper metal by direct-injection enthalpimetry (194). The results obtained were within 3% of certified data and coefficients of variation were <5%.

Although "spectro–electro" analysis has received considerable attention, the incorporation of an optical detector into a calorimetric cell has never been seriously pursued. Weisz et al. (195) have designed just such a device capable of batch or flow calorimetry. The determination of sulfide and thiosulfate ion by virtue of their catalytic effect on the iodine–azide reaction has been used to evaluate the instrument. Three reagent-addition procedures were used to introduce sulfide ion into the flow system: direct pumping of sulfide solution, syringe addition of hydrogen sulfide (diluted with nitrogen), and a nitrogen sweep over the head-space of an H_2S generating cell. In each case, analyte recoveries were poor, ranging from 58 to 156% for calorimetric data. Optical data, derived from the decrease in transmittance caused by nitrogen bubble formation, was equally unsatisfactory. Non-linear calibration graphs resulted from all experiments. In view of the satisfactory precision and accuracy obtained with other cited applications of this reaction, it must be concluded that instrumental design flaws are the primary cause of inaccuracies in these data.

6.11.6. Iodide-Catalyzed Cerium (IV)/Arsenic (III) and Manganese (III)/Arsenic (III) Reactions

The calorimetric applications of these reactions are presented in Table 6.44.

TABLE 6.44. Calorimetric Applications of the Iodide-Catalyzed Reactions between Arsenic (III) and Cerium (IV) or Manganese (III)

Analyte(s)	Mode[a]	Indicator	Detection Limit (DL); Range (R)	Reference	Text Section Reference
Aliphatic secondary amines, alone and in mixtures	T	Ce^{4+}/As^{3+}	0.1–5 μmol (R)	196	6.11.6a
Iodide	F	Ce^{4+}/As^{3+}	0.01–10 μmol L^{-1} (R), 0.15 ng (DL)	197	6.11.6b
Ag^{+}, Hg^{2+}, Pd^{2+}	T	Mn^{3+}/As^{3+}	0.5–500 μg Ag^{+}, Hg^{2+} (R) 0.2–500 μg Pd^{2+} (R)	198	6.11.6c

[a]T = titration; F = flow.

6.11.6a. Secondary Aliphatic Amines

The inhibition by silver ion of the iodide ion-catalyzed reaction of cerium (IV) with arsenic (III) has been used indirectly to determine some aliphatic secondary amines (196). The amines are first converted to the silver diakyldithiocarbamates by reaction with carbon disulfide and silver nitrate, followed by extraction into chloroform. Interference with primary amines can be avoided by introduction of 2-ethylhexaldehyde to form the corresponding imines. The silver ion associated with the derivatized secondary amines is then stripped with nitric acid and determined by titration with iodide to a thermochemical endpoint indicated by the catalyzed redox reaction. Diethyl-, dipropyl-, dibutyl-, ethylpropyl-, and ethylbutylamine were determined in the range 0.2–5 μmol with 97–101% recoveries and 2–4% precision rsd. Data for the determination of a secondary amine in the presence of primary and tertiary amines are presented in Table 6.45.

6.11.6b. Iodide

The combination of calorimetric flow technology and catalyzed reaction chemistry has produced the lowest detection limit reported to date for a thermochemical method not involving a biochemical process. Elvecrog and Carr (197) report

TABLE 6.45. Determination of Secondary Amines in Mixtures of Primary, Secondary, and Tertiary Amines

		Taken (μmol)	Secondary Amine (μmol)	Recovery (%)	rsd ($n = 5$) (%)
Mixture 1:	ethylamine	550			
	diethylamine	1.20	1.18	98.3	±3.0
	triethylamine	200			
Mixture 2:	butylamine	840			
	dibutylamine	1.64	1.62	98.8	±3.0
	tributylamine	1000			

Source: Ref. 196, by permission of Pergamon Press.

that 0.01–10 μmol L^{-1} of iodide ion can be determined in samples as small as 120 μL by monitoring the temperature changes associated with the injection of iodide solutions into a flowing stream containing a cerium (IV)/arsenic (III) solution. This represents a lowest limit of detection of 0.15 ng of iodide ion. The reagent concentrations were chosen so that the rate of reaction was represented by

$$-d[\text{Ce (IV)}]/dt = k\,[\text{Ce(IV)}][\text{I}^-] \qquad (6.86)$$

that is, the reaction kinetics were pseudo-zero order with respect to arsenic (III) concentration, and [at constant cerium (IV) concentration] pseudo-first order with respect to iodide ion concentration. Possible interfering species include copper (I), tin (II), and chromium (III). Strong oxidants will prereact with arsenic (III) and will not interfere at low concentrations. The report details the strong dependence of peak shape (and therefore analytical precision) on column packing, sample volume, and concentration and flow rate. Each of these parameters should be optimized for the best results. These considerations are discussed in more detail in Chapter 4.

6.11.6c. Silver (I), Mercury (II), and Palladium (II)

The manganese (III)–arsenic (III) reaction $2Mn^{3+} + As^{3+} = 2Mn^{2+} + As^{5+}$ is catalyzed by iodide ion and can therefore be used as an indicator reaction for the determination of iodide precipitants, if the associated solubility product is such that the free iodide level remains below the catalytic threshold concentration prior to the equivalence point. Kiba and Furosawa (198) have described a thermometric, catalyzed endpoint, titration procedure for the determination of 0.5–500 μg of silver (I), mercury (II), and 0.2–500 μg palladium (II) based on this chemistry. Phosphoric acid must be added to the reaction mixture at a concentration 20 times greater than that of manganese (III) in order to prevent the formation of manganese dioxide. The insolubility of lead (II), bismuth (III), and thallium (I) iodides is not sufficient to allow a quantitative determination of these ions. Indeed, their presence does not adversely affect the analytical data for silver (I), mercury (II), and palladium (II). Cyanide and thiosulfate ions interfere quantitatively with the determination of palladium (II) and obscure the mercury (II) endpoint, but do not interfere with the determination of silver (I). The only apparent advantage of this procedure versus the corresponding cerium (IV)–arsenic (III) method is that it is applicable to solutions containing phosphate, which interferes with the latter by precipitating cerium phosphate.

6.11.7. Hydrogen Peroxide Decomposition/Oxidation Reactions

The exothermic transition-metal-catalyzed decomposition and oxidation reactions of hydrogen peroxide are perhaps the most widely used catalyzed reactions in the analytical sector. Recent calorimetric applications of these reactions are summarized in Table 6.46. As is apparent from the list of applications, the reaction is by no means selectively catalyzed and, therefore, with few exceptions, it is used to determine metal ion concentration in simple solutions.

Kiss (199) has described a catalyzed endpoint procedure in which several chelating agents are determined by thermometric titration with copper (II) or iron (III). The oxidation of resorcinol, pyrocatechol, hydroquinone, or dimethyl-*p*-phenylenediamine, catalyzed by copper (II), provides the exothermic heat change at the endpoint. Iron (III) can be used as the titrant with the latter two reactions. Extension of the procedure to the determination of metal ions requires a back- or substitution titration in which the titrant consumes excess complexing agent or displaces a complex-bound analyte, respectively. In a subsequent report, the same author has shown that many of the same species can be determined by back-titration of excess complexing agent with manganese (II) to an endpoint indicated by the simple decomposition of hydrogen peroxide if ethanolamine is used as the solvent (200).

In a technique termed the "catalytic–kinetic difference method," the analyte (a catalyst) and a solution of known catalyst concentration are introduced simultaneously into identical indicator reaction solutions (201). The difference in thermal response between the two reacting solutions is monitored as the analytical signal. By use of such a technique, Pantel and Weisz (201) have determined microgram quantities of copper (II); the exothermic autodecomposition of hydrogen peroxide in ammoniacal solution was used as the indicator reaction. The

TABLE 6.46. Calorimetric Applications of the Catalyzed Oxidation and Decomposition Reactions of Hydrogen Peroxide

Analyte(s)	Indicator(s)	Mode[a]	Range (R); Detection Limit (DL)	Reference	Text Section Reference	Notes
EDTA, CyDTA, NTA, alkaline earths, Cu^{2+}, Fe^{3+}, Ni^{2+}, Co^{2+}, Zn^{2+}, Sn^{2+}, Mn^{2+}, Al^{3+}, Bi^{3+}, Cr^{3+}, Sb^{3+}, In^{3+}, Ga^{3+}, Th^{4+}	H_2O_2/resorcinol or catechol catalyzed by Cu^{2+}, H_2O_2/hydroquinone or dimethyl-*p*-phenylenediamine catalyzed by Cu^{2+} and Fe^{2+}	T	mg–μg (R)	199	6.11.7	Direct, back, or substitution titrations used
EDTA, DCTA, NTA, alkaline earths, Zn^{2+}, Cd^{2+}, Mn^{2+}, Pb^{2+}, Hg^{2+}, Tl^{+}, Al^{3+}, Ga^{3+}, Cu^{2+}, In^{3+}, Th^{4+}	H_2O_2	T	0.1 μg-4 mg (R)	200	6.11.7	
Cu^{2+}	H_2O_2	I	6–23 μg mL^{-1} (R)	201	6.11.7	
Mo^{6+}, W^{6+}	H_2O_2/I^- or $S_2O_3^{2-}$	I	Mo^{6+}, 2×10^{-2} mg L^{-1} with I^-, 3×10^{-3} mg L^{-1} with $S_2O_3^{2-}$, 0.06 mg L^{-1} with rubeanic acid (DL) W^{6+}, 10^{-2}–8×10^{-2} mg L^{-1} (R)	202	6.11.7	

Co^{2+}	H_2O_2/2-aminopyridine	I	0.01 μg mL^{-1} (DL)	203	6.11.7	10.0 mol L^{-1} H_2O_2, 1×10^{-5} mol L^{-1}, 2-aminopyridine. Interference from Fe, Mn, Cu, and Ni removed by extraction with 1-(2-pyridylazo)-2-napthol into $CHCl_3$
Co^{2+}	H_2O_2/2-6-diaminopyridine	I	5×10^{-9} mol L^{-1} (DL)	204	—	
Mn^{2+}	H_2O_2/1,10-phenanthroline	I	—	205	—	pH 9.5, 2.5 mol L^{-1} H_2O_2, 1×10^{-3} mol L^{-1} 1,10-phenanthroline. Ag, Cu, Co, Ni, Mg, Cu, Sr, Ba, Zn, Cd, Hg, Al, Cr, Pb, V, W, and Mo do not interfere. Fe removed by extraction
CN^-, Cu^{2+}	H_2O_2	I	3–80 μg CN^- (R)	206	6.11.7	CN^- determination based on inhibition time to onset of Cu catalyzed decomposition of H_2O_2
Fe^{3+}, Zr^{4+}, Th^{4+}, V^{6+}, Mo^{6+}, W^{6+}	H_2O_2/I^-/ascorbic acid	I		207	6.11.7	Determination based on "Landolt Effect"
Hydrazobenzene	H_2O_2/V^{5+}	I	30—70 μg (R)	208	6.11.7	

[a]T = titration; I = injection.

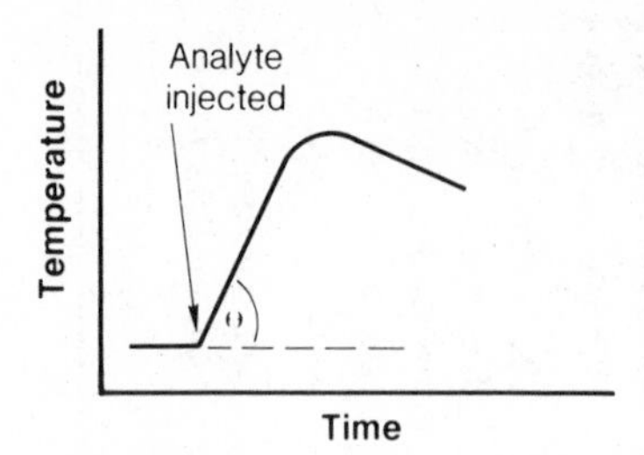

Figure 6.18. Typical enthalpogram for the determination of copper (II) by the "catalytic–kinetic difference" method (201) (by permission of Elsevier Science Publishers). Indicator solution: 5 mL aqueous ammonia (2 mol L^{-1}) and 2 mL hydrogen peroxide made up to 27 mL with water. Analyte: 3 mL copper (II) solutions.

tangent of the angle θ, shown in Fig. 6.18, was shown empirically to be linearly related to the square of the copper (II) concentration over a limited range. Within this range, analytical recoveries range from 97% to 103%.

The oxidation reactions of iodide, thiosulfate, and rubeanic acid [dithiooxamide, $H_2(NHCS)_2$] with hydrogen peroxide have been evaluated as indicator reactions for the determination of molybdenum (VI) and tungsten (VI) by direct-injection enthalpimetry (202). All three reactions produced acceptable data after optimization of reagent concentrations to linearize the response with respect to the concentration of the catalyst. Significantly, the tungsten-catalyzed oxidation of rubeanic acid was shown to be relatively selective with the alkaline earths, cobalt (II), nickel (II), lead (II), manganese (II), titanium (IV) ions showing no interference at the ppm level. Interference was observed from copper (II), which forms an oxidizable complex with rubeanic acid.

Copper (II), in ammoniacal solution, will catalyze both the oxidation of cyanide to cyanate and the autodecomposition of hydrogen peroxide. In the presence of both cyanide and hydrogen peroxide it has been observed that the autodecomposition reaction is delayed for a time period that is linearly related to the concentration of cyanide. By use of this principle, a calorimetric method has been developed for the determination of microgram amounts of cyanide (206). The reaction is initiated by injection of 0.5 mL of hydrogen peroxide into a preequilibrated solution of copper (II), ammonia, and cyanide ion. As the "retardation time" is related to the concentration of both cyanide and copper (II) ions, the latter must remain constant in all experiments. Obviously, the level of copper (II) can be adjusted to alter the sensitivity of the measurement. At the 20 μg level, retardation times between 5 and 1000 s were observed for amounts of cyanide in the range 6–30 μg. Typical enthalpograms are shown in Fig. 6.19. The heat effect associated with the oxidation of cyanide (during the retardation period) is insignificant because of the low level of cyanide in solution. The slope of the decomposition reaction signal, as manifested by the angle θ, is proportional to the concentration of copper (II) ion in solution and can be used analytically at constant cyanide levels. Analytical recoveries for the determination of cyanide ranged from 89.8 to 103.4%.

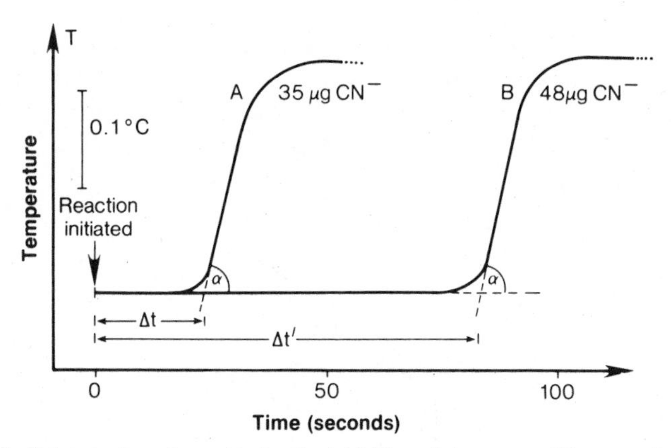

Figure 6.19. Determination of cyanide by the inhibition of the copper (II) − catalyzed autodecomposition of hydrogen peroxide (206): 40 mg copper (II), 5 mL hydrogen peroxide, 0.4 mol L^{-1} ammonia (by permission of Elsevier Science Publishers).

Gaal et al. (207) have described an enthalpimetric determination of iron (III), thorium (IV), zirconium (IV), vanadium (V), molybdenum (VI), and tungsten (VI) by use of the "Landolt effect." The principle of the method is best illustrated by reference to the series of enthalpograms shown in Fig. 6.20. Hydrogen peroxide is injected into a solution containing the analyte, iodide, and ascorbic acid. As long as ascorbic acid remains in solution, the temperature change associated with the following concurrent reactions is observed:

$$2I^- + H_2O_2 + 2H^+ = I_2 + 2H_2O \tag{6.87}$$

$$I_2 + \text{ascorbic acid} = 2I^- + \text{dehydroascorbic acid} \tag{6.88}$$

as depicted by region A of the enthalpogram. When the ascorbic acid is exhausted, iodine will appear in solution and the reaction represented by Eq. (6.87) will proceed to completion. The resulting change in enthalpy will produce an inflection point on the enthalpogram. The time Δt from reaction initiation to the appearance of iodine (the Landolt effect) is proportional to the concentration of analyte (catalyst) at constant iodide, ascorbic acid, and hydrogen peroxide concentration. Specifically, a linear relationship exists between the reciprocal of Δt and the catalyst concentration (209). This technique does have the advantage that the iodide concentration remains essentially undepleted during first part of the experiment (region A), resulting in a linear enthalpogram and a well-defined inflection point. This is reflected in favorable accuracy and precision data as shown in Table 6.47.

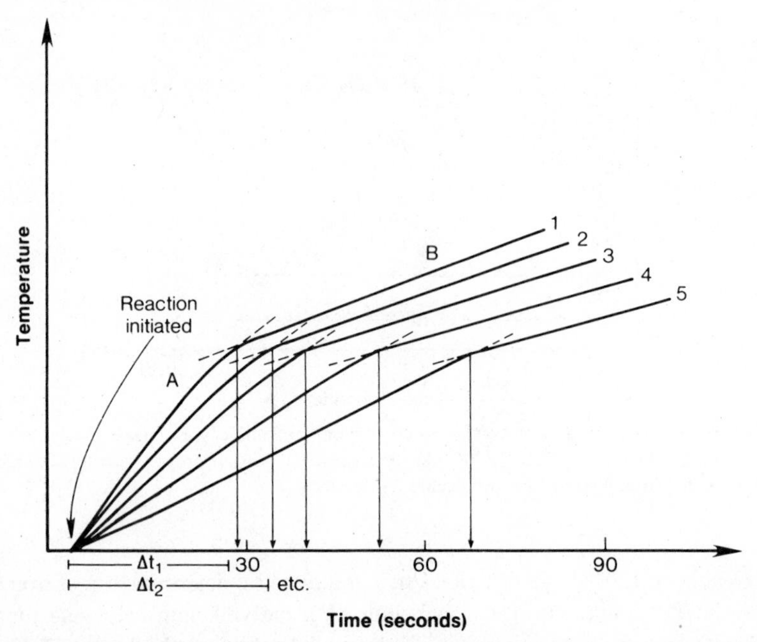

Figure 6.20 "Landolt effect" enthalpograms obtained for the determination of molybdenum by use of the hydrogen peroxide–iodide–ascorbic acid indicator reaction (207). Curves 1–5 represent the determination of 9, 7, 5, 3, and 1 mg L^{-1} molybdenum, respectively (by permission of Elsevier Science Publishers).

Although not strictly a kinetic method of analysis, a simple enthalpimetric determination of hydrazobenzene has been developed by use of a catalytic reaction (208). In the method 100 μl of a 5×10^{-4} mol L^{-1} vanadium (V) solution was injected into an ethanolic mixture of hydrazobenzene and hydrogen peroxide. The exothermic oxidation reaction to form azobenzene is complete within 1–1.5 min under these conditions. Calibrative quantitation of the temperature pulse results in the determination of milligram amounts of analyte with an accuracy and precision of about 1%.

6.12. GASEOUS REAGENTS

In 1969 Zambonin and Jordan described "gas enthalpimetry" as "an equilibrium heat pulse" method applicable to reactions between gases and liquids (210). In this pioneering study, the analytical potential of calorimetric gas analysis was evaluated by characterizing the thermochemistry of the aquation and Bronsted

TABLE 6.47. Enthalpimetric Determination of Transition Metals by Use of the "Landolt Effect"

Ions	pH	Useful Range (μg mL^{-1})	Taken (μg mL^{-1})	Found[a] (μg mL^{-1})	Error (%)	rsd (%)
Fe^{3+}	3.6	0.1–1.0	0.419	0.408	−2.6	2.2
			0.628	0.643	+2.4	3.8
Th^{4+}	3.6	100–700	209.0	206.0	−1.4	0.2
			419.0	410.0	−2.2	2.1
Zr^{4+}	1.6	10–100	42.0	42.7	+1.7	1.7
			69.9	71.3	+2.0	2.6
V^{5+}	4.5	4–44	19.0	19.4	+2.1	0.2
			28.6	29.1	+1.7	0.7
Mo^{6+}	4.5	1–10	4.19	4.10	−2.2	0.9
			6.28	6.32	+0.6	1.0
W^{6+}	4.5	10–100	41.9	40.7	−2.9	0.6
			62.9	64.5	+2.5	0.2

Source: Ref. 207, by permission of Elsevier Science Publishers.

[a]Four determinations.

acid–base reactions of carbon dioxide, sulfur dioxide, nitrogen dioxide, and dinitrogen tetroxide. From an experimental viewpoint, the main advantage of injecting a gas into a liquid is that the gas has a negligible heat capacity. Consequently, temperature mismatch between reagent and analyte becomes less important, unless a temperature-dependent equilibrium exists in the gas phase.

Although conceived as a method for the determination of gaseous analytes, gas enthalpimetry has been applied primarily to the determination of analytes in solution by calibration of heat effects associated with the injection of gaseous reagents as shown in Table 6.48.

6.12.1. Unsaturated Compounds

The thermochemistry of the hydrogenation reactions of unsaturated organic compounds has recently been reviewed (221, 222). Under the appropriate conditions, these reactions are generally quantitative, fast, and have relatively large enthalpy changes. For example, in a stirred slurry of palladium catalyst, the hydrogenation of several micromoles of unsaturate is usually complete in less than 10 s and

TABLE 6.48. Application of Gaseous Reagents to Analytical Calorimetry

Analyte(s)	Reaction Medium	Reagent(s)	Mode[a]	Detection Limit (DL); Range (R)	Reference	Text Section Reference
			Unsaturated Organic Compounds			
Non-aromatic alkenes	Hexane	H_2/Pd catalyst	I	1.4–37.5 μmol (R)	211	6.12.1a
Fatty acids	Hexane or acetic acid	H_2/Pd catalyst	I	0.6–28.6 μmol (R)	212	6.12.1b
Fatty acids, methyl esters, and triglycerides	Hexane	H_2/Pd catalyst	I	1.3–4.0 mg (R)	213	6.12.1b
Allylic and vinylic compounds	Hexane or cyclohexane	H_2/Pd catalyst	I	26.4–184 μmol (R)	214	6.12.1c
Linear dienes and cyclic trienes	Hexane	H_2/Pd	I	7.3 nmol-52 μmol (R)	215	6.12.1d
Various olefins	Hexane	H_2Pd	I	0.07–3.0 μmol (R)	216	
			Metals			
Cu^{2+}, Cd^{2+}, Ag^{+}, Fe^{3+}, Hg^{2+}, Bi^{3+}, Tl^{+}, and Pb^{2+}	Water	H_2S	I	20–1170 ppm	217	6.12.2
			Oxidants			
I_2, IO_3^-, IO_4^-, and BrO_3^-	Water	SO_2	I	15–50 μmol	218, 219	6.12.2
			Gaseous Bases			
Ammonia and aliphatic amines	Air or nitrogen	HCl	T	0.03–0.04 mol L^{-1}	220	6.12.3

evolves between 109 and 126 kJ mol^{-1} of heat (221). These reactions have been used to develop direct-injection enthalpimetric methods termed "hydrogen enthalpimetry" for the determination of a variety of unsaturated compounds (211–216). The calorimeter design used in these experiments is shown in Fig. 6.21.

In the method (211) about 25 mL of hexane and 600 mg of 5% palladium catalyst are introduced into the calorimeter chamber which is then sealed with the aid of a sealing compound. After the sealant has set (0.5–1 h), hydrogen is allowed to flow into the calorimeter via the inlet needle inserted through a hard rubber septum; the hydrogen pressure is typically maintained at about 20 psi throughout the experiment. The cell is then allowed to reach thermal equilibrium following the rapid rise in temperature caused by the activation of the catalyst, and then the unsaturate (20–80 μL) is introduced via the microburet or a microsyringe. The ensuing temperature change is monitored and calibrated versus the temperature change obtained for a known standard sample of the analyte. This procedure, with a few modifications, applies to all the methods described in Sections 6.12.1a–6.12.1d. Rogers reports that the reaction times for most hydrogenation reactions are fast enough, in the presence of an excess of catalyst, that the shape of the temperature–time curve is limited by thermistor response time and the heat-transfer properties of the calorimeter (221). Some typical enthalpograms obtained from the hydrogenation of unsaturated fatty acids are shown

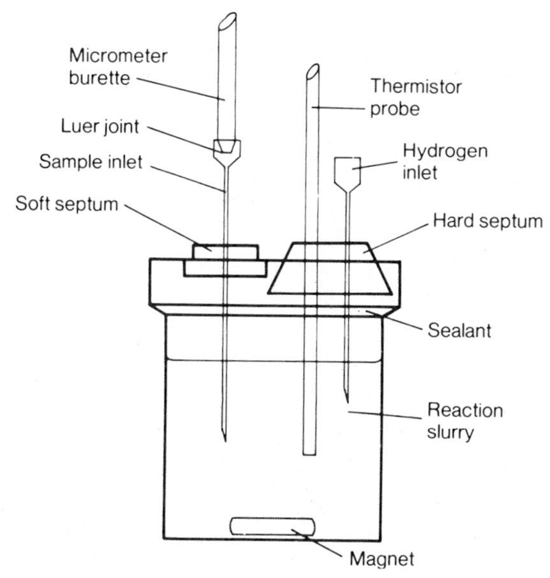

Figure 6.21. Hydrogen calorimeter (213) (by permission of Pergamon Press).

in Fig. 6.22. Kinetic lag has been observed in enthalpograms obtained for the hydrogenation of sterically hindered molecules such as those containing tetra-substituted double bonds. Interference from thermal events associated with the adsorption onto or desorption from the catalyst surface tend to be small because the enthalpy changes of these interactions are equal and opposite in sign and therefore cancel in the long term. Catalyst degradation may be a significant factor with impure samples; however, this has not proved to be a problem with relatively pure samples. Typical errors are in the vicinity of 2–3% in the range from 200 nmol to 150 μmol of analyte, rising to 10–16% in the 10–200-nmol region. The principal advantage of "hydrogen enthalpimetry," compared with chromatographic methods of determining unsaturation, is sample throughput. Samples can be analyzed calorimetrically at a rate of one per minute or less versus 20–30 min for a chromatographic analysis of a complex sample. On the other hand, the chromatographic data will resolve individual unsaturates; hydrogen enthalpimetry will only provide a measure of the total level of unsaturates.

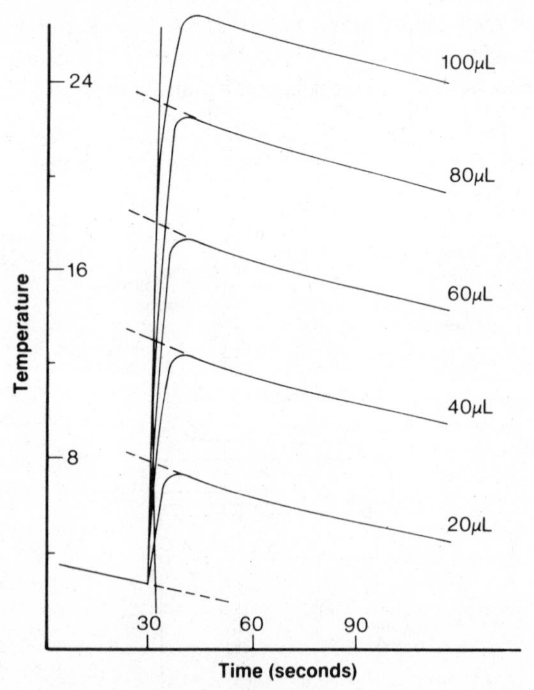

Figure 6.22. Enthalpograms obtained for the hydrogenation of unsaturated fatty acids in the presence of a palladium catalyst. Unsaturate concentration ranges from 0.6 to 30 μ mol (212) (by permission of Academic Press).

6.12.1a. Non-Aromatic Alkenes

The determination of micromole quantities of cyclopentene, 2-pentene, cyclohexene, heptene, methylallylbenzene, and hexene has been achieved with a precision better than 3% rsd (211). Precision deteriorated significantly (~10%) at analyte levels less than 2 μmol.

6.12.1b. Fatty Acids, Methyl Esters, and Triglycerides

Rogers and Sasiela (212) have determined micromole quantities of oleic, elaidic, and linoleic acids, methyl linolenate, and linolenyl alcohol by "hydrogen enthalpimetry." No selectivity is obtained between double bonds in a mixture of acids; however, the method does discriminate against aromatic compounds which are not hydrogenated under the conditions already described in Section 6.12.1. With a few exceptions, the enthalpy of hydrogenation increases linearly with the number of double bonds in an unsaturated molecule. For example, the ratio of the hydrogenation enthalpies for oleic, linoleic, and linolenic acids is 1:2:3 (212). Accordingly, in the absence of thermochemical data, quantitation is usually achieved by constructing a calibration curve for each unsaturate based on data obtained from samples of known composition. However, it has been shown that the hydrogenation enthalpy *per double bond* is consistent for many unsaturated fatty acids at -125.5 kJ mol^{-1} (213). Accordingly, it is possible to determine lipids based on a comparison of the enthalpy of hydrogenation for each lipid with that of a universal standard. The total unsaturation of any one compound can then be expressed in terms of an equivalent mass (or number of moles) of an arbitrary standard. This procedure has been evaluated for the determination of oleic and linoleic acids, methyl linoleate, methyl linolenate, trilinolein, and trilinolenin (213). The results indicated that the fatty acid residues do indeed behave independently, that is, the signal obtained for triolein was three times that obtained for methyl oleate, etc. Some typical analytical data are shown in Table 6.49.

6.12.1c. Allylic and Vinylic Unsaturates

It has been observed that allylbenzene, allylcyclohexane, allylcyclopentane, methallylbenzene, vinylcyclohexane, and vinylcyclopentane all add 1 mol of hydrogen across the isolated double bond without perturbing the unsaturation in the ring (214). This reaction, which is fast and quantitative, has been used for the direct-injection enthalpimetric determination of micromole quantities of these compounds with a relative precision better than 3%. Abnormally high results obtained for the hydrogenation of vinyl benzene by the same procedure are

TABLE 6.49. Determination of Some Common Lipids by Hydrogen Enthalpimetry

Lipid	N	mg Taken	db[a]	Equiv. mg Found	Equiv. (mg/db)	σ (mg)
Oleic acid	9	4.03	1	4.07	4.07	0.12
Linoleic acid	7	1.80	2	3.55	1.78	0.03
Methyl linoleate	15	3.19	2	6.66	3.33	0.16
Methyl linolenate	9	1.94	3	5.80	1.93	0.08
Trilinolein	12	2.15	2	4.16	2.08	0.08
Trilinolenin	9	1.31	3	4.02	1.34	0.10

Source: Ref. 213, by permission of Marcel Dekker Inc.

[a]db = double bonds per fatty acid residue.

attributed to either partial ring hydrogenation or polymerization. Clearly, conjugation of the vinyl group with the aromatic ring significantly affects the hydrogenation chemistry.

6.12.1d. High-Sensitivity Determinations

The detection limits of hydrogen enthalpimetry have been improved by replacing the conventional Wheatstone bridge circuit with an enforced linear resistance bridge incorporating an operational amplifier based on field-effect transistors (215). This adaptation results in a 20-fold increase in sensitivity. Typical data are presented in Table 6.50. Reduction of the calorimeter cell volume from 40 to 5 mL and use of a positive-temperature or Curie-point thermistor as the temperature transducer lower the detection limit even further to the nanomole region with acceptable precision (216). For example, 11.7–35.2 nmol of oleic acid was determined with a relative precision ranging from 1.9 to 3.2%.

6.12.2. Hydrogen Sulfide and Sulfur Dioxide as Gaseous Thermochemical Reagents

Gaseous hydrogen sulfide has been used as a precipitant for the enthalpimetric determination of several metal cations (217). In the procedure, hydrogen sulfide is slowly injected into acidified solutions of the cations at a rate of 750 mL min^{-1}. In most cases, well-defined linear titration enthalpograms were obtained. However, volumetric calculation was subject to error because of variations in gas dissolution rate as a function of pH and temperature. Consequently, the total heat pulse was used as the analytical signal and quantitation was performed by means of a family of calibration graphs. The pH of the solution is a critical

TABLE 6.50. The Determination of Various Olefins by Direct-Injection Enthalpimetry with an Enforced Linear Resistance Bridge

Compound	Amount (μmol)	rsd (%)[a,b]
1-Hexene	0.62–2.48	4.9 (30)
Cyclohexene	0.91–3.04	3.7 (14)
Cyclooctene	0.47–1.06	7.8 (15)
Cyclododecene	0.43–1.36	9.8 (18)
1,3-Cyclooctadiene	0.72–2.41	6.5 (18)
1,5-Cyclooctadiene	0.65–1.98	6.5 (17)
1,5,9-Cyclododecatriene	0.33–1.64	3.9 (17)
1,5,9-Trimethylcyclododecatreine	0.25–1.98	7.8 (16)
Bicyclo-[4,3,0]-nona-3,7-diene	0.75–2.26	3.0 (14)
5-Methylene-2-norbornene	0.50–1.49	4.2 (14)
Dicyclopentadiene	0.07–0.52	6.4 (12)

Source: Ref. 215, by permission Plenum Publishing Corp.

[a]Number of experiments in parentheses.

[b]The relative standard deviation was obtained by calculation of the deviation of a set of results from the "best" straight line (least squares) representing temperature change versus analyte concentration.

parameter in determining the rate and extent of precipitation of all the species examined. The optimum conditions for each cation and the corresponding detection limits of the technique are shown in Table 6.51.

The procedure can be extended to the determination of (a) copper (II) in the presence of lead (II), silver (I), nickel (II), cobalt (II), and zinc (II) or (b) mercury (II) in the presence of silver (I), lead (II), and zinc (II). Silver (I) and lead (II) can be removed by addition of chloride and sulfate ion. The precipitation of cobalt (II), nickel (II), and zinc (II) can be inhibited by manipulation of pH.

Bark and Prachuabpaibul (218, 219) have used a similar procedure to determine iodine, iodate, bromate, and periodate by injection of gaseous sulfur dioxide into acidified solutions of these reagents. Each oxidant was determined at the 2 μmol level with a reproducibility not less than 1%.

6.12.3. Gas-Phase Titration of Ammonia and Aliphatic Amines

The calorimetric titration of gaseous analytes with gaseous reagents is feasible using very simple apparatus if the analytical reaction is fast and quantitative. Duffield and Hume (220) have shown that hydrogen chloride will titrate ammonia, methylamine, dimethylamine, trimethylamine, ethylamine, diethylamine,

TABLE 6.51. Conditions for the Precipitation of Various Metal Cations by Gaseous Hydrogen Sulfide

Cations	Precipitation Medium	Detection Limit[a] (ppm)
Cu^{2+}	0.5 mol L^{-1} HCl	20
Cd^{2+}	0.05 mol L^{-1} HCl	75
Bi^{3+}	0.5 mol L^{-1} HCl/HNO_3	70
Hg^{2+}	0.5 mol L^{-1} HCl	33
Pb^{2+}	0.5 mol L^{-1} CH_3COOH	70
Ag^{+}	0.5 mol L^{-1} H_2SO_4	33
Fe^{3+}	Acetate buffer, pH 6	33
Tl^{+}	Borax buffer, pH 9	130

Source: Ref. 217.

[a]Arbitrarily defined as the amount of cation which, when dissolved in 15 mL of the appropriate solvent, will give a pen deflection of 25 mm at a sensitivity of 0.5 mV fsd. (This figure will also depend on the voltage supplied to the bridge, not quoted in the report.

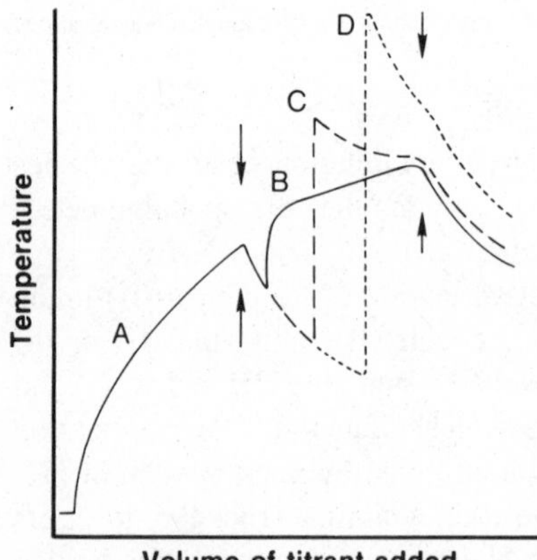

Figure 6.23. Typical titration enthalpograms for gaseous ammonia or aliphatic amines with hydrochloric acid (220) (by permission of Marcel Dekker Inc.). Curve A: direct titration of a single sample. Curves B, C, and D: examples of consecutive addition method in which the second sample is added 15%, 40%, or 70% after the first endpoint. Arrows indicate the endpoint.

and *n*-butylamine in the gas phase to an acceptable endpoint. The reaction cell, a three-necked, 1 L flask surrounded by polystyrene insulation, is first flushed with air or nitrogen; the sample (0.03–0.04 mol L^{-1}) is then added by syringe or buret. With the stirrer activated, the gaseous titrant (0.03–0.04 mol L^{-1}) is added from a syringe buret and the heat change is monitored in the usual fashion. A consecutive sample addition procedure, illustrated in Fig. 6.23, was developed to increase sample throughput; this technique also produced more accurate data than single analyte titrations. For five replicate experiments, the consecutive addition procedure gave 0.0 and −0.8% errors for the titration 5 and 10 mL ammonia samples, respectively. The corresponding reproducibility figures were 1.4 and 2.1% rsd. The non-linearity of the enthalpograms was attributed to heat transfer to the equipment.

6.13. MISCELLANEOUS APPLICATIONS

6.13.1. Water

The calorimetric determination of small amounts of water by monitoring the enthalpy change (-67.4 kJ mol^{-1} of water) associated with the Karl Fischer reaction was first reported in 1966 (223). This reaction has recently been used to determine the water content of acetone, milk powder, vitamin C tablets (224), crystal hydrates (225), and antibiotics (226). Small samples (10–15 mg) of liquid or solid samples were added to 200 mL of concentrated Karl Fischer reagent in the calorimeter cell. Liquids were injected by syringe, solids were simply added through a funnel. The effect of side reactions for a particular sample was determined by addition of the sample to exhausted reagent in a separate experiment. Blank responses were then subtracted from the analytical signal and the water content determined by comparison with data obtained with known standards. The redox reaction between ascorbic acid and iodine (in the Karl Fischer reagent) must be subtracted from the vitamin C tablet data. As the reagent is in considerable excess, samples can be successfully introduced into the cell at short intervals until the reagent is exhausted.

A different procedure involving the reaction of lithium hydride with water has been used to determine the water content of lubricating oils where interference from additives precludes the use of the Karl Fischer reaction (227). Water contents less than or equal to 0.01% can be determined by adding solid hydride to a solution of the sample dissolved in water-free xylene.

6.13.2. Hydrogen Peroxide

Several reactions have been evaluated for application to the determination of hydrogen peroxide by direct-injection enthalpimetry (228). The thermochemistry of some redox reactions involving hydrogen peroxide is presented in Table 6.52. Clearly, the oxidation of iron (II) will provide the greatest sensitivity. The choice of reagent will depend on potential interferences in the presence of strong redox reagents, for example, interaction with peroxide stabilizing agents. Titration with sodium hypobromite is also possible (see Section 6.13.3c).

TABLE 6.52. Enthalpies of Reaction for Several Redox Reactions of Hydrogen Peroxide

Reaction	Titration Medium (mol L^{-1} H_2SO_4)	ΔH_R (kJ mol^{-1})
$5H_2O_2 + 2MnO_4^- + 6H^+ = 2Mn^{2+} + 5O_2 + 8H_2O$	1	−160.6
$H_2O_2 + 2Ce^{4+} = 2Ce^{3+} + 2H^+ + O_2$	1	−99.6
$2I^- + H_2O_2 + 2H^+ = I_2 + 2H_2O$	0.05	−194.1
$2Fe^{2+} + H_2O_2 + 2H^+ = 2Fe^{3+} + 2H_2O$	1	−246.9

Source: Ref. 228.

6.13.3. Thermochemical Reagents

6.13.3a. Iodine Monochloride

Iodine monochloride can be used as a general purpose oxidimetric thermochemical titrant (229). In order to simplify the enthalpograms it is recommended that titrations be performed in the presence of excess mercury (II) which prevents the oxidation of iodide formed in the primary analytical reaction (see Section 6.7.6). Under these conditions the enthalpogram has one inflection point. The list of compounds titrated and the corresponding stoichiometries are presented in Table 6.53. A titration medium of 2 mol L^{-1} hydrochloric acid is recommended for optimum results.

6.13.3b. N-Bromosuccinimide

N-Bromosuccinimide is a wide-ranging analytical reagent, engaging in oxidation, substitution, and addition reactions with a variety of inorganic and organic compounds. Bark and Prachuabpaibul (230) have evaluated this reagent as a thermochemical titrant for the compounds listed in Table 6.54. In the analyte range 5–50 μmol, recoveries ranged from 97% to 102%. Typical redox reaction enthalpies are of the order of −250 kJ mol^{-1}. The principal disadvantage of this reagent is its limited stability; solutions (0.07 mol L^{-1}) stored at 4°C in the dark will remain stable for about 3 days.

6.13.3c. Sodium Hypobromite

The application of sodium hypobromite as an oxidimetric calorimetric reagent has received considerable attention in the literature (231–234). A summary of

TABLE 6.53. Applications of Iodine Monochloride as a Thermochemical Titrant

Analyte	Stoichiometry (Reagent/Analyte)	Type of Enthalpogram[a]
Hydrazine sulfate	2:1	Well defined
Hydroquinone	2:1	Considerable curvature at endpoint, prohibitive above 5 mg analyte
Phenylhydrazine hydrochloride	2:1	Slight curvature at endpoint
Potassium antimonyl tartrate [antimony (III)]	1:1	Well defined
Sodium arsenite [arsenic (III)]	1:1	Well defined
Semicarbazide hydrochloride	1:1	Considerable curvature at endpoint
Ascorbic acid	1:1	Well defined
Thiosemicarbazide	1:1	Excessive enthalpogram curvature prohibits extrapolation

Source: Ref. 229.

[a]All titrations performed in the presence of excess mercury (II); all enthalpograms exhibit one endpoint.

applications and relevant conditions is presented in Table 6.55. Although hypobromite will engage in redox reactions with a large number of compounds in alkaline solution the major drawbacks to the use of this reagent are its relatively poor stability and the non-stoichiometric nature of its reactions. Rusz-Szorad and Kalasz, in their series of reports on thermometric titrations with sodium hydrobromite (232–234), claim that if care is taken to avoid the formation of bromate during hypobromite preparation, the decomposition of the reagent can be minimized. Nonetheless, they recommend daily reagent standardization. As can be seen in Table 6.55, direct titration is usually not practicable due to enthalpogram curvature caused by slow kinetics and/or incomplete equilibria. In the cases where direct titrations do result in reproducible endpoints, the reaction is often non-stoichiometric and quantitation must be achieved by the use of empirical factors obtained from the titration standards. When direct titration is

TABLE 6.54. Applications of *N*-Bromosuccinimide as a Thermochemical Titrant

Analyte	Titration Medium	Stoichiometry (Reagent/Analyte)	Type of Enthalpogram[a]
$NaAsO_2$	0.1–2 mol L^{-1} HCl or 0.01–0.05 mol L^{-1} NaOH	1:1	1
NH_4SCN	$NaHCO_3$, pH 7–8	4:1	3
KSCN	$NaHCO_3$, pH 7–8	4:1	1
$SbO{\cdot}C_4H_6O_6$ (tartar emetic)	1–3 mol L^{-1} HCl	1:3	1
NaI	2 mol L^{-1} HCl	1:2	2
Hydroquinone	0.1–1 mol L^{-1} HCl	1:1	1
Thiourea	$NaHCO_3$, pH 7–8	4:1	1
Ascorbic acid	0.1–1.0 mol L^{-1} HCl	1:1	1
$N_2H_5HSO_4$	1–2 mol L^{-1} HCl	2:1	1
$N_2H_4{\cdot}H_2O$	1–2 mol L^{-1} HCl	2:1	1
Monomethylhydrazine	1–2 mol L^{-1} HCl	2:1	1
Unsymmetrical dimethylhydrazine	1–2 mol L^{-1} HCl	2:1	4
4-Nitrophenylhydrazine	1–2 mol L^{-1} HCl	2:1	1
2,4-Dinitrophenyl-hydrazine	1–2 mol L^{-1} HCl	2:1	1
Semicarbazide	1–2 mol L^{-1} HCl	2:1	1
1-Phenylsemicarbazide	1–2 mol L^{-1} HCl	1:1	1
Isonicotinic acid hydrazide	1–2 mol L^{-1} HCl	2:1	1

Source: Ref. 230.

[a]Type 1—single end-point; no end-point extrapolation required.

Type 2—Two end-points; the first is well-defined and corresponds to the formation of iodine, the second is curved and is due to the formation of ICl_2^- or IBr_2^-.

Type 3—two distinct slopes; the first represents the oxidation of thiocyanate to sulfate and succinimide, the second is caused by the slow reaction of bromine with ammonium ion to form nitrogen.

Type 4—two inflection points; the first corresponds to a stoichiometry of 2:1 (reagent/analyte), the second is unassigned.

TABLE 6.55. Applications of Sodium Hyprobromite as a Thermochemical Titrant

Analyte	Mode[a]	Medium	Reaction	ΔH_R (kJ mol^{-1})	Reference	Notes
Nitrogen-Containing Compounds						
Ammonia	I	0.05–0.5 mol L^{-1} NaOH		−741.0	231	
	T	pH 9.5–10.5		−270.7	232, 233	Non-stoichiometric
Urea	T	0.5 mol L^{-1} NaOH		−281.2	232, 233	Non-stoichiometric
Hydrazine	T	pH 7.0–7.3, sodium bicarbonate		−245.2	232, 233	Non-stoichiometric
Glycine, Cystine	I	0.5 mol L^{-1} NaOH		−808, −2634	231	Evidence of sulfur oxidation
Metal Ions						
Chromium (III)	T	2.5 mol L^{-1} NaOH		—	232, 234	Indirect titration with arsenic (III)
Manganese (III)	T	7.5 mol L^{-1} NaOH		—	232, 234	Indirect titration with I^-
Anions						
Iodide	T	0.1–1.5 mol L^{-1} NaOH		−879	232, 234	Non-stoichiometric
Thiosulfate	T	0.5 mol L^{-1} NaOH		−38.4	232, 234	Direct or indirect titration
Thiocyanate	T	0.5–1.5 mol L^{-1} NaOH		−379.5	232, 234	Direct or indirect titration
Sulfite	T	> 0.5 mol L^{-1} NaOH		—	232, 234	Indirect titration with arsenic (III)
Hydrogen Peroxide	T	0.5 mol L^{-1} NaOH		—	232, 234	Indirect titration with arsenic (III)

[a] T = titration; I = injection.

not possible, sodium arsenite or potassium iodide are the recommended back-titration reagents.

Malingerova and Malinger (231) propose that the problem of reagent stability, poor kinetics, and incomplete reactions can be avoided by the injection of a large excess of hypobromite in a direct-injection enthalpimetry procedure. In a review of potential direct-injection enthalpimetry applications of this reagent, the determination of amino acids, phenols, and aromatic amines is suggested in addition to those compounds listed in Table 6.55. The enthalpy assignments from the titration (232) and direct-injection enthalpimetry (231) experiments are distinctly different even though the experimental conditions are apparently the same. For example, the enthalpy of reaction of hypobromite with ammonium salts under alkaline conditions is cited as -741 kJ mol^{-1} (231) and -270 kJ mol^{-1} (232). In general, the enthalpy assignments from direct-injection enthalpimetry data are much higher than those calculated from titration data for hypobromite reactions with nitrogen-containing compounds. As the differences are too large to be explained by mass action effects of the excess reagent, it seems likely that the reaction stoichiometries and products may vary with the amount of excess hypobromite by analogy with the complex amine chlorination reactions of hypochlorite (235, 236).

REFERENCES

1. L. S. Bark and S. M. Bark, *Thermometric Titrimetry,* Pergamon Press, Oxford, 1969.
2. H. J. V. Tyrrell and A. E. Beezer, *Thermometric Titrimetry,* Chapman and Hall, London, 1968.
3. G. A. Vaughan, *Thermometric and Enthalpimetric Titrimetry,* Van Nostrand Reinhold, London, 1973.
4. J. Barthel, *Thermometric Titrations,* Wiley-Interscience, New York, 1975.
5. J. Jordan, J. K. Grime, D. H. Waugh, C. D. Miller, H. M. Cullis, and D. Lohr, *Anal. Chem.,* **48**, 427A (1976).
6. L. S. Bark, P. Bate, and J. K. Grime, "Thermometric and Enthalpimetric Titrimetry," in *Selected Annual Reviews of the Analytical Sciences,* L. S. Bark, Ed., Society for Analytical Chemistry, London, 1972, Vol. 2.
7. J. Jordan, "Thermometric Enthalpy Titrations," in *Treatise in Analytical Chemistry,* I. M. Kolthoff and P. J. Elving, Eds., Wiley-Interscience, New York, 1968, Part I, Vol. 8.
8. R. M. Izatt, E. H. Redd and J. J. Christensen, *Thermochim. Acta* **64**, 355 (1983).
9. L. D. Hansen, R. M. Izatt, and J. J. Christensen, "Applications of Thermometric Titrimetry to Analytical Chemistry," in *Treatise on Titrimetry,* J. Jordan, Ed., Marcel Dekker, New York, 1974, Vol. 2.
10. L. S. Bark, *J. Therm. Anal.* **12**, 266 (1977).

11. J. Jordan, J. D. Stutts, and W. J. Brattlie, *Enthalpimetric Analysis, Thermochemical Titrations and Related Methods,* NBS Spec. Publ. 580, Proceedings of the Workshop on the State-of-the-Art of Thermal Analysis, U.S. GPO, Washington, DC, 1980.
12. L. S. Bark, *J. Therm. Anal.* **21**, 119 (1981).
13. J. J. Christensen and R. M. Izatt, "Thermochemistry in Inorganic Solution Chemistry," in *Physical Methods in Advanced Inorganic Chemistry,* H. A. O. Hill and P. Day, Eds., Wiley-Interscience, New York, 1968.
14. E. J. Greenhow, *Chem. Rev.* **77**, 835 (1977).
15. A. P. Chagas, O. E. S. Godinho, and J. L. M. Costa, *Talanta* **24**, 593 (1977).
16. N. Kiba and T. Takeuchi, *Talanta* **20**, 875 (1973).
17. R. Geyer, J. Birkhahn, and G. Eppert, *Chem. Tech. (Leipzig)* **26**, 775 (1974).
18. N. Kiba and T. Takeuchi, *Anal. Chim. Acta* **66**, 75 (1973).
19. N. Kiba and T. Takeuchi, *J. Inorg. Nucl. Chem.* **36**, 847 (1974).
20. N. Kiba and T. Takeuchi, *Bull. Chem. Soc. Jpn.* **46**, 3086 (1973).
21. N. Kiba and T. Takeuchi, *J. Inorg. Nucl. Chem.* **37**, 159 (1975).
22. N. D. Jespersen, *J. Inorg. Nucl. Chem.* **35**, 3873 (1973).
23. R. S. Schifreen, C. S. Miller, and P. W. Carr, *Anal. Chem.* **51**, 278 (1979).
24. G. Dube and F. M. Kimmerle, *Anal. Chem.* **47**, 285 (1975).
25. P. Marik-Korda, *J. Therm. Anal.* **12**, 291 (1977).
26. P. Marik-Korda, "Iodometric Determination of Sulfate by Direct Thermometric Method," in *Proceedings of the First European Symposium on Thermal Analysis,* Heyden, London 1976, p. 143.
27. F. Trischler, *Fresenius' Z. Anal. Chem.* **300**, 288 (1980).
28. N. Kiba, M. Furusawa, and T. Takeuchi, *J. Inorg. Nucl. Chem.* **38**, 1385 (1976).
29. M. B. Williams and J. Janata, *Talanta* **17**, 548 (1970).
30. L. S. Bark and A. E. Nya, *Anal. Chim. Acta* **87**, 473 (1976).
31. P. Marik-Korda, "Use of Non-Selective Reagents for the Simultaneous Determination of Iodide and Bromide by Direct Thermometry," in *Therm. Anal., Proc. Int. Conf., 4th,* I. Buzas, Ed., Heyden, London, 1975, Vol. 1, p. 795.
32. L. S. Bark and A. E. Nya, *J. Therm. Anal.* **12**, 277 (1977).
33. L. S. Bark and A. E. Nya, *Thermochim. Acta* **23**, 321 (1978).
34. P. Marik-Korda, *Talanta* **20**, 569 (1973).
35. G. Ewin and J. O. Hill, "A Thermometric Titrimetric Study of the Complexation of Alkaline Earth Metals by Linear Polycarboxylic Acids," in *Proc. 2nd Aust. Thermodyn. Conf.,* 1981, p. 422.
36. J. Rusz-Szorad and A. Halasz, *Hung. Sci. Instrum.* **32**, 25 (1975).
37. K. Doering, *Thermochim. Acta* **19**, 385 (1977).
38. J. K. Grime, A. D. Campbell, and A. H. Yahaya, *Anal. Chim. Acta* **110**, 139 (1979).
39. K. Doi and M. Tanaka, *Anal. Chim. Acta* **71**, 464 (1974).
40. K. Doi, *Anal. Chim. Acta* **74**, 357 (1975).
41. S-C. Hsin and T-S. Huang, *J. Chin. Chem. Soc. (Taipei)* **21**, 25 (1974).
42. P. Boudeville, J. L. Burgot, and Y. Chauvel, *Thermochim. Acta* **43**, 313 (1981).
43. K. Doi, *Talanta* **25**, 97 (1978).

44. K. Doi, *Talanta* **27**, 859 (1980).
45. M. Zamek, *Coll. Czech. Chem. Commun.* **41**, 3754 (1976).
46. R. Vidal and L. M. Mukherjee, *Talanta* **21**, 303 (1974).
47. J. L. Burgot, *Talanta* **25**, 233 (1978).
48. M. A. Bernard and J. L. Burgot, *Anal. Chem.* **51**, 2122 (1979).
49. J. L. Burgot and M. A. Bernard, *Analusis* **8**, 305 (1980).
50. V. Cerda, E. Casassas, F. Borrull, and M. Esteban, *Thermochim. Acta* **55**, 1 (1982).
51. A. Izquierdo and J. Carrasco, *Talanta* **28**, 341 (1981).
52. J. Lumbriarres, C. Mongay, and V. Cerda, *Analusis* **8**, 62 (1980).
53. R. M. Blanco and J. Palomas, *Quim. Anal.* **31**, 97 (1977).
54. V. Cerda, E. Cassassas, and F. Garcia Montelongo, *Thermochim. Acta* **47**, 343 (1981).
55. D. S. Sabde and R. B. Kharat, *J. Indian Chem. Soc.* **58**, 1125 (1981).
56. D. S. Sabde and R. B. Kharat, *J. Indian Chem. Soc.* **58**, 191 (1981).
57. E. J. Forman and D. N. Hume, *Talanta* **11**, 129 (1964).
58. R. J. N. Harries, *Talanta* **15**, 1345 (1968).
59. J. J. Christensen, R. M. Izatt, D. P. Wrathall, and L. D. Hansen, *J. Chem. Soc. A*, 1212 (1969).
60. J. J. Christensen, D. P. Wrathall, R. M. Izatt, and D. O. Tolman, *J. Phys. Chem.* **71**, 3001 (1967).
61. J. J. Christensen, R. M. Izatt, and L. D. Hansen, *J. Am. Chem. Soc.* **89**, 213 (1967).
62. P. Marik-Korda and E. Eckhart, *J. Therm. Anal.* **17**, 171 (1979).
63. R. Volf, M. Stastny, J. Vulterin, and M. Waldman, *Chem. Prum.* **26**, 648 (1976).
64. D. Jeffries and J. Fresco, *J. Chem. Educ.* **51**, 545 (1974).
65. J. Vulterin, M. Stastny, R. Volf, and B. Mueller, *Chem. Prum.* **29**, 83 (1979).
66. M. Stastny, J. Vulterin, R. Volf, and P. Marsolek, *Chem. Zvesti* **33**, 410 (1979).
67. F. Trischler, *Fresenius' Z. Anal. Chem.* **292**, 141 (1978).
68. J. Vulterin and P. Straka, *Sb. Vys. Sk. Chem.-Technol. Praze, Anal. Chem.* **H11**, 189 (1976).
69. L. S. Bark and P. Prachuabpaibul, *Fresenius' Z. Anal. Chem.* **280**, 373 (1976).
70. L. S. Bark, D. Edwards, and P. Prachuabpaibul, *Proc. Soc. Anal. Chem.* **11**, 170 (1974).
71. L. S. Bark and P. Prachuabpaibul, *Anal. Chim. Acta* **72**, 196 (1974).
72. L. S. Bark and D. Edwards, *Fresenius' Z. Anal. Chem.* **272**, 202 (1974).
73. W. A. de Oliveira and A. A. Rodella, *Talanta* **26**, 965 (1979).
74. R. Volf, M. Stastny, J. Vulterin, and M. Waldman, *Chem. Prum.* **28**, 513 (1978).
75. L. S. Bark and P. Prachuabpaibul, *Anal. Chim. Acta* **84**, 207 (1976).
76. J. Vulterin, P. Straka, M. Stastny, and R. Volf, *Cesk. Farm.* **24**, 10 (1975).
77. J. Vulterin, P. Straka, M. Stastny, and R. Volf, *Chem. Prum.* **24**, 618 (1974).
78. J. D. Lamb, J. E. King, J. J. Christensen, and R. M. Izatt, *Anal. Chem.* **53**, 2127 (1981).
79. R. Volf, M. Stastny, N. Jagrova, and J. Vulterin, *Chem. Prum.* **29**, 132 (1979).
80. L. S. Bark and P. Bate, *Analyst* (London) **98**, 103 (1973).

81. K. Y. Khalaf and T. W. Gilbert, *Analyst* (London)**103**, 623 (1978).
82. E. A. Lewis, J. Barkley, and T. St. Pierre, *Macromolecules* **14**, 546 (1981).
83. K. Y. Khalaf, P. MacCarthy, and T. W. Gilbert, *Geoderma* **14**, 319 (1975).
84. G. C. Krescheck and W. A. Hargraves, *J. Coll. Interfac. Sci.* **48**, 481 (1974).
85. G. C. Kresheck and W. A. Hargraves, *J. Coll. Interfac. Sci.* **83**, 1 (1981).
86. J. K. Grime and J. D. Wernery (unpublished work).
87. L. S. Bark and J. K. Grime, *Anal. Chim. Acta* **64**, 276 (1973).
88. L. S. Bark and O. Ladipo, *Analyst (London)* **101**, 203 (1976).
89. P. Marcu and I. Grecu, *Pharmazie* **31**, 167 (1976).
90. L. S. Bark and L. Kershaw, *Anal. Proc. (London)* **18**, 307 (1981).
91. L. S. Bark and L. Kershaw, *J. Therm. Anal.* **18**, 371 (1980).
92. F. Trischler and J. Friss, *Acta Pharm. Hung.* **51**, 115 (1981).
93. J. L. Burgot, *Talanta* **25**, 339 (1978).
94. J. L. Burgot, *Ann. Pharm. Fr.* **37**, 125 (1979).
95. L. S. Bark and J. K. Grime, *Analyst* (London) **98**, 452 (1973).
96. E. J. Greenhow and L. E. Spencer, *Anal. Chem.* **47**, 1384 (1975).
97. T. Fujie and M. Ogawa, *Yakugaku Zasshi* **94**, 593 (1974).
98. L. S. Bark and J. K. Grime, *Analyst (London)* **99**, 38 (1974).
99. L. S. Bark and L. Kershaw, *Analyst (London)* **100**, 873 (1975).
100. P. Coassolo, M. Sarrazin, J. C. Sari, and C. Briand, *Biochem. Pharmacol.* **27**, 2787 (1978).
101. M. Otagiri, G. E. Hardee, and J. H. Perrin, *Biochem. Pharmacol.* **27**, 1401 (1978).
102. K. Y. Khalaf, P. MacCarthy, and T. W. Gilbert, *Geoderma* **14**, 331 (1975).
103. E. M. Perdue, *Geochim. Cosmochim. Acta* **42**, 1351 (1978).
104. E. M. Perdue, J. H. Reuter, and M. Ghosal, *Geochim. Cosmochim. Acta* **44**, 1841 (1980).
105. O. Talibudeen, K. W. T. Goulding, B. S. Edwards, and B. S. Minter, *Lab. Prac.* **26**, 952 (1977).
106. R. Bezman, *J. Catalysis* **68**, 242 (1981).
107. D. J. Eatough, S. Salim, R.M. Izatt, J. J. Christensen, and L. D. Hansen, *Anal. Chem.* **46**, 126 (1974).
108. R. J. Kvitek, P. W. Carr, and J. F. Evans, *Anal. Chim. Acta* **124**, 229 (1981).
109. J. K. Grime and E. D. Sexton, *Anal. Chem.* **54**, 902 (1982).
110. G. Steinberg, *Chemtech* 730 (1981).
111. Yu. O. Begak and Yu. N. Kukushkin, *Zh. Prikl. Chim. (Leningrad)* **47**, 1391 (1974).
112. I. Sajo and J. Brandstetr, *Thermochim. Acta* **37**, 325 (1980).
113. J. Brandstetr, J. Huleja, I. Sajo, and H. Strauss, *Fresenius'* Z. *Anal. Chem.* **304**, 385 (1980).
114. Z. Jedrasik, *Pr. Osr. Badaw.-Rozwoj. Przetwornikow Obrazu, (Warsaw)* **2**, 58 (1981).
115. E. VanDalen and L. G. Ward, *Anal. Chem.* **45**, 2248 (1973).
116. R. Magrone, R. R. Jean, P. Saccone, and R. V. Weber, *Anal. Chim. Acta* **102**, 233 (1978).
117. Y. Bodard, R. R. Jean, and R. V. Weber, *Light Met.*, 171 (1979).

118. C. Yoshimura and H. Karakawa, *Kinki Daigaku Rikogakubu Kenkyu Hokoku* **17**, 45 (1982).
119. F. Trischler and K. Doering, *Silikattechnik* **25**, 265 (1974).
120. F. Trischler, *Hung. Sci. Instrum.* **28**, 57 (1973).
121. A. Oelschlaeger and I. Herrmann, *Silikattechnik* **25**, 123 (1974).
122. A. Oelschlaeger and I. Herrmann, *Silikattechnik* **25**, 94 (1974).
123. A. Oelschlaeger and I. Herrmann, *Silikattechnik* **27**, 350 (1976).
124. E. Farkas, F. Tamas, and F. Wittmann, *Silikattechnik* **29**, 195 (1978).
125. J. Brandstetr, D. Funke, P. Rovnanikova, and I. Sajo, *Silikattechnik* **31**, 47 (1980).
126. K. Doering, *Therm. Anal.*, Heyden, London, 1974, Vol. 3, p. 543. *Proc. Int. Conf., 4th.*
127. K. Doering, *Thermochim. Acta* **8**, 485 (1974).
128. K. Doering, *Fresenius' Z. Anal. Chem.* **269**, 288 (1974).
129. K. Doering, *Talanta* **21**, 312 (1974).
130. K. Doering, *Hung. Sci. Instrum.* **28**, 37 (1974).
131. I. Sajo and A. Sipos, *Silikattechnik* **26**, 225 (1975).
132. J. Brandstetr, J. Huleja, and S. Honzova, *Slevarenstvi* **25**, 426 (1977).
133. K. Doering, *Fresenius Z. Anal. Chem.* **276**, 297 (1975).
134. K. Doering, *Anal. Chim. Acta* **80**, 192 (1975).
135. M. Czaklosz and J. Nowicka, *Chem. Anal. (Warsaw)* **25**, 1115 (1980).
136. A. Halasz, K. Polyak, and J. Rusz-Szorad, *Magy. Kem. Lapja.* **29**,341 (1974), **29**, 345 (1974).
137. A. Halasz, J. Rusz-Szorad, and M. Ifcsics, *Hung. Sci. Instrum.* **28**, 41 (1974).
138. G. Peuschel and F. Hagedorn, *Fresinius' Z. Anal. Chem.* **277**, 177 (1975).
139. R. Weber, G. Blanc, G. Peuschel, and F. Hagedorn, *Anal. Chim. Acta* **86**, 79 (1976).
140. R. Weber, *Adv. Autom. Anal., Technicon Int. Cong., 7th,* Mediad, Tarrytown, 1977, Vol. 2, p. 357.
141. L. L. Wall and C. W. Gehrke, *Adv. Autom. Anal., Technicon Int. Cong., 7th,* Mediad, Tarrytown, 1977, Vol. 2, p. 355.
142. A. A. Ankawi and L. S. Bark, *J. Therm. Anal.* **12**, 285 (1977).
143. L. Staeudel, A. Stille, and H. Woehrmann, *GIT Fachz. Lab.* **23**, 291 (1979).
144. L. S. Bark, D. Griffin, and P. Prachuabpaibul, *Analyst (London)* **101**, 306 (1976).
145. F. Hagedorn, *GIT Fachz. Lab.* **22**, 498 (1978); **22**, 500 (1978).
146. P. Dupont, *Am. Nutr. Aliment.* **32**, 905 (1978).
147. F. G. Bodewig, *Anal. Chem.* **46**, 454 (1974).
148. M. Zamek and F. Strafelda, *Coll. Czech. Chem. Commun.* **40**, 1888 (1975).
149. M. Mike, *Publ. Hung. Mining Res. Inst.* **16**, 141 (1973).
150. M. Mike, *Hung. Sci. Instrum.* **28**, 47 (1974).
151. M. A. H. Al-Gifri and L. S. Bark, *Proc. Anal. Div. Chem. Soc.* **13**, 359 (1976).
152. M. A. H. Al-Gifri and L. S. Bark, *Proc. Eur. Symp. Therm. Anal., 1st,* Heyden, London, 1976, p. 149.
153. J. Brandstetr and S. Stastnik, *Kniznice Odb. Ved. Spisu Vys. Uceni Tech. Brne B.* **B-78**, 77 (1977).
154. R. Sell, *Kali Steinsalz* **7**, 395 (1979).

155. F. Strafelda, J. Huleja, and R. Volf, *Sb. Vys. Sk. Chem.-Technol. Praze, Anal. Chem.* **H9**, 177 (1973).
156. R. Volf, F. Strafelda, and J. Huleja, *Sb. Vys. Sk. Chem.-Technol. Praze, Anal. Chem.* **H10**, 99 (1974).
157. F. Millero, S. R. Schrager, and L. D.Hansen, *Limnol. Oceanogr.* **19**, 711 (1974).
158. L. D. Hansen, L. D. Whiting, D. J. Eatough, T. E. Jensen, and R. M. Izatt, *Anal. Chem.* **48**, 634 (1976).
159. D. J. Eatough, T. Major, J. Ryder, M. Hill, N. F. Mangelson, N. L. Eatough, and L. D. Hansen, *Atmos. Environ.* **12**, 263 (1978).
160. D. J. Eatough, N. L. Eatough, M. W. Hill, N. F. Mangelson, J. Ryder, and L. D. Hansen, *Atmos. Environ.* **13**, 489 (1979).
161. L. D. Hansen, B. E. Richter, and D. J. Eatough, *Anal. Chem.* **49**, 1779 (1977).
162. D. J. Eatough, L. D. Hansen, R. M. Izatt, and N. F. Mangelson, *Proc. IMR Symposium 8th,* NBS Spec. Publ. 464, U.S. GPO, Washington, D.C., 1977, p. 643.
163. E. J. Greenhow, *Chem. Ind. (London),* 422 (1972).
164. E. J. Greenhow, *Chem. Ind. (London),* 466 (1972).
165. E. J. Greenhow and L. E. Spencer, *Analyst (London)* **98**, 90 (1973).
166. E. J. Greenhow and L. H. Loo, *Analyst (London)* **99**, 360 (1974).
167. E. J. Greenhow, R. Hargitt, and A. A. Shafi, *Angew. Makromol. Chem.* **48**, 55 (1975).
168. E. J. Greenhow and A. A. Shafi, *Angew. Makromol. Chem.* **53**, 187 (1976).
169. D. A. Castle and E. J. Greenhow, *Inst. Pet.* 1p 75–015, 15 (1975).
170. E. J. Greenhow and A. A. Shafi, *Proc. Anal. Div. Chem. Soc.* **12**, 286 (1975).
171. E. J. Greenhow and A. Nadjafi, *Anal. Chim. Acta* **109**, 129 (1979).
172. E. J. Greenhow and L. E. Spencer, *Analyst (London)* **98**, 485 (1973).
173. E. J. Greenhow and A. A. Shafi, *Talanta* **23**, 73 (1976).
174. E. J. Greenhow and L. E. Spencer, *Analyst (London)* **98**, 81 (1973).
175. E. J. Greenhow and L. E. Spencer, *Analyst (London)* **98**, 98 (1973).
176. E. J. Greenhow and R. Hargitt, *Proc. Anal. Div. Chem. Soc.* **10**, 276 (1973).
177. E. J. Greenhow and A. A. Shafi, *Analyst (London)* **101**, 421 (1976).
178. E. J. Greenhow, A. Nadjafi, and L. A. Dajer de Torrijos, *Analyst (London)* **103**, 411 (1978).
179. L. A. Dajer de Torrijos and E. J. Greenhow, *Proc. Anal. Div. Chem. Soc.* **16**, 7 (1979).
180. E. J. Greenhow and L. A. Dajer de Torrijos, *Analyst (London)* **104**, 801 (1979).
181. E. J. Greenhow and L. E. Spencer, *Analyst (London)* **99**, 82 (1974).
182. E. J. Greenhow and L. E. Spencer, *Analyst (London)* **101**, 777 (1976).
183. E. J. Greenhow and L. E. Spencer, *Analyst (London)* **100**, 747 (1975).
184. G. L. Jeyaraj and E. J. Greenhow, *Proc. Anal. Div. Chem. Soc.* **19**, 326 (1982).
185. G. A. Vaughan and J. J. Swithenbank, *Analyst* **90**, 594 (1965).
186. E. J. Greenhow, *Chem. Ind.* (London) 456 (1974).
187. N. Kiba, Y. Sawada, and M. Furusawa, *Talanta* **29**, 416 (1982).
188. E. J. Greenhow and L. E. Spencer, *Talanta* **24**, 201 (1977).
189. F. F. Gaal, B. D. Arramovic, and V. J. Vajgand, *Microchem. J.* **27**, 231 (1982).

190. E. J. Greenhow, *Analyst* (London) **102**, 504 (1977).
191. F. Feigl and V. Anger, *Spot Tests in Inorganic Analysis*, 6th ed., Elsevier, Amsterdam, 1972, p. 437.
192. N. Kiba, T. Suto, M. Furusawa, and T. Takeuchi, *Therm. Anal., Proc. Int. Conf. 5th,* H. Chihara, Ed., Heyden, London, 1977, p. 38.
193. N. Kiba, T. Suto, and M. Furasawa, *Talanta* **28**, 115 (1981).
194. N. Kiba, N. Masahiro, and M. Furusawa, *Talanta* **27**, 1090 (1980).
195. H. Weisz, W. Meiners, and F. Guenter, *Anal. Chim. Acta* **107**, 301 (1979).
196. N. Kiba, Y. Suzuki, and M. Furusawa, *Talanta* **28**, 691 (1981).
197. J. M. Elvecrog and P. W. Carr, *Anal. Chim. Acta* **121**, 135 (1980).
198. N. Kiba and M. Furusawa, *Anal. Chim. Acta* **98**, 343 (1978).
199. T. F. A. Kiss, *Mikrochim. Acta,* 847 (1973).
200. T. F. A. Kiss, *Mikrochim. Acta,* 471 (1975).
201. S. Pantel and H. Weisz, *Anal. Chim. Acta* **68**, 311 (1974).
202. R. Feys, J. Devynck, and B. Tremillon, *Talanta* **22**, 17 (1975).
203. R. P. Panteler, L. D. Alfimova, A.M. Bulgakova, and I. V. Pulyaeva, *Zh. Anal. Khim.* **30**, 946 (1975).
204. R. P. Panteler, L. D. Alfimova, and A. M. Bulgakova, *Ukr. Khim. Zh.* **41**, 801 (1975).
205. R. P. Panteler, L. D. Alfimova, and A. M. Bulgakova, *Zh. Anal. Khim.* **30**, 1584 (1975).
206. H. Weisz, S. Pantel, and W. Meiners, *Anal. Chim. Acta* **82**, 145 (1976).
207. F. F. Gaal, V. I. Soros, and V. J. Vajgand, *Anal. Chim. Acta* **84**, 127 (1976).
208. F. Trischler, *J. Therm. Anal.* **16**, 119 (1979).
209. G. Svehla, *Analyst (London)* **94**, 513 (1969).
210. P. G. Zambonin and J. Jordan, *Anal. Chem.* **41**, 437 (1969).
211. D. W. Rogers and R. J. Sasiela, *Talanta* **20**, 232 (1973).
212. D. W. Rogers and R. J. Sasiela, *Anal. Biochem.* **56**, 460 (1973).
213. D. W. Rogers and M. K. Adeniran, *Anal. Lett.* **12**, B11, 1149 (1979).
214. D. W. Rogers and R. J. Sasiela, *Mikrochim. Acta,* 33 (1973).
215. L. A. Williams, B. Howard, and D. W. Rogers, *Anal. Calorim.* **3**, 207 (1974).
216. D. W. Rogers and A. Goldberg, *Anal. Calorim.* **4**, 125 (1974).
217. L. S. Bark and P. Prachuabpaibul, *Fresenius' Z. Anal. Chem.* **283**, 293 (1977).
218. L. S. Bark and P. Prachuabpaibul, *Proc. Eur. Therm. Anal.*, 1st, Heyden, London, 1976, p. 145.
219. L. S. Bark and P. Prachuabpaibul, *Fresenius' Z. Anal. Chem.* **282**, 201 (1976).
220. B. J. Duffield and D. N. Hume, *Anal. Lett.* **7**, 681 (1974).
221. D. W. Rogers, *Am. Lab.* **12**, 18 (1980).
222. D. W. Rogers, *Am. Lab.* **14**, 15 (1982).
223. J. C. Wasilewski and C. D. Miller, *Anal. Chem.* **38**, 1750 (1966).
224. P. Marik-Korda, "Water Determination by DIE Method," *Therm. Anal., Proc. Int. Conf. Therm. Anal., 6th*, Birkhauser, Basel, 1980, Vol. 1, pp. 529.
225. P. Marik-Korda, "Determination of the Water Content of Some Crystal Hydrates by a Modified Direct Enthalpimetric (DIE) Method, "*Czech. Conf. Calorimetry,*

(Lect. Short Commun.), 1st, Inst. Inorg. Chem. Czech. Acad. Sci., Prague, 1977, p. C1211.

226. K. Szoke and P. Marik-Korda, *Acta Pharm. Hung.* **51**, 111 (1981).
227. Y. Kuriya, M. Nakayama, T. Nishishita, T. Takazawa, Y. Itoo, and S. Tsutsui, *Sekiyu Gakkai Shi* **20**, 573 (1977).
228. J. Brandstetr and P. Sapakova, *Coll. Czech. Chem. Commun.* **38**, 2249 (1973).
229. L. S. Bark and J. K. Grime, *Talanta* **22**, 443 (1975).
230. L. S. Bark and P. Prachuabpaibul, *Anal. Chim. Acta* **87**, 505 (1976).
231. N. Malingerova and M. Malinger, *J. Therm. Anal.* **13**, 149 (1978).
232. J. Rusz-Szorad and A. Halasz, *Acta. Chim. Acad. Sci. Hung.* **99**, 215 (1979).
233. J. Rusz-Szorad and A. Halasz, *Hung. Sci. Instrum.* **40**, 19 (1977).
234. J. Rusz-Szorad and A. Halasz, *Hung. Sci. Instrum.* **42**, 21 (1978); **42**, 25 (1978).
235. J. M. Antelo, F. Arce, F. Barbadillo, J. Casado, and A. Varela, *Environ. Sci. Technol.* **15**, 912 (1981).
236. W. H. Dennis, Jr., L. A. Hull, and D. H. Rosenblatt, *J. Org. Chem.* **32**, 3783 (1967).

CHAPTER

7

APPLICATION OF SOLUTION CALORIMETRY TO BIOCHEMICAL AND CLINICAL ANALYSES

J. Keith Grime

The Procter & Gamble Company
Ivorydale Technical Center
Cincinnati, Ohio

A survey of calorimetric literature during the past decade reveals a considerable emphasis on biochemical and clinical chemistry. In fact, it can be argued that the major motivation for the development of new calorimetric instrument technology, data collection and treatment methods, and methodological innovations during this period is a direct result of the demands placed on the technique by biochemical and clinical chemistry. Examples include the increased focus on flow methods to increase sample throughput, the advent of small-cell technology for sample-limited (e.g., physiological fluid) determinations, and the addition of kinetic direct-injection methodology to accommodate the measurement of rate-controlled (primarily enzymatic) reactions, to mention just a few.

The plethora of research reports resulting from this activity has allowed the evaluation of calorimetric methods as a measurement tool in almost every aspect of biochemical and clinical analysis from the fundamental to the applied. This chapter will be limited to the discussion of the theoretical principles, methodology, and application of calorimetric methods for the study and analysis of biochemical and/or clinical systems. Instrumentally, the emphasis is on isoperibol technology, a reflection of its popularity in this area of research. However, the application of heat-conduction calorimetry to biochemical analyses is also discussed in some detail. Differences in the interpretation of isoperibol and heat-conduction calorimetry data have been delineated as appropriate with each application area.

The design of an analytical experiment based on enzyme-catalyzed reactions is governed primarily by the prevailing kinetics. Similarly, the equations describing the analytical signal are described as much by fundamental enzyme kinetics as by calorimetric theory. Accordingly, this chapter has been classified primarily according to application rather than instrumental approach. The relevant enzyme kinetic theory is developed in each case and subsequently incorporated into the working equations of calorimetry.

The reader is referred to the following reviews, monographs, and books for further information on biochemical calorimetry (1–16).

7.1. THERMOMETRIC ENTHALPY TITRATION OF PROTEINS

Prior to the advent of commercially available stable enzyme preparations, the contribution of calorimetric measurements to clinical analysis was necessarily limited by the inherent non-selectivity of the detection system. The principal exception is the area of protein analysis, which has been the subject of many calorimetric studies.

7.1.1. The Determination of Serum Protein

12-Phosphotungstic acid (PTA), an efective anionic precipitant of proteins, is often used to remove proteins from analytical solutions in which proteins would represent an interference. Since the reaction is stoichiometrically related to the number of positively charged sites in the protein molecule (17), which is in turn proportional to the number of basic amino acids present, the enthalpy of precipitation can be used as the basis for the quantitative determination of the protein by thermometric enthalpy titration (18):

$$H_nP^{n+} + \frac{n}{3}\,PTA^{3-} = H_nP(PTA)\frac{n}{3}\,(s) \qquad (7.1.)$$

where H_nP^{n+} is the totally protonated form of the protein existing under acid conditions.

The shape of the enthalpogram is dependent on pH, the viscosity of the solution, and the rate of stirring as shown in Figs. 7.1–7.3, which represent the titration of bovine serum albumin with PTA. At low pH (<2), the first endpoint has been attributed to the reaction

$$H_nPCl_x^{(n-x)+} + \frac{(n-x)}{3}\,PTA^{3-} = H_nPCl_x(PTA)_{(n-x)/3} \qquad (7.2.)$$

where chloride ion has been incorporated to provide electroneutrality. The second and final endpoint corresponds to the reaction

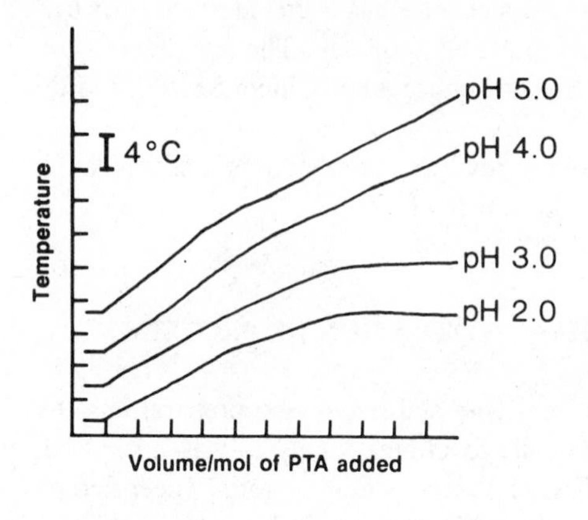

Figure 7.1. The effect of pH on the shape of the enthalpogram obtained for the titration of bovine serum albumin with 12-phosphotungstic acid (PTA) (from Ref. 17, by permission of Plenum Publishing Corp.).

$$H_nPCl_x(PTA)_{(n-x)/3} + \frac{x}{3}PTA^{3-} = H_nP(PTA)_{n/3} + xCl^- \quad (7.3.)$$

and is used as the basis for the determination of total protein. In most calorimetric experiments the heat generated by the mechanical stirring of the solution imparts an effectively constant exothermic background noise to the titration curve of any particular species. Typical concentration changes of the analyte do not usually result in a significant change in the contribution to the noise from this source. An exception to this behavior are species which substantially effect the viscosity of the analytical solution. This is particularly evident with the precipitation titration of proteins as shown in Fig. 7.2. The viscosity of the solution increases with increasing protein concentration resulting in convex curvature of the enthalpogram in the region prior to the first endpoint. The assignment of this phenomenon to increased stirring heats is confirmed by the data shown in Fig. 7.3 in

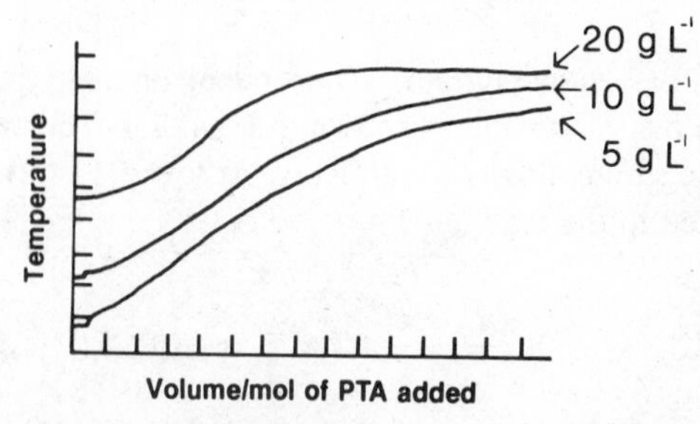

Figure 7.2. The effect of protein concentration on the shape of the enthalpogram obtained for the titration of bovine serum albumin with PTA (from Ref. 18, by permission of the American Chemical Society).

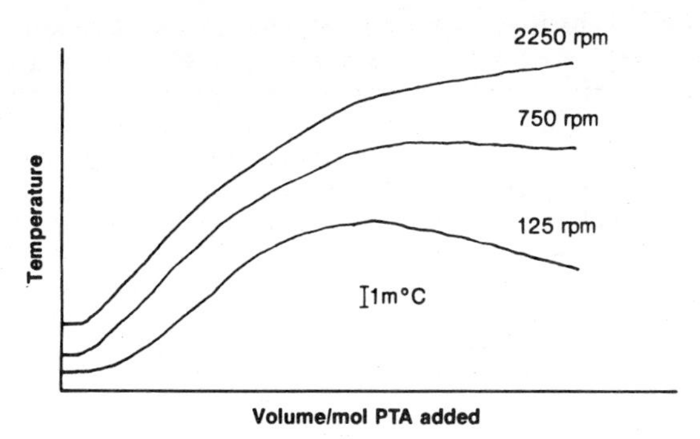

Figure 7.3. The effect of stirring rate on the shape of the enthalpogram obtained for the titration of bovine serum albumin with PTA (from Ref. 18, by permission of the American Chemical Society).

which the posttitration slope is increased relative to the initial (baseline) slope with increased stirring rate.

The apparent stoichiometry of Eq. (7.1) is dependent on both the concentration of the protein and the rate of titrant addition as shown in Tables 7.1 and 7.2. Specifically, the apparent stoichiometry increases at low protein concentration and at higher rates of titrant addition. Both of these effects can be attributed to spurious (i.e., non-stoichiometric) endpoints resulting from slow reaction kinetics. As with any linear titration procedure, positive endpoint errors result if the titration reaction rate approaches or is less than the rate of addition of reagent. In practice, concentrations of protein between 1 and 10 g of protein per liter can be determined with a precision of 0.3–0.8% if titration rates are less than or equal to 7.66 μeq s^{-1}. The optimum titration rate is 1.4 μeq s^{-1}. The application

TABLE 7.1. Thermometric Titration of Bovine Serum Albumin with 12-Phosphotungstic Acid—Effect of Protein Concentration on the Reaction Stoichiometry and Precision

BSA Conc. (g L^{-1})	BSA Conc. Taken (meq L^{-1})	Titratable Groups Found per 100,000 g of Protein	% rsd
0.50	0.735	164.4	2.2
1.00	1.47	146.6	1.5
2.00	2.94	145.2	0.2
5.00	7.35	144.0	0.8
10.00	14.7	145.9	0.6
20.00	29.4	154.6	1.8

Source: Ref. 18, by permission of the American Chemical Society.

TABLE 7.2. Effect of the Rate of Titrant Addition on the Apparent Stoichiometry of the Reaction between 12-Phosphotungstic Acid and Bovine Serum Albumin (BSA)

		Titratable Groups	
Rate of Addition (μeq/PTA s^{-1})	BSA Conc. (μeq L^{-1})	Found Per 100,000 g Protein	Percentage Range[a]
0.56	1.250	144.3	0.3
1.41	1.250	154.5	0.3
0.56	6.321	144.5	0.0
1.41	6.321	145.2	0.3
3.06	6.321	149.2	0.8
7.66	6.321	151.6	0.7

Source: Ref. 18, by permission of the American Chemical Society.

[a]Relative range of duplicate measurements.

of this calorimetric procedure to the determination of protein in serum samples produces acceptable data when compared with the classical Kjeldahl and biuret procedures (Fig. 7.4) with correlation coefficients of 0.982 and 0.990, respectively. No significant interferences were observed on species present in serum which would precipitate PTA, for example, bilirubin, NH_4^+, and the amino

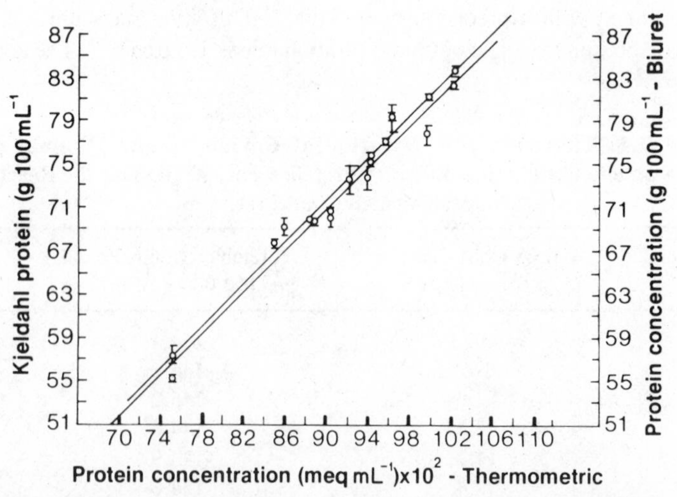

Figure 7.4. Correlation of enthalpimetric analysis data of total serum protein with Kjeldahl nitrogen and biuret colorimetry. (From Ref. 18, by permission of the American Chemical Society).

acids, the presumption being that at the range of dilution examined the solubility products are not exceeded. (See Table 7.3.)

7.1.2. Proton Dissociation

The unique dependence of a calorimetric titration curve on the enthalpy of reaction ΔH_R, as well as the molar free energy ΔG_R, sometimes allows the serial determination of species not possible with a comparable potentiometric titration. The titration of boric acid is often used as an example of this characteristic (19). Similar arguments can be used to explain the differences between the potentiometric and thermometric titration curves of ovalbumin (20) with sodium hydroxide as shown in Fig. 7.5. Three quite discernible endpoints, corresponding to the deprotonation of carboxyl, imidazole, and amino functionalities are observed for the thermometric titration, illustrating the difference in enthalpies between each of these neutralization reactions. The potentiometric titration curve, on the other hand, lacks definition.

Potentiometric and calorimetric data in combination can, in some instances, be a powerful tool for the characterization of the proton ionization chemistry of

TABLE 7.3. Precision Data for the Thermometric Enthalpy Titration of Serum Protein with 12-Phosphotungstic Acid[a]

Sample Number	Protein Found[b] (meq)	Percentage Range[c]
1	99.4	0.2
2	90.2	0.1
3	81.6	2.2[d]
4	95.9	0.4
5	90.5	0.2
6	90.4	0.2
7	83.8	0.6
8	76.7	0.3
9	85.4	0.3
10	75.0	0.8
Average		0.3[e]

Source: Ref. 18, by permission of the American Chemical Society.

[a]Ten individual 1-mL samples of human serum diluted to 30 mL with 0.1 mol L^{-1} hydrochloric acid.

[b]Amount of protein found in units of equivalents of reactive units per 1 mL of sample serum based on the standardization of PTA versus cesium.

[c]Range relative to the protein concentration determined in duplicate runs.

[d]Value rejected on the basis of a test at the 99% confidence interval.

[e]Average of the above results.

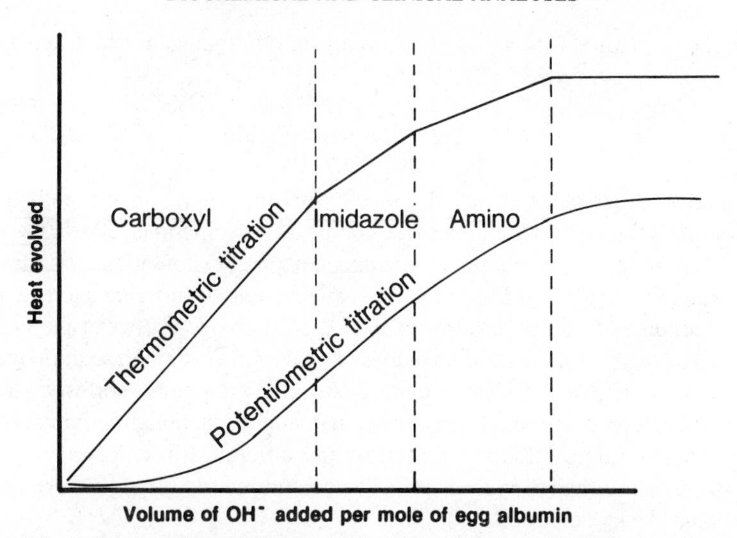

Figure 7.5. Comparison of thermometric enthalpy titration and potentiometric titration curves for the titration of protonated egg albumin with sodium hydroxide (from Ref. 20, by permission of Centre National de la Recherche Scientifique).

proteins. A good example is the acid–base titration of insulin described by Izatt et al. (21). Figure 7.6 shows the thermometric and potentiometric titration curves for the titration of 1.7×10^{-4} mol L^{-1} insulin dimer with 1.7×10^{-1} mol L^{-1} hydrochloric acid, determined independently. The data analysis of these curves is based on the assumption that the protein consists of a series of related functional groups, and that the members within each series have identical pK and ΔH_R values (22). The enthalpy and pH values determined at preset intervals on the titration curves are thus combined to provide a unique test for the presence of thermodynamically equivalent functional groups. It can be shown (21) that the heat change at any instant, $q_{R,p}$, can be represented by the equation

$$q_{R,p} = nm_A\Delta H_R - \frac{q_{R,p}}{K_{eq}a_{H^+,p}} \tag{7.4}$$

where n is the number of equivalent functional groups in the protein, $a_{H^+,p}$ is the corresponding hydrogen ion activity at point p, and m_A is the total number of moles of protein in solution. K_{eq} is the equilibrium constant for the protonation of that particular set of identical functionalities. A plot of $q_{R,p}$ versus $q_{R,p}/a_{H^+,p}$ will therefore be linear with a slope of $-1/K_{eq}$ and an intercept of $nm_A\Delta H_R$. This equation will be valid for each set of equivalent functional groups.

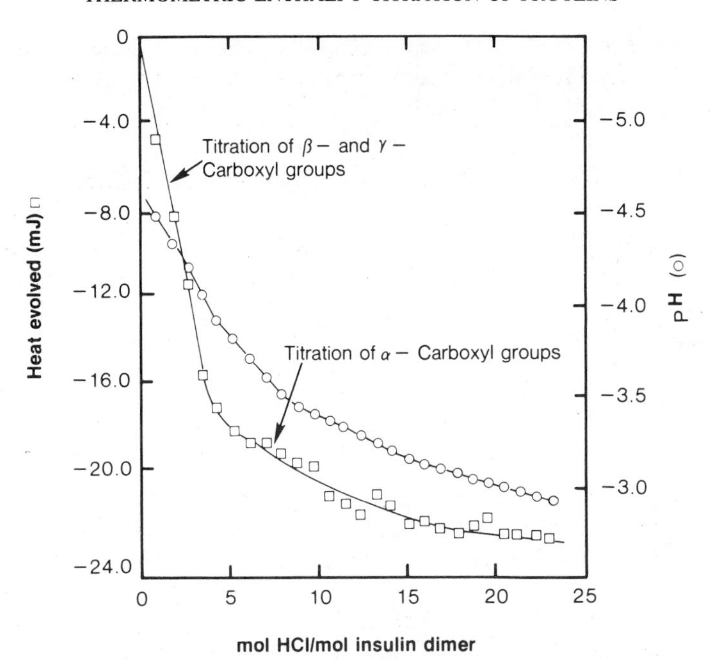

Figure 7.6. Comparison of calorimetric and pH titration data for the titration of insulin dimer (0.1735 mmol L^{-1}) with 0.1766 mol L^{-1} hydrochloric acid (from Ref. 33, by permission of Elsevier Science Publishers).

Accordingly, in just two experiments, a complete thermodynamic and stoichiometric analysis of the protonation chemistry of a protein can be obtained.

ΔH_R and pK data are available directly from the slope and midpoints of the calorimetric and potentiometric titration curves, respectively. Similarly, n can be obtained from the inflection points of either curve (21, 23). This approach does not, however, provide a check for the presence of functional groups with atypical thermodynamic parameters.

A least squares analysis as described in Chapter 5 would also be an appropriate method of data analysis. To date, the application of this technique has not been reported for protein proton ionization data. The incorporation of a low-thermal-mass microelectrode into a calorimetric cell would clearly represent the ideal instrumental design for the generation of complementary potentiometric and calorimetric data. Experimental titration calorimeters, containing pH and reference electrodes, have been described (24, 25). Commercial equipment with this facility is available from Tronac Inc. In practice the success of a simultaneous potentiometric and calorimetric experiment will depend on the relative response time of the electrode with respect to the titration rate and the response time of the thermistor (~0.5 s). In the case of a slow response time, it may be necessary to adapt the experimental design to incremental addition. A programmable buret system clearly facilitates this operation.

7.1.3. The Interaction of Proteins with Surfactants

The binding of surfactants to proteins can also be conveniently studied by calorimetric techniques. The incorporation of a surfactant into the experiment adds an extra dimension to data interpretation since the enthalpy contribution of micellar changes must be delineated from the protein-binding chemistry. This can be achieved empirically by comparisons of the enthalpograms obtained for the titration of the surfactant into a buffer solution in the presence and absence of the binding agent, in this instance, a protein. A hypothetical example is shown in Fig. 7.7. Curve A represents a titration "blank" in which micellar effects are evaluated. In region 1, demicellization, that is, the dissociation of concentrated micelles in the titrant to give monomers in the reaction cell, will generate a slope $q_{1/n}$, where n is the number of moles of surfactant added. At x the critical micelle concentration (cmc) is attained and the enthalpy change beyond that point will be generated by the dilution of concentrated micelles since, by definition, surfactant monomers cannot form in this region. Curve B resents an identical experiment in which an interacting protein has been added to the solution. Region 3 now contains an additional process, the binding of the protein to the surfactant monomers, as well as the demicellization reaction. Since the contributing enthalpies are additive, the enthalpy associated with protein–surfactant binding, q_B/n, is given by,

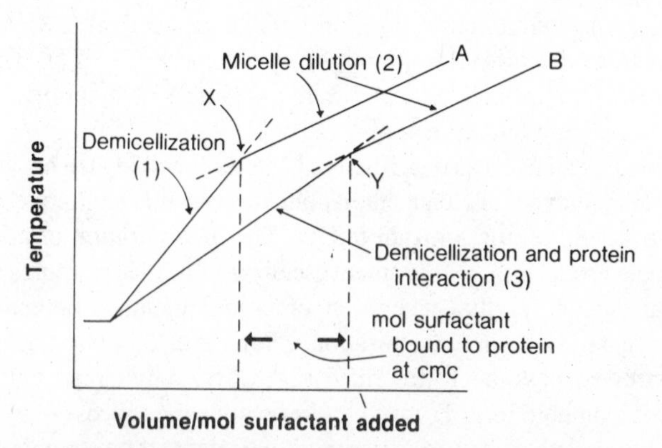

Figure 7.7. Hypothetical enthalpograms for the titration of a surfactant into (A) buffer solution and (B) buffer solution and an interacting protein. x is the critical micelle (CMC) in the absence of a binding protein. y is the apparent CMC in the presence of an interacting protein.

$$q_B/n = q_1/n - q_3/n \tag{7.5.}$$

where q_3/n is the slope in region 3.

Equation (7.5) can be written

$$\Delta H_B = \Delta H_1 - \Delta H_3 \tag{7.6.}$$

where ΔH_B is the molar enthalpy of protein–surfactant binding. The stoichiometry of the binding reaction at the CMC is given by the shift in CMC caused by the presence of the protein as shown in Fig. 7.7. This type of data manipulation has been used to study the binding of Triton nonionic surfactants **1** to bovine serum albumin (26) and the interaction of block polypeptides with anionic and cationic surfactants (27). Other enthalpimetric investigations of fundamental protein chemistry include studies on the adsorption of proteins by membranes (28) and the reaction of metal ions with proteins (29).

$$CH_3-C(CH_3)_2-CH_2-CH(CH_3)_2-C_6H_4-O(CH_2-O)_nCH_2-OH$$

1

7.2. THE CALORIMETRIC DETERMINATION OF CALCIUM AND MAGNESIUM IONS IN SERUM

The selectivity problem associated with the application of calorimetry to clinical determinations can sometimes be overcome by manipulation of the reaction thermodynamics and/or kinetics by a judicious choice of reagent and experimental variables, for example, pH.

An example of this principle is the determination of calcium and magnesium

TABLE 7.4. Thermodynamic Parameters for the Reaction of Ca^{2+} and Mg^{2+} with EDTA

Analyte	ΔH_R (kJ mol^{-1})	ΔH_R in THAM Buffer at pH 8.0 (kJ mol^{-1})	log K_{eq}
Ca^{2+}	−25.2	−75.6	10.6
Mg^{2+}	16.8	−33.6	8.7

Source: Refs. 30 and 31.

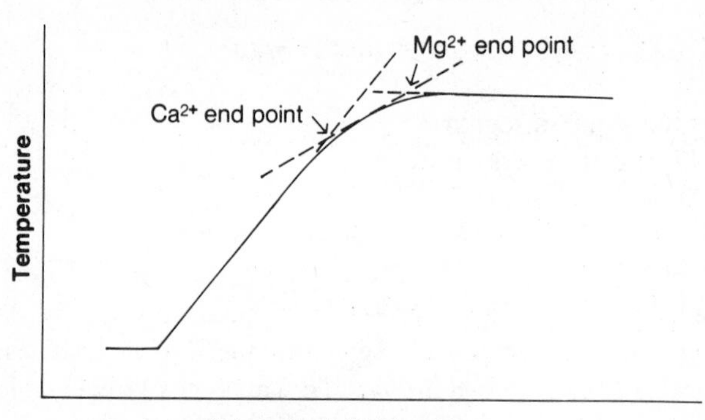

Figure 7.8. Typical enthalpogram for the titration of 1 mL of serum containing 2.5 mmol L^{-1} Ca^{2+} and 1.3 mmol L^{-1} Mg^{2+} in THAM buffer (pH 8.0) with EDTA (from Ref. 30, by permission of the American Association for Clinical Chemistry).

ions in serum by enthalpimetric titration with ethylenediaminetetraacetate, EDTA (30). The principle of the method becomes apparent upon inspection of the relevant thermodynamic parameters shown in Table 7.4. The thermodynamics of the reactions are such that the calcium ions will titrate quantitatively prior to the reaction of magnesium ions. Moreover, the significant difference in enthalpies will facilitate endpoint location. In the absence of any other concurrent reactions, the titration enthalpogram would be composed of an exotherm, followed by an endotherm (31). However, it has been shown (30) that at the concentration levels of calcium and magnesium ions in serum, 2.5 mmol L^{-1} and 1.0 mmol L^{-1}, respectively, the optimum pH for enthalpogram linearity is 8.0; slow reaction kinetics prohibit a higher pH level. In the nature of the EDTA complexation process, a concurrent buffer protonation will occur at this pH. In THAM buffer the *total* enthalpy change for the titration of both ions will be dominated by amine protonation (-50.4 kJ mol^{-1}) and a double exotherm will result. A typical enthalpogram for a clinical serum sample is shown in Fig. 7.8. Correlation data with an atomic spectroscopy procedure is shown in Table 7.5. The precision of the enthalpimetric titration is determined primarily by the reproducibility of the titrant delivery system since a typical volume necessary to reach the endpoint is only 0.02 mL. Sodium sulfide is the recommended masking agent for the removal of potential interference by iron, zinc, and copper ions.

7.3. THE ELUCIDATION OF METAL ION BINDING AND FUSION MECHANISMS IN PHOSPHOLIPID MEMBRANES

Phospholipid membranes, for example, phosphatidylserine vesicles, are often used as models to mimic the behavior of physiological membranes. Membrane

TABLE 7.5. Correlation Data for the Determination of Ca^{2+} and Mg^{2+} in Sera by Thermometric Enthalpy Titration (TET) and Atomic Absorption Spectrometry

Sample Number	Ca^{2+} Determination (mmol L^{-1})		Mg^{2+} Determination (mmol L^{-1})	
	AA	TET	AA	TET
1	2.11	1.99	0.88	0.88
2	2.30	2.27	0.79	0.81
3	2.07	1.98	0.82	0.80
4	2.13	2.09	0.77	0.82
5	2.36	2.23	0.82	0.84
6	2.31	2.23	0.81	0.83
7	2.24	2.27	0.82	0.81
8	2.20	2.29	0.80	0.83
9	2.19	2.18	0.78	0.79
10	2.32	2.34	0.71	0.78
11	2.41	2.29	0.81	0.88
12	3.08	3.08	0.65	0.73

Standard Deviation of Difference = ± 0.06 (Ca^{2+}) ± 0.04 (Mg^{2+})

Correlation coefficient = 0.98

Source: Ref. 30.

fusion and rupture processes which occur in the presence of divalent cations can be identified by observing the attendant heat changes (32). The role of calcium ion in this complex sequence of events can be elucidated by superimposition of calorimetric and potentiometric (ion-selective electrode) data as shown in Fig. 7.9. Four distinct regions of the enthalpogram can be attributed to four different processes with the aid of corresponding calcium ion activity data (33). The calorimetric data were obtained in a continuous titration, the potentiometric data from a stepwise addition procedure.

The primary correlation point is arrowed in Fig. 7.9. At this point, a sharp increase in exothermic response is accompanied by a corresponding decrease in free calcium ion activity. The behavior is consistent with a model in which the vesicle framework ruptures due to calcium-ion-induced crystallization of the lipid acyl chains. The breakdown of the vesicle produces a sudden release of non-complexed phospholipid into solution decreasing the calcium ion activity. This model is strengthened by three other observations: (1) the solution becomes turbid at this point; (2) cessation of the titration in region 3 results in an ongoing evolution of heat which does not occur in regions 1 or 2; and (3) the enthalpy

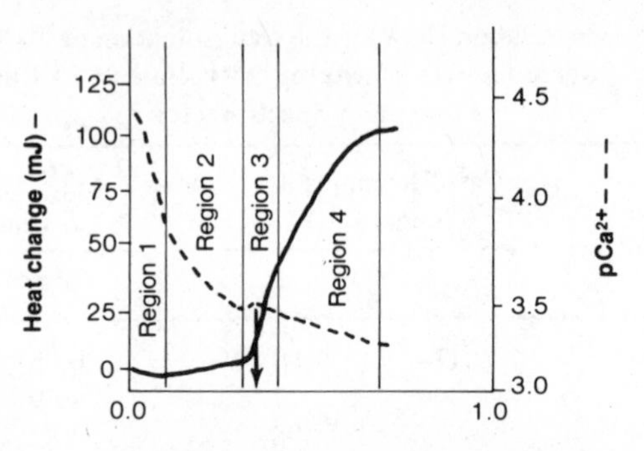

Figure 7.9. Comparison of calorimetric and pH data for the titration of phosphatidylserine vesicles with $CaCl_2$. Calorimetric titration performed in THAM buffer, pH 7.4 (from Ref. 33, by permission of Elsevier Science Publishers).

change measured in regions 3 and 4 is identical to that measured by differential scanning calorimetry for acyl chain crystallization in the absence of Ca^{2+}.

7.4. BIOCHEMICAL CALORIMETRY—ANALYTICAL CONSIDERATIONS

At first glance, the combination of enzyme chemistry with a calorimetric measurement is an ideal fit because each technology minimizes a limitation of the other. The primary disadvantage of calorimetric methods for the analysis of complex solutions is a total lack of selectivity in the detection system which will sum the heat changes of the reactions occurring *in toto*; this problem is largely circumvented by the inherent selectivity of an enzyme-catalyzed reaction. Similarly, one of the drawbacks of "enzymatic analysis" *per se* is that the analytical enzymatic reaction does not always produce a measurable specie. Secondary reagents, often a secondary enzyme, must be added solely for the purpose of producing an analytical signal by catalyzing the consumption of primary reaction product to produce protons or a chromophore, for example. Since each enzyme system has its own prerequisites of pH, co-substrate, ionic strength, activators, etc., the analytical solution can become complex, increasing the chance for interference. Moreover, a compromise must usually be made between the optimum conditions for each enzyme-substrate reaction.

As most chemical or biochemical reactions are accompanied by a measurable enthalpy change, it is usually possible to monitor the *primary* enzymatic reaction calorimetrically, thereby simplifying an enzymatic analysis considerably. Moreover, the tolerance of the thermistor to the physical properties of the solution, for example, the presence of insolubles, turbidity, or color, can in some instances allow a determination of an analyte in a physiological liquid without sample pretreatment.

7.4.1. Chemical and Biochemical Amplification of Enthalpy Changes

Physiological enzymes or substrates are often present in small concentrations. Accordingly, enthalpy amplification procedures are frequently used to increase the sensitivity of the calorimetric determination. The simplest and most convenient method is "buffer amplification." All enzyme reactions are carried out under well-defined conditions of pH and temperature. The requirement of a buffer affords an opportunity to utilize any attendant protonation or deprotonation reactions to amplify the *total* enthalpy change associated with an enzyme-catalyzed reaction without introducing secondary reagents. The concurrent buffer reaction will be stoichiometrically and kinetically linked to the primary enzymatic reaction and it will therefore be a true representation of its extent and rate. The choice of buffer for a calorimetric experiment will be dependent on both the pH and sensitivity required since protonation/deprotonation enthalpies can differ by factors of 2 or 3. The two buffers most commonly used for analytical methods in the physiological pH range (6–8) are tris-hydroxymethylaminomethane (THAM) and phosphate. In the clinical enzyme area, phosphate buffers are often avoided because of the possibility of inhibition of enzyme activity by complexation of essential polyvalent cation activators, for example, Ca^{2+}, Mg^{2+}, Fe^{3+}. McGlothlin and Jordan have determined the thermodynamic parameters associated with the protonation/deprotonation of 19 buffer systems compatible with enzyme-catalyzed reactions in the pH ranges 6–10.5 (34). The enthalpy changes and pK_a associated with each system are represented in Table 7.6. In each case the enthalpy and pK data refer to reactions of the type,

$$RNH_3^+ + H_2O = RNH_2 + H_3O^+ \tag{7.7}$$

The enthalpy of protonation can therefore be obtained by reversing the sign of the enthalpy data shown.

As the large majority of enzyme-catalyzed reactions have their optimal activity in the pH range 7–9, THAM buffer is by far the most often reported thermochemical amplification buffer. There are noteable exceptions as discussed in Section 7.5.5.

TABLE 7.6. Enthalpy of Reaction Data for a Series of "Biological Buffer" Systems at 25°C

Trivial Name	Formula	pK_a	ΔH_R (kJ mol^{-1})
MES	$O\ \ \ N^{+}HCH_2CH_2SO_3^{-}$	6.08	12.68
bis–tris	$(HOCH_2)_3CN^{+}H(CH_2CH_2OH)_2$	6.41	29.25
ACES	$H_2NCOCH_2N^{+}H_2CH_2CH_2SO_3^{-}$	6.65	30.12
ADA	$H_2NCOCH_2N^{+}H\begin{smallmatrix}CH_2COO^{-}\\CH_2COO^{-}\end{smallmatrix}$	6.75	11.51
MOPS	$O\ \ \ N^{+}HCH_2CH_2CH_2SO_3^{-}$	6.76	19.0
PIPES	$NaO_3SCH_2CH_2N\ \ \ N^{+}HCH_2CH_2SO_3^{-}$	6.79	8.70
BES	$(HOCH_2CH_2)_2N^{+}HCH_2CH_2SO_3^{-}$	6.92	23.10
HEPES	$HOCH_2CH_2HN^{+}\ \ \ NCH_2CH_2SO_3^{-}$	7.24	16.40
TES	$(HOCH_2)_3CN^{+}H_2CH_2CH_2SO_3^{-}$	7.34	29.25
Ethyl glycinate	$H_3N^{+}CH_2COOCH_2CH_3$	7.57	46.32
Glycinamide	$H_3N^{+}CH_2CONH_2$	7.73	44.77
HEPPS	$HOCH_2CH_2H^{+}N\ \ \ NCH_2CH_2CH_2SO_3^{-}$	7.82	17.95
Tricine	$(HOCH_2)_3CN^{+}H_2CH_2COO^{-}$	8.00	30.50
THAM	$(HOCH_2)_3CN^{+}H_3$	8.03	47.28
Glycylglycine	$H_3N^{+}CH_2CONHCH_2COO^{-}$	8.21	43.72
Bicine	$(HOCH_2CH_2)_2N^{+}HCH_2COO^{-}$	8.31	26.23
TAPS	$(HOCH_2)_3CN^{+}H_2CH_2CH_2CH_2SO_3^{-}$	8.34	40.12
N,N-Dimethyl-glycine	$(CH_3)_2N^{+}HCH_2COO^{-}$	9.95	31.51
CAPS	$N^{+}H_2CH_2CH_2CH_2SO_3^{-}$	10.35	48.53

Source: Ref. 34, by permission of Marcel Dekker, Inc.

If the primary enzymatic reaction does not produce or consume protons and extra sensitivity is required, the remaining options are to "trap" one of the products by *in situ* derivatization or resort to the incorporation of a secondary enzyme system that will catalyze the consumption of one of the products. This, in turn, will increase the enthalpy change produced per mole of substrate consumed.

7.5. THE DETERMINATION OF ENZYME ACTIVITY

An enzyme is a true catalyst in that its role is to accelerate the rate of a reaction without itself undergoing any permanent change. The final thermodynamic equilibrium position of the reaction is fixed regardless of the nature of the catalyst or indeed in the absence of it. An enzyme, therefore, does not affect the equilibrium position of a reaction, merely the time it takes to achieve that equilibrium.

Because only a very small mass of catalytically active enzyme is typically required to significantly accelerate the rate of reaction of its optimum substrate and the fact that the molar mass of an enzyme is often very large (and in many cases uncertain), the most practical way to characterize an enzyme is via its catalytic ability.

7.5.1. Definition of Enzyme Activity

There are currently two accepted units of enzyme activity. An *International Unit (IU)* is that amount of enzyme that will catalyze the transformation of 1×10^{-6} mol of substrate per minute under defined conditions of pH, temperature, and, if zero-order kinetics do not prevail, substrate concentration.

Recently, an SI equivalent unit is being reported; the *katal* is that amount of enzyme that will catalyze the transformation of 1 mol of substrate per second. Therefore, IU = 16.67 nkatals.

7.5.2. Kinetic Considerations

The fundamental equation describing the rate of most enzyme-catalyzed reactions is

$$v = \frac{V_{\max}[S_0]}{K_m + [S_0]} \tag{7.8}$$

where v is the initial rate of reaction, $V_{\max}$ is the maximum velocity, $[S_0]$ is the initial substrate concentration, and K_m is the Michaelis constant (see Section

7.8). The steady-state derivation of Eq. (7.8) is based on the classical enzyme–substrate intermediate mechanism,

$$E + S \underset{k_{-1}}{\overset{k_{+1}}{\rightleftharpoons}} ES \xrightarrow{k_{+2}} E + P \tag{7.9}$$

for a one-substrate system. It assumes that the rate of formation of ES equals its rate of dissociation *in any direction*. Importantly, the derivation does not impose any restrictions on the relative magnitudes of k_{+2} and k_{-1}. V_{max} is in fact equal to $k_{+2}[E_{tot}]$, where $[E_{tot}]$ is the total concentration of enzyme.

Equation (7.8), although derived from a simple mechanistic model, is valid for many more complex mechanisms. Adaptation to other mechanisms merely requires the incorporation of additional terms into the parameters V_{max} and K_m. The characteristics of the equation are represented graphically in Fig. 7.10. The plot takes the form of a rectangular hyperbola with asymptotes of $[S_0] = -K_m$ and $v = V_{max}$.

For a two-substrate mechanism, the nature of the rate equation will depend on the prevailing mechanism, for example, random order, ping-pong, sequential, etc. However, the equation usually takes a form similar to

$$v = \frac{V_{max}}{1 + \dfrac{K_{m(S_0)_1}}{[S_0]_1} + \dfrac{K_{m(S_0)_2}}{[S_0]_2}} \tag{7.10}$$

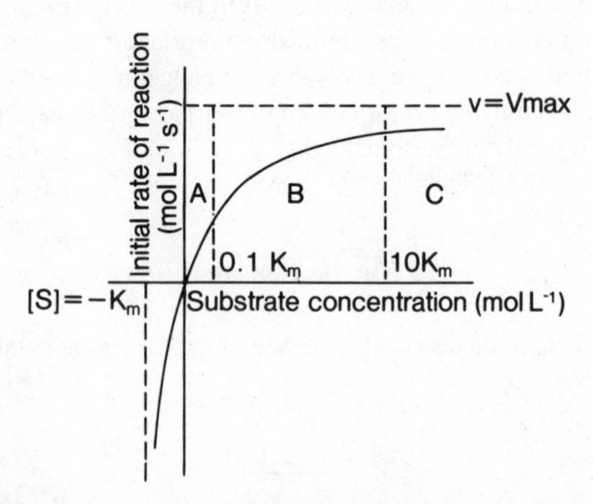

Figure 7.10. Graphical representation of the Michaelis–Menten equation [Eq. (7.8)]. (A) First-order region, $v = V_{max}/K_m$; (B) fractional-order region $v = V_{max}[S_0]/(K_m + [S_0])$; (C) Zero order, $v = V_{max}$.

A detailed discussion of the characteristics of this equation is beyond the scope of this text, and the reader is referred elsewhere for more detail (35). In order to design an experiment for the determination of enzyme activity, the experimental conditions should (ideally) be such that the rate of reaction is limited by the concentration of the enzyme only. The overall rate of reaction at any substrate concentration is actually governed by the breakdown of the ES intermediate to products. Therefore,

$$v = k_{+2} [ES] \tag{7.11}$$

When the substrate concentration is large enough ($>>K_m$), the active sites of the enzyme becomes "saturated" with substrate and the enzyme is functioning at its maximum capacity. As all the enzyme is then in the form of ES, its concentration is rate limiting and the conditions necessary for accurate enzyme activity determination prevail. Under these conditions Eq. (7.11) becomes

$$v = k_{+2}[E_{tot}] = V_{max} \tag{7.12}$$

In general, the "rule of thumb" for the reaction kinetics to be pseudo-zero order with respect to substrate concentration is that $[S_0]$ should be equal to at least $10K_m$ as shown in Fig. 7.10 (see Section 7.5.3a).

Equations (7.11) and (7.12) only apply to a one-substrate system; the kinetics of a two-substrate reaction are obviously more complex. Whenever experimental conditions permit, Eq. (7.10) can be simplified by the use of saturating amounts of both substrates, that is, $[S_0]_1 >> K_{m(S1)}$ and $[S_0]_2 >> K_{m(S2)}$. The denominator then becomes unity and the reaction can be considered pseudo-zero order with respect to each substrate. This often may not be possible because of limited substrate solubility or substrate inhibition at elevated concentrations. In such instances the concentration of the rate-limiting substrate should be specified with the assigned enzyme activity. If K_m is known or can be easily determined (see Section 7.8), V_{max} can then be calculated from Eq. (7.10).

7.5.3. Theoretical and Methodological Principles of Calorimetric Enzyme Activity Determinations

The most common method of determining enzyme activity is by "kinetic direct-injection enthalpimetry," an isoperibol technique which is simply a time-resolved variant of direct-injection enthalpimetry (36).

The essential parameters of an isoperibol calorimetric measurement are given by

$$\Delta T = \frac{\sum n_i \Delta H_i}{C_P} \tag{7.13}$$

where i concurrent reactions are occurring each with an enthalpy of reaction ΔH; n represents the number of moles of product at equilibrium, and C_P is the heat capacity of the calorimeter and its contents. Similarly, Eq. (7.13) can be written in a time-resolved form,

$$\frac{\Delta T}{\Delta t} = \frac{\sum n_i \Delta H_i}{\Delta t \; C_P} \tag{7.14}$$

for a kinetic measurement. For an enzyme-catalyzed reaction, the rate of reaction with pseudo-zero-order kinetics prevailing is given by Eq. (7.12), which can be rewritten in the form

$$v = \frac{d\,[\mathrm{S_0}]}{dt} = \frac{\Delta n_P}{\Delta t} = k_{+2}[\mathrm{E_{tot}}]V \qquad (\mathrm{mol\ s^{-1}}) \tag{7.15}$$

where V is the volume of the solution (L) and n_P is the number of moles of product at time t. The dimensions of k_{+2} and $[\mathrm{E_{tot}}]$ are $\mathrm{s^{-1}}$ and $\mathrm{mol\ L^{-1}}$, respectively. Substitution of Eq. (7.15) into Eq. (7.14) and rearrangment gives

$$\Delta T = \frac{k_{+2}[\mathrm{E_{tot}}]V\Delta H_R}{C_P}\,\Delta t \tag{7.16}$$

Accordingly, a plot of ΔT versus Δt will be linear with a slope equal to $k_{+2}[\mathrm{E_{tot}}]V\Delta H_R/C_P$. The salient features of a calorimetric determination of enzyme activity are conveyed by Eq. (7.16). The sensitivity of the measurement, as determined by the slope of the response per mole of substrate consumed in unit time is directly proportional to the total enthalpy of reaction and inversely proportional to the heat capacity of the system. Substituting Eq. (7.12) into (7.16) we obtain

$$\Delta T = \frac{V_{\mathrm{max}}\Delta H_R V \Delta t}{C_P} \tag{7.17}$$

If V_{max} is expressed in dimensions of $\mathrm{mol\ L^{-1}\ s^{-1}}$, then $V_{\mathrm{max}}V$ is equal to the enzyme activity (EA) in katals ($\mathrm{mol\ s^{-1}}$). Rearrangement gives the working equation of "enzymatic enthalpimetry," that is,

$$EA = \frac{\Delta T}{\Delta t} \frac{C_P}{\Delta H_R} \qquad (\text{mol s}^{-1}) \tag{7.18}$$

In terms of a corresponding heat change Eq. (7.18) can be written

$$EA = \frac{\Delta q_R}{\Delta t} \frac{1}{\Delta H_R} \qquad (\text{mol s}^{-1}) \tag{7.19}$$

Conversion to International Units can be achieved by use of the appropriate factor (Section 7.5.1).

The methodological principles of a kinetic direct-injection experiment for the determination of *EA* are shown schematically in Fig. 7.11. A small volume (0.1–1.0 mL) of substrate solution at a concentration necessary to produce a pseudo-zero-order reaction rate is injected into the preequilibrated reaction cell containing a buffered solution of the enzyme and any co-factors (co-substrates, activators, etc.). The linear temperature–time response is monitored for an appropriate period, typically several minutes. Depending on time constraints, calibration can either be performed after the reaction has gone to completion, during which time the response will asymptotically approach the postreaction baseline, or calibration can be achieved as the reaction proceeds as shown in Fig. 7.11. Following a

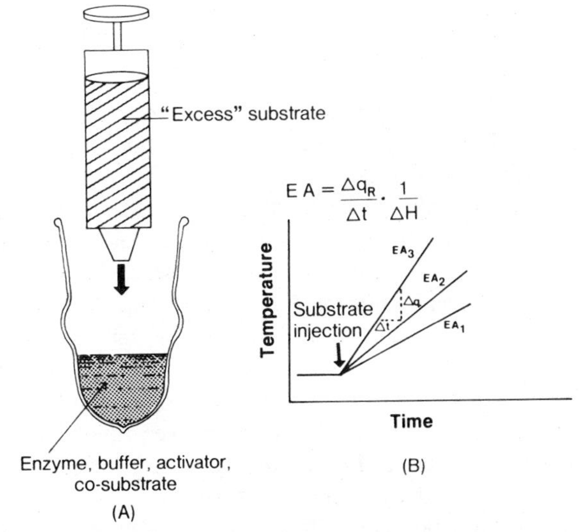

Figure 7.11. The determination of enzyme activity by isoperibol calorimetry: (A) Reagent mixing sequence. (B) Response curve for a series of enzyme activity levels $EA_3 > EA_2 > EA_1$. Initial substrate concentration constant for all experiments. t_c is the activation time for the Joule heating circuitry.

zero offset, the rate of heat change $\Delta q_R/\Delta t$ is calibrated in the usual manner by activation of Joule heating circuitry. The latter procedure demands that zero-order kinetics prevail throughout the measurement and calibration period. The enzyme activity can then be calculated by insertion of $\Delta q_R/\Delta t$ into Eq. (7.19).

Knowledge of the *total* enthalpy of reaction, including attendant buffer reactions, is required for calculation of enzyme activity from Eq. (7.19). If this datum is not tabulated, as is often the case for biochemical reactions, it can be determined in a substrate limiting experiment as described in Section 7.6.

Caloric calibration or the determination of reaction enthalpies can be circumvented by the preparation of a calibration graph of calorimetric response (recorder deflection) versus the activity of a series of enzyme standards. Clearly, *all* experimental conditions must remain constant in order to maintain the integrity of a purely calibrative approach. The most likely interference in the clinical sector is a difference in heat capacity between sample and standard caused by protein content or insolubles. Standard addition procedures can be used to minimize matrix effects. The limit of detection for the determination of enzyme activity by the isoperibol injection approach is about 1 IU, given a maximum specimen volume of 1 mL. The microcell technology described in Chapter 3 is an important advance in the application of calorimetry to clinical analysis since it allows a much smaller sample. A 0.1-mL serum sample could be used in a 2-mL cell. Similarly, sensitivity could be increased by about a factor of 10 if the same 1-mL sample was used in the small cell.

Heat-conduction calorimeters employ a different sample–reagent mixing mechanism. The calorimeter cell consists of two discrete compartments, for example, concentric cylinders (Fig. 7.12A), which initially separate small volumes (1 mL) of the substrate and the buffered enzyme solution. The reaction is initiated by mechanical rotation of the cell through 360° after an initial equilibration period; this type of reagent mixing is usually called "batch addition." Some instruments incorporate a matched reference cell that can be used to monitor a reagent "blank" response, thereby minimizing thermal effects due to dilution and mixing. The measurement principle is also different in that the reaction vessel is placed in a constant-temperature heat sink (typically an aluminum block) between two thermopile systems. The heat flow through the thermopiles dq/dt is then proportional to the temperature difference between the reaction vessel and the heat sink, which in turn is related to the potential difference generated across the thermopiles. The critical difference between the heat-conduction and the isoperibol designs, in terms of data interpretation, is that the heat-conduction device, by definition, produces a derivative signal in contrast to the integral output of an isoperibol calorimeter. The *absolute* value of the heat-conduction instrument output is therefore directly related to the *rate* of heat change and hence to the enzyme activity, that is,

$$\mathbf{E}_c = \varepsilon_c \Delta \frac{q}{\Delta t} \tag{7.20}$$

therefore,

$$\mathbf{E}_c = \varepsilon_c (EA) \Delta H_R \tag{7.21}$$

where **E** is the thermopile voltage and ε is the energy equivalent or calibration constant of the calorimeter as defined in Chapter 3. The energy equivalent is determined to compensate for the inefficiency of the thermopile system and the residual heat not transferred from the reaction vessel.

Calibration is achieved in exactly the same manner as an isoperibol experiment by passage of current through a calibration resistor or by measuring the heat effect associated with a reaction of known enthalpy (see Chapter 3). The shape of the recorder output will depend on the relative concentrations of substrate and enzyme as shown in Fig. 7.12B. A transitory peak, for example, peak EA_3, implies that a steady-state (pseudo-zero-order) condition is not achieved, or if it is, only for a short period. As the enzyme activity decreases, at the same substrate concentration, the zero-order region is elongated. If curve EA_3 did

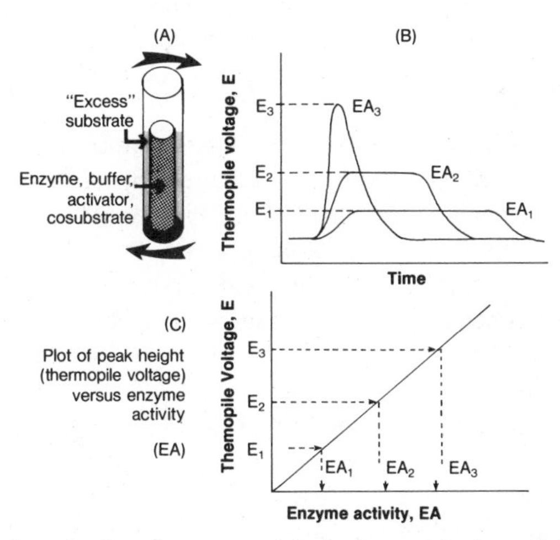

Figure 7.12. The determination of enzyme activity by heat conduction calorimetry: (A) Reagent mixing sequence. (B) Response curves for a series of enzyme activity levels, $EA_3 > EA_2 > EA_1$. (C) Corresponding analytical calibration plot of peak height (peak thermopile voltage) versus enzyme activity, assuming pseudo-zero-order kinetics prevail in each experiment.

pertain to a non-substrate limiting rate of reaction, a plot of peak height pversus enzyme activity (or volume of sample) would be linear, as shown in Fig. 7.12C.

A second factor affecting the peak shape is the response time of the detection system. As discussed in Chapter 3, the response time of a heat-conduction device is longer than a detection system based on a thermistor by at least two orders of magnitude. This characteristic places more emphasis on the achievement of zero-order kinetics for a known period of time. This time period should significantly exceed the response time of the instrument. As the response time for any particular instrument is known, calculations of the type shown in Section 7.5.3a can identify the minimum amount of substrate necessary to maintain a linear rate for the appropriate time period. Heat-conduction calorimeters require extensive correction procedures when used to monitor reactions taking place over an extended period of time. Conventionally, those corrections are made by mathematical functions that compensate primarily for conductive heat loss. Without such corrections, the measured temperature change per unit time is not a true representation of the heat change occurring within the calorimeter reaction vessel. Therefore, calibration standards are essential for accurate analytical data. A simulation method, based on a computer program, has been developed to reconstruct the ideal voltage–time curve which would result if no time lag or heat loss occurred. These curves, known as "thermogenesis curves," have been incorporated into the software of the Prosen–Berger batch microcalorimeter (37). A series of voltage–time curves and the subsequent "thermogenesis curves" for the reduction of pyruvate catalyzed by lactate dehydrogenase (38, 39) are shown in Figs. 7.13A and 7.13B. The authors claim that this procedure increases accuracy and sensitivity (39).

The choice of instrument for enzyme activity determinations will be governed by sensitivity and/or sample throughput requirements. In terms of sensitivity, both heat-conduction and low volume (<2 mL) isoperibol batch instruments will determine mIU levels of activity (assuming ΔH_R >40 kJ mol^{-1}). If sample throughput is the dominant factor, a flow calorimeter is clearly the best choice; in this case, the isoperibol version is the most sensitive. The reader should consult Table 3.2 for sensitivity specifications of the various instrument types.

7.5.3a. Optimization of Substrate Concentration

The optimum amount of substrate used in a direct-injection or batch addition experiment is determined by several criteria, all of which will affect the accuracy and precision of an enzyme activity determination. The most obvious prerequisite is that the rate of reaction is described by Eq. (7.12), that is, the reaction rate is determined by the concentration of enzyme and is independent of substrate concentration. This situation is desirable from two experimental viewpoints:

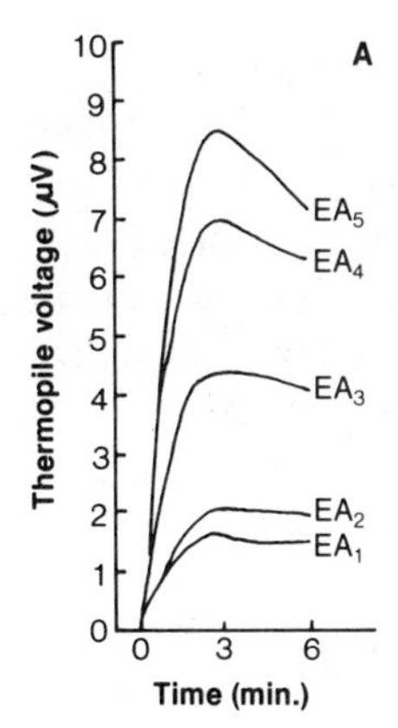

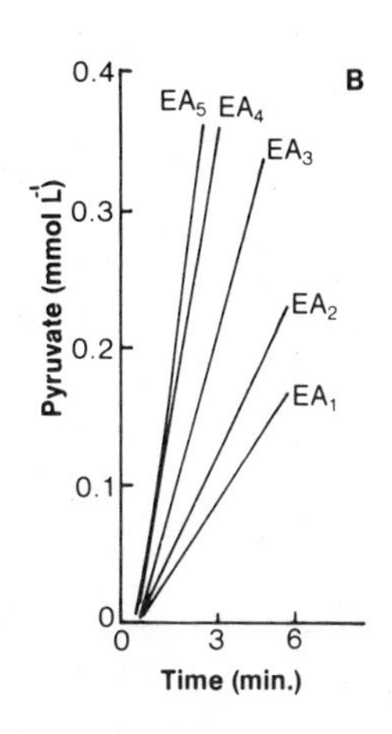

Figure 7.13. (A) A series of voltage–time curves recorded for the lactate-dehydrogenase-catalyzed reduction of pyruvate, $EA_1 < EA_2 < EA_3 < EA_4 < EA_5$. (B) Corresponding thermogenesis curves, software-corrected for non-chemical heat effects. The corrected slope is a direct representation of the rate of pyruvate reduction (from Ref. 38, by permission of the American Association for Clinical Chemistry).

1. It will ensure that small variations in substrate concentration will not affect the rate assignment. Moreover, under conditions where the substrate concentration is non-rate limiting, it need not be accurately measured or specified with the result.
2. The reaction rate will be linear for a significant period of time, reducing extrapolation error and, therefore, increasing precision.

In general, the approximation that the substrate concentration should be greater than $10K_m$ will ensure that the above requirements are met. The amount of substrate required can be determined empirically by measuring the reaction rate at a series of increasing substrate concentrations while keeping the enzyme concentration constant and equal to the maximum activity anticipated in sample analysis. An experimental plot of v versus $[S_0]$, similar to Fig. 7.10, will allow an approximate determination of the value of $[S_0]$ consistent with V_{max}.

Closer inspection of the characteristics of Eq. (7.8) show the effect of substrate concentration on analytical accuracy and precision. An initial substrate concentration of $5K_m$ will result in a reaction rate equal to $5/6 V_{max}$ or $0.83V_{max}$. Doubling the substrate concentration to $10K_m$ will in fact only raise the reaction rate to $10/11 V_{max}$ or $0.91V_{max}$, that is, a 9% increase and still not equal to the absolute value of V_{max}. In practice, of course, attainment of V_{max} is not practicable at realistic substrate concentrations since v approaches V_{max} asymptotically. This characteristic is shown in Table 7.7.

The relatively high cost of biochemical reagents makes the choice of the higher substrate concentration questionable, since, ostensibly, it does not meet the primary kinetic requirement for maximal reaction velocity. However, the effect on analytical precision may be the overriding factor. This can be seen by use of an integrated version of Eq. (7.8), namely,

$$V_{max}t = [S_0] - [S_t] + K_m \ln \frac{[S_0]}{[S_t]} \tag{7.22}$$

where $[S_0]$ is the initial substrate concentration and $[S_t]$ is the substrate concentration at any time t. For a derivation of this equation see Section 7.8.

Whatever the *initial* substrate concentration, its value will decrease in magnitude as the reaction proceeds. For the purposes of calculation, it will be assumed that a 1% increase in substrate concentration during the period of measurement is acceptable. Equation (7.22) will allow the calculation of the time lapse before that 1% decrease occurs or, in other words, the time period in which the reaction time will be effectively constant. A sample calculation comparison between $[S_0] = 5K_m$ and $[S_0] = 10K_m$ is presented below:

TABLE 7.7. Relationship Between Substrate Concentration and Reaction Rate According to Eq. (7.8)

[S]	v
$1000K_m$	$0.999V_{max}$
$100K_m$	$0.99V_{max}$
$10K_m$	$0.91V_{max}$
K_m	$0.5V_{max}$
$0.1K_m$	$0.091V_{max}$
$0.01K_m$	$0.01V_{max}$

$[S_0] = 5K_m$
Initial reaction rate $= 0.833V_{max}$
Reaction rate after 1% decrease in $[S_0] = 0.825V_{max}$

Equation (7.8) can be rearranged to give

$$\frac{V_{max}}{v} - 1 = \frac{K_m}{[S_t]} \tag{7.23}$$

therefore, when $v = 0.825V_{max}$, $[S_t] = 4.71K_m$.
Rearranging Eq. (7.22) gives

$$t = \frac{[S_0]}{V_{max}} - \frac{[S_t]}{V_{max}} + \frac{K_m}{V_{max}} \ln \frac{[S_0]}{[S_t]} \tag{7.24}$$

$$t = \frac{5K_m}{V_{max}} - \frac{4.71K_m}{V_{max}} + \frac{K_m}{V_{max}} \ln \frac{5K_m}{4.71K_m} \tag{7.25}$$

$$t = 0.29 \frac{K_m}{V_{max}} + 0.06 \frac{K_m}{V_{max}} \tag{7.26}$$

or

$$t = 0.35 \frac{K_m}{V_{max}} \tag{7.27}$$

The dimensions of K_m and V_{max} are mol L^{-1} and s^{-1}, respectively. Following the identical procedure for $[S_0] = 10K_m$ gives

$$t = 1.01 \frac{K_m}{V_{max}} \tag{7.28}$$

The conclusion from the calculation is that doubling the substrate concentration will in practice treble the time for which the reaction rate remains effectively linear. Whether or not this is desirable will, of course, depend on the relative magnitudes of K_m and V_{max}. Nonetheless, use of this valuable equation removes the guesswork from the experimental design. As will be seen later, similar calculations can be used to optimize substrate determination experiments (Section 7.6).

An additional constraint on the magnitude of the substrate concentration arises from the design of a kinetic direct-injection enthalpimetry experiment. Typically, the volume of reagent injected is at least 10 times less than the volume of solution in the reaction cell. Consequently, if a substrate concentration of $10K_m$ is required in the cell, its concentration in the syringe may be of the order of $100K_m$. The temperature change associated with the dilution of large concentrations of substrate can lead to spurious reaction rate assignments in an isoperibol instrument where a small, but finite, heat transfer between the cell and its environment occurs. For example, positive (exothermic) deviations from anticipated reaction curve gradients have been observed for substrate injections producing an initial endothermic dilution response (40). This is caused by the cell and its contents striving to reequilibrate with the surrounding thermostat bath after the initial perturbation from thermal equilibrium. In summary, the minimum amount of substrate consistent with the maintenance of analytical precision and accuracy should be used in a calorimetric experiment.

From a practical standpoint, the most common impediment to the balance of the above parameters are limited substrate solubility and "substrate inhibition."

7.5.4. Fixed-Time Kinetic Injection Enthalpimetry

The sensitivity of kinetic injection enthalpimetry can be increased by a factor of 10 by recourse to a classical fixed-time approach in which the unreacted or residual substrate is determined after a fixed incubation period with the sample enzyme solution (41). The analytical signal in this approach is the temperature change associated with the complete reaction of the residual substrate with the injection of an excess of *reagent* enzyme at the end of the precisely timed incubation period. Inevitably, the reagent enzyme and analyte enzyme are identical. The concentration of the reagent enzyme is, however, much larger and should be sufficient to catalyze the consumption of the residual substrate in a short-time period, for example, 5–10 min. The methodological principles of this approach are shown schematically in Fig. 7.14.

The sample enzyme is introduced quantitatively into the reaction cell and diluted to an appropriate volume with buffer. The substrate and reagent enzyme solutions are loaded into separate syringes; only the substrate solution volume must be known precisely. The entire syringe and cell assemblies are then allowed to reach thermal equilibrium, AB. The substrate solution (0.5 mL) is then injected and the stirrer is left activated for several seconds to homogenize the solution. After incubation for about 25 min, the stirrer is reactivated to establish a baseline. Precisely 30 min after the injection of the substrate (C), the reagent enzyme solution is injected and the attendant heat effect q_t is monitored. After the establishment of a suitable postreaction baseline (D), the calibration factor q_{CAL} is determined by Joule heating *in situ*.

The calculation of enzyme activity (*EA*) can be rationalized as follows:

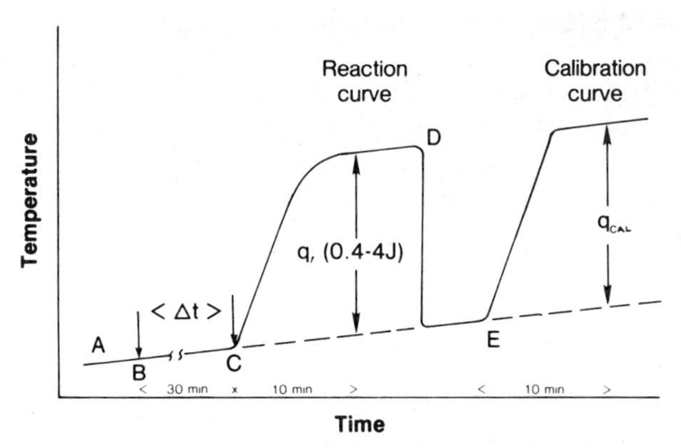

Figure 7.14. Fixed-time kinetic enthalpimetry sequence of events: AB, Isothermal baseline indicates sample and buffer at thermal equilibrium; B, precise amount of substrate injected into sample/buffer solution, stirrer activated for 10 s; BC, incubation period Δt stirrer activated after ca. 25 min., baseline recorded for 4 min; C, reagent enzyme injected; CD, temperature change recorded as residual substrate is consumed; D, zero offset; E, Joule heating calibration sequence intiated (from Ref. 41, by permission of Elsevier Science Publishers).

$$EA = \frac{\Delta n}{\Delta t} = \frac{(n_0 - n_t)}{\Delta t} \tag{7.29}$$

where n_0 and n_t are the total (initial) and residual number of moles of substrate respectively, and Δt is the incubation period. Also,

$$n_t = \frac{q_t}{\Delta H_R} \tag{7.3}$$

where ΔH_R is the overall enthalpy of reaction associated with the reagent enzyme–substrate reaction. Therefore, substituting Eq. (7.30) into Eq. (7.29) we obtain

$$EA = \frac{[n_0 - (q_t/\Delta H_R)]}{\Delta \text{t}} \tag{7.31}$$

Equation (7.31) represents the working equation of fixed-time kinetic injection enthalpimetry. n_0 is known, q_t and Δt can be determined from Fig. 7.14 and ΔH_R can be determined in a separate experiment (see Section 7.6).

The increase in sensitivity obtained versus the direct-injection (derivative) approach can best be illustrated by considering the determination of an enzyme activity of 0.1 IU by both the fixed-time and derivative methods. Substitution of 0.1 IU and a typical enthalpy change of 40 kJ mol^{-1} into Eq. (7.19) gives a gradient, dq/dt, of 8×10^{-5} J s^{-1} or ca. 5 mJ min^{-1}. Over a typical measurement period for a derivative procedure, say 5 min, a total heat change of 25 mJ would be generated, which is below the detection limit for conventional isoperibol calorimeters with reaction cell volumes greater than 10 mL.

Now we can consider the fixed-time approach to the same determination of a 0.1 IU enzyme sample. If pseudo-zero-order kinetics prevail, 3.0×10^{-6} mol of substrate would be consumed in a 30-min incubation period at this level of enzyme activity. If this substrate consumption represented a minimum of 20% decrease in the initial amount of substrate to maintain analytical precision, n_0 and n_t would have values of 1.5×10^{-5} mol and 1.2×10^{-5} mol, respectively. Using the same enthalpy change of 40 kJ mol^{-1}, the consumption of n_t catalyzed by the injection of reagent enzyme would generate an enthalpy change of 480 mJ, that is, well within the detection limit of isoperibol equipment. This increase in sensitivity is obtained without a significant decrease in sample throughput, since the thermal equilibration procedures, which dominate the lapsed experiment time, are incorporated into the incubation period. The optimum amount of substrate chosen for the determination is governed by three factors:

1. The residual amount of substrate n_t must be large enough to determine enthalpimetrically.
2. Δn or $n_0 - n_t$ should be large enough to ensure that it does not approach the imprecision of the determination, that is, it should be at least twice the standard deviation of the method.
3. Δn should be small enough that pseudo-zero-order kinetics prevail throughout the incubation period.

The latter two limitations are clearly conflicting and require a compromise. The error in allowing a 20% decrease in substrate concentration can be calculated using the integrated rate equation as illustrated in Section 7.5.3.

The fixed-time approach to kinetic injection enthalpimetry has the disadvantages of any two-point kinetic method, regardless of the detection method. Nonlinearity of reaction rate, initial lag phases, and non-enzymatic reactions will, if undetected, cause significant errors. Moreover, the calculation of enzyme activity by this approach involves the subtraction of two relatively large quantities, n_0 and n_t. This arithmetic operation significantly affects the precision of the data if n_0 is not adjusted commensurate with the level of enzyme activity. The reason for this adjustment is that a small decrease in enzyme activity integrated over a 30-min incubation period will have a deleterious effect on the signal-to-noise ratio if n_0 remains constant, because under these conditions n_0 approaches n_t.

Fixed-time kinetic enthalpimetry is, in practice, limited to the determination of small levels of enzyme activity (0.06–1 IU). At activities above 1 IU the amount of substrate required to maintain pseudo-zero-order kinetics over extended incubation periods becomes excessive. Moreover, the procedure has no inherent advantage over the derivative approach at this level of sample activity.

7.5.5. Selected Applications

Both variants of cholinesterase (ChE)—serum or butyrylcholinesterase and erythrocyte or acetycholinesterase—have been the subject of several calorimetric studies. The generalized reaction catalyzed by these enzymes is represented by

$$RC(=O)—O(CH_2)_2N^+(CH_3)_3X^- + H_2O \overset{ChE}{=} R—C(=O)—OH + HO(CH_2)_2N^+(CH_3)_3X^- \tag{7.32}$$

where R = CH_3 for acetylcholine and C_3H_7 for butyrylcholine, the optimum substrates for acetyl ChE and butyrylChE, respectively. X^- is typically chloride or bromide ion. The activity of both enzymes is maximized in the pH range 7–8. Accordingly, the reaction is usually monitored in the presence of THAM buffer. Indeed, the concomitant protonation of THAM represents the dominant calorimetric signal for either hydrolysis. The enthalpy of enzymatic hydrolysis per se is small and endothermic. Sturtevant (42) has reported the hydrolysis of acetylcholine by acetylChE to have an enthalpy of + 1.17 kJ mol^{-1}. In a separate study the enthalpy of reaction for the hydrolysis of butyrylcholine has been reported as $+1.6 \pm 0.5$ kJ mol^{-1} (40). The incorporation of the protonation enthalpy into the calorimetric determination of ChE activity is therefore a classic example of "buffer amplification" producing a 30-fold increase in the enthalpy change per mole of substrate examined. The physiological significance of both enzymes is that the normal level (2–5 IU mL^{-1}) in serum is considerably diminished (up to 70%) by malfunctions of the liver and in cases of organophosphate pesticide poisoning. The additive effect of the protonation enthalpy is therefore critical to the success of the calorimetric assay of relatively small abnormal levels of ChE activity.

Precision data for the determination of serum (butyryl) ChE by isoperibol calorimetry have been compared with an established spectrophotometric procedure (40) in which the residual substrate is converted to its hydroxamic acid derivative and complexed with Fe (III) to form a chromophore (43). As shown in Table 7.8, the correlation is favorable. Some of the problems associated with the introduction of serum into a calorimetric experiment have been detailed in a report concerned with the determination of acetylChE by heat conduction calorimetry (44). The principal effects are shown in Fig. 7.15, which represents the signals observed when different volumes of serum, diluted to 1 mL with THAM buffer, are mixed with 1 mL of buffer. The most obvious feature is that the dilution of serum per se results in a significant endothermic response at serum volumes greater than 0.6 mL. The authors also report an exothermic response for the mixing of equal volumes of buffer as is evident from the recorder trace labeled 0 in Fig. 7.15. This response is difficult to rationalize as the concentration of the buffer in both cells is the same in this particular experiment. Presumably the signal can be attributed to an instrumental artifact. Additionally, the overall enthalpy change observed over the reaction period for the enzyme assay, ca. 12 min, contains a contribution from the non-enzymatic hydrolysis of acetylcholine. All of these effects, which would constitute a complex error propagation in the analysis, can be effectively removed by subtraction of the signal from a reference cell in which the appropriate volume of serum, diluted with buffer, is added to buffer alone. However, as the authors point out, a compromise between the

TABLE 7.8. Comparison of Enthalpimetric and Colorimetric Data for the Determination of Cholinesterase Activity in Aqueous Solutions and Reference Sera

Sample	Activity ($IU\ mL^{-1}$) Colorimetry[a]	Precision[b] % rsd	Activity ($IU\ mL^{-1}$) Enthalpimetry	Precision[b] % rsd	Error[c]
Aqueous Solutions					
ca. 3 $IU\ mL^{-1}$	3.34	1.6 (7)	3.30	2.5 (6)	−1.2
ca. 4 $IU\ mL^{-1}$	4.07	2.2 (7)	4.02	1.5 (6)	−1.2
ca. 5 $IU\ mL^{-1}$	5.61	1.2 (6)	5.57	1.9 (6)	−0.7
Welcomtrol II reference sera	3.80	2.3 (15)	3.84	1.4 (9)	+1.0

Source: Ref. 40, by permission of Elsevier Science Publishers.

[a]Based on the method in Ref. 43.

[b]Number of experiments in parentheses.

[c]Difference between enthalpimetric and colorimetric data expressed as percentage.

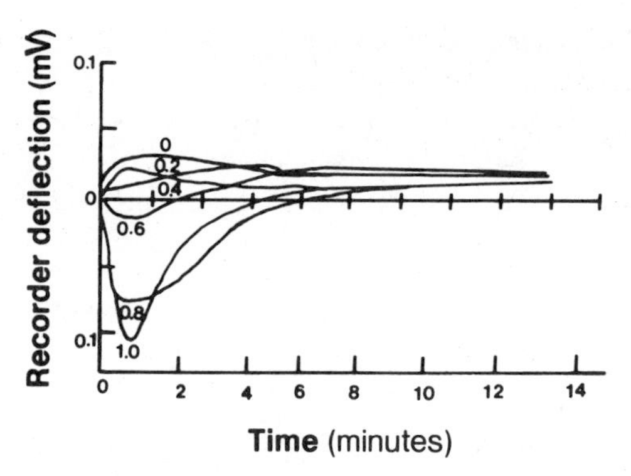

Figure 7.15. Heat-conduction microcalorimeter response curves for the reaction of serum with acetylcholine: 0–0.8 mL of pooled serum diluted to 1.0 mL with buffer, then mixed with 1 mL acetylcholine. (from Ref. 44, by permission of the American Association for Clinical Chemistry).

length of thermal equilibration periods and sample throughput exists. Thermal equilibration periods can be decreased and efficiency increased by leaving the reference cell empty and precalibrating the spurious thermal effects. Data from these preliminary experiments can be used for subsequent subtraction data or, even better, for establishing the experimental conditions in which interference effects are minimized. In the report cited, 0.4 mL of serum was identified as the sample volume producing minimal thermal interference.

The problems mentioned here illustrate the pitfalls for the inexperienced in the application of calorimetry to complex solution analysis. Such effects are commonplace; however, they can be easily circumvented or corrected by well-designed experiments following exhaustive preliminary "blank" response tests. The need for this type of correction procedure is perhaps the primary motivation for the relatively recent introduction of computerized data collection systems into research calorimeters. In another ChE study the correlation between spectrophotometric and calorimetric procedures for the determination of ChE activity was determined (45). The salient result in this report was that spectrophotometric assay with acetylthiocholine (which utilized a secondary reaction with the chromagen, dithiobisnitrobenzoic acid) did not require a secondary reagent either with the same substrate, $r = 0.976$, or with acetylcholine, when r was determined to be 0.900. In this instance, the often quoted convenience of calorimetrically monitoring the primary enzymatic reaction is clearly an advantage in circumventing interference.

For the most part, a relatively large enthalpy of protonation/deprotonation is a desirable feature of a buffer system in the design of a calorimetric enzyme assay if the buffer reaction enthalpy is the dominant heat effect or if its sign is the same as the enzymatic reaction enthalpy. However, if the enthalpies of the enzyme-catalyzed reaction and the buffer reaction are opposite and their magnitude approximately equal, very low sensitivity will obtained. Such is the case when THAM is used to buffer the lactate-dehydrogenase- (LDH) catalyzed reduction of pyruvate (39). The relevant reactions are represented by

$$\Delta H_1 \qquad \text{pyruvate} + \text{NADH} + \text{H}^+ \overset{\text{LDH}}{=} \text{lactate} + \text{NAD}^+ \tag{7.33}$$

$$\Delta H_2 \qquad \text{buffer} - \text{H}^+ = \text{buffer} + \text{H}^+ \tag{7.34}$$

$$\Delta H_R = \Delta H_1 + \Delta H_2 \text{pyruvate} + \text{NADH} + \text{buffer} - \text{H}^+ \overset{\text{LDH}}{=} \text{lactate} + \text{NAD}^+ + \text{buffer} \tag{7.35}$$

The reduction of pyruvate in the presence of nicotinamide adenine dinucleotide (reduced form) involves the deprotonation of the buffer. The overall enthalpy change ΔH_R will be determined by the addition of the reduction and deprotonation enthalpies, that is, $\Delta H_1 + \Delta H_2$. The effect of the choice of buffer on the magnitude of the enthalpy change per mole of pyruvate consumed is shown by reference to Table 7.9. A sensitivity increase of approximately threefold is effected for the determination of LDH activity by the use of phosphate buffer which has a smaller subtractive endothermic effect on the exothermic reduction reaction.

Tian–Calvet heat-conduction calorimetry has been used to determine serum aldolase activity via the measurement of fructose phosphate degradation (46). Significant reagent dilution enthalpy correction was required to achieve a response directly proportional to aldolase activity. In a direct clinical application of this

TABLE 7.9. Enthalpies of Reaction for the LDH-Catalyzed Reduction of Pyruvate and the Deprotonation of THAM and Phosphate Buffers

Buffer	Overall Reaction Enthalpy, ΔH_R (kJ mol^{-1})	Deprotonation Enthalpy, ΔH_2 (kJ mol^{-1})	Pyruvate Reduction Enthalpy, ΔH_1 (kJ mol^{-1})
Phosphate	−47.32	+14.77	−62.09
THAM	−15.35	+47.07	−62.42

Source: Ref. 39, by permission of the American Association for Clinical Chemistry.

procedure (47), the ratio of serum aldolase activity toward fructose-1,6-diphosphate and fructose-1-monophosphate was correlated with the incidence of carcinomas and the method proposed as a routine carcinoma screening procedure.

The enzyme peroxidase has no physiological significance per se. Nonetheless, it is an important enzyme analytically in that it can be used as a coupling reagent to consume hydrogen peroxide produced in a primary enzymatic reaction, for example, reactions catalyzed by glucose oxidase, amino acid oxidase, cholesterol oxidase, etc. The mechanism of peroxidase activity involves an electron donor (AH_2) in addition to the principal substrate, hydrogen peroxide (S) as shown in

$$E + S \underset{k_2}{\overset{k_1}{\rightleftharpoons}} ES \tag{7.36}$$

$$ES + AH_2 \xrightarrow{k_4} E + P \tag{7.37}$$

One of the problems associated with the determination of peroxidase activity is the attainment of pseudo-zero-order kinetics. It has been shown (48) that if the hydrogen peroxide concentration $[H_2O_2]$ is such that $k_1[H_2O_2] >> k_4[AH_2]$, where $[AH_2]$ is the donor concentration, then the rate of reaction can be represented by $k_1 = k_4[AH_2]$ and the kinetics are zero order if $[H_2O_2] << [AH_2]$. This situation is complicated by the fact that peroxidase activity is inhibited by elevated peroxide concentrations. The choice of the electron donor is therefore a complex issue, since the concurrent fulfillment of the conditions $k_1[H_2O_2] >> k_4[AH_2]$ and $[H_2O_2] << [AH_2]$ implies that $k_1 >> k_4$. The number of substrates which meet these requirements *and* produce an analytically measurable species, for example, a chromophore, are limited. In practice, most of the reported procedures do not achieve zero-order conditions and require a statement of substrate concentration in the final result.

The choice of a secondary substrate can be widenened and simplified by the use of calorimetric detection, thereby avoiding the use of unnecessarily complex molecules selected primarily on the basis of unsaturation and hence chromogenic activity. In particular, the oxidation of iodine ion

$$2H^+ + H_2O_2 + 2I^- \rightarrow I_2 + 2H_2O \tag{7.38}$$

has been utilized to determine horseradish peroxidase activity in the range 2–80 IU (49). The overall enthalpy of this reaction at pH 3.9 (sodium acetate/acetic acid buffer) was determined to be -123.7 ± 0.7 kJ mol^{-1}. The concentration of both substrates was optimized to minimize the effect of small variations in

TABLE 7.10. Selected Applications—Calorimetric Determination of Enzyme Activity

Enzyme	Substrate	Matrix	Type of Calorimetry	Reference
Adenosine triphosphatase	Adenosine triphosphate	Fibroplast tissue	Heat conduction	51
Aldolase	Fructose phosphate	Serum	Heat conduction	46
Aldolase	Fructose 1,6-diphosphate/fructose-1-monophosphate	Serum	Heat conduction	47
Cholinesterase	Acetylcholine	Serum	Heat conduction	44
Cholinesterase	Acetylcholine/acetylthiocholine	Water	Heat conduction	45
Cholinesterase	Butyrylcholine	Serum	Isoperibol	40
Cholinesterase	Butyrylcholine	Serum	Isoperibol (fixed time)	41
Cyclic nucleotide phosphodiesterase	Cyclic adenosine monophosphate	Water	Heat conduction	52
Hexokinase	Mg^{2+}/adenosine triphosphate/glucose	Water	Isoperibol	53
Lactate dehydrogenase	Pyruvate/reduced nicotine adenosine dinucleotide	Human epidermis	Heat conduction	50
Lactate dehydrogenase	Pyruvate/reduced nicotine adenosine dinucleotide	Water	Heat conduction	38
Peroxidase	Hydrogen peroxide/iodine ion	Water	Isoperibol	49
Phosphonate esterase	4-Nitrophenyl phenylphosphonate	Water	Heat conduction	54
Proteinases trypsin, pepsin, pronase	p-Tosyl-L-arginine methyl ester, casein, hemoglobin	Water	Heat conduction	55

concentration on reaction rate. Zero-order conditions were not achieved with respect to either substrate.

The tolerance of calorimetric techniques to the opacity of the sample matrix opens up the possibility of the direct determination of enzyme activity in untreated tissue specimens. For example, sections of human epidermis stretched over a small frame with gauze windows can be introduced into a calorimetric cell (50). The problems of tissue damage and the effect this may have on the interpretation of *in vitro* data (as with tissue homogenates) is avoided with this procedure. This principle has been validated with the determination of lactate dehydrogenase activity in human epidermis. The correlation of reaction enthalpy with a pure enzymatic reaction in homogeneous solution is quite remarkable. A value of -46.9 kJ mol^{-1} was obtained for the tissue reaction, which compares with -47.2 kJ mol^{-1} determined for the reaction in homogeneous solution. Twelve-milligram samples of tissue exhibited 22.1 ± 2.3 mIU of LDH activity per milligram wet weight of epidermis. Heat-conduction instruments are required to achieve this level of sensitivity. A list of selected applications of calorimetry to the determination of enzyme activity is summarized in Table 7.10.

7.6. THE DETERMINATION OF SUBSTRATES

There are two methodological approaches to the determination of substrate concentration by use of an enzyme-catalyzed reaction, namely, the kinetic (or initial-slope) procedure and the thermodynamic (or equilibrium) method.

7.6.1. Kinetic (Initial-Slope) Approach

The basic principles of the kinetic approach to substrate determination are evident from further inspection of Eq. (7.8). If $[S_0] << K_m$, then Eq. (7.8) becomes

$$v = \frac{V_{max}}{K_m}[S_0] \tag{7.39}$$

that is, the *initial* reaction rate shows a first-order dependence on the *initial* substrate concentration $[S_0]$, *at constant enzyme concentration*. Under these conditions, V_{max}/K_m becomes a constant analogous to a pseudo-first-order rate constant with the dimensions of time^{-1} if V_{max} and K_m are given the units of mol L^{-1} s^{-1} and mol L^{-1}, respectively.

The reaction kinetics can therefore be expressed in the form of a conventional pseudo-first-order rate equation, that is,

$$n_{P,t} = [S_0]V\left[1 - exp\left(\frac{-V_{max}\,t}{K_m}\right)\right] \tag{7.40}$$

where $n_{P,t}$ is the number of moles of product formed at time t and V is the volume of solution. The heat change associated with the reaction is, therefore,

$$q_t = [S_0]V\Delta H_R\left[1 - \exp\left(\frac{-V_{max}\,t}{K_m}\right)\right] \tag{7.41}$$

Since $\Delta T = q_t/C_P$, the temperature change at time t is

$$\Delta T = \frac{\Delta H_R[S_0]\,V}{C_P}\left[1 - \exp\left(\frac{-V_{max}\,t}{K_m}\right)\right] \tag{7.42}$$

If an initial rate measurement is made, that is, $t \rightarrow 0$, then Eq. (7.42) simplifies to

$$\Delta T = \frac{\Delta H_R[S_0]\,V}{C_P}\,\frac{V_{max}\,t}{K_m} \tag{7.43}$$

Equation (7.43) embodies the principles of an initial-slope technique for a calorimetric experiment. A plot of ΔT versus time will be linear with a slope determined by $[S_0]$. Clearly, this is only valid if the enzyme concentration is invariant so that V_{max} becomes a constant. By definition, the rate measurement must be made as a tangential extrapolation of the *initial* rate as shown in Fig. 7.16A. The resultant calibration graph, Fig. 7.16B, obtained for a series of substrate standards whose concentration brackets the anticipated sample concentration, is then used to determine the concentration of substrate in the sample. The term V_{max}/K_m in fact will determine the range of substrate concentrations amenable to this procedure, since its magnitude controls the slope in Fig. 7.16A. As a "rule of thumb," the first-order form of Eq. (7.8) begins to dominate as $[S_0]$ becomes less than $0.1K_m$. If $[S_0] > 0.1K_m$, adherence to Eq. (7.39) can be achieved by dilution of the sample. If, on the other hand, $[S_0] < 0.1K_m$, such that the rate is too slow to be precisely measured, v can be increased by addition of more enzyme which in turn will increase the magnitude of V_{max}.

The initial response delays in heat-conduction instruments, which can be of the order of several minutes, precludes the application of initial-slope extrapolations for this type of instrument unless "thermogenesis curves" are generated to compensate for such effects as described in Section 7.5.3.

The design of the isoperibol experiment in which one reagent is injected into a solution of the other contained in the calorimeter cell introduces an unavoidable

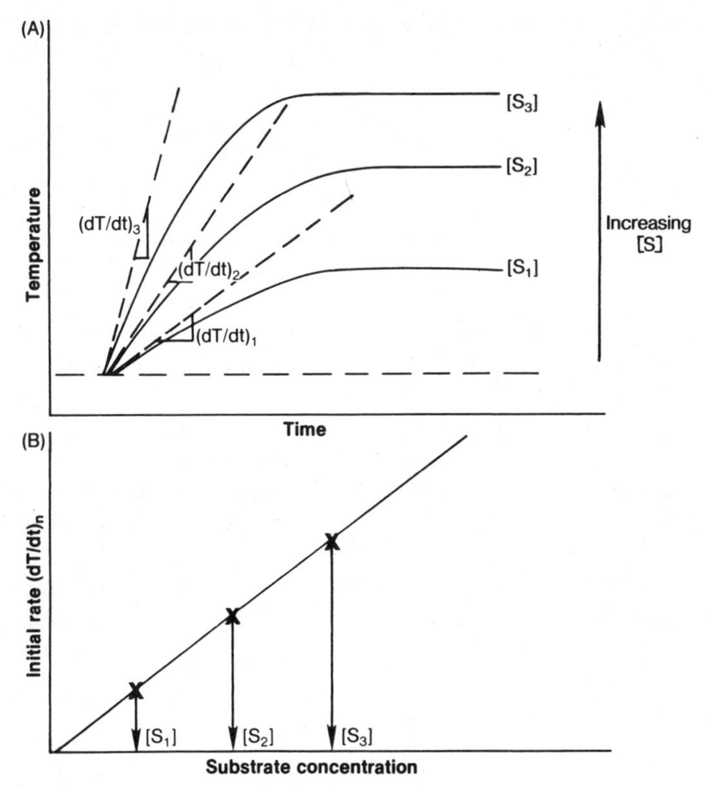

Figure 7.16. Determination of substrate concentration by the initial rate extrapolation procedure (isoperibol calorimeter): (A) extrapolation procedure, $[S_3]>[S_2]>[S_1]$; (B) corresponding analytical calibration graph.

imprecision into the determination. When first-order kinetics prevail, the reaction rate will be linearly dependent on both the substrate and enzyme concentration, the latter determining the value of V_{max} in Eq. (7.39). The variation in enzyme concentration caused by irreproducibility of the volume introduced into the cell by the syringe will be the dominant source of experimental imprecision even with gravimetrically calibrated syringes. This source of error can be eliminated by the use of the equilibrium approach to substrate determination.

7.6.2. Equilibrium ("End-Point") Approach

The equilibrium approach to the determination of substrate concentration is based on the measurement of the integral temperature change ΔT associated with the complete conversion of the substrate to products. Adopting the terminology of biochemical reactions and including an analytical solution volume term V, Eq. (7.13) can be rewritten,

$$[S_0] = \frac{\Delta T C_P}{\Delta H_R V} = \frac{q_{tot}}{\Delta H_R V} \tag{7.44}$$

for a 1:1 reaction going to completion. In an isoperibol experiment, an "excess" of enzyme, that is, an amount sufficient to bring about the complete consumption of the substrate in an acceptable time period (~10–20 min) is injected into a buffered solution of the substrate and the ensuing temperature change is monitored. Since the amount of enzyme added will affect only the rate of the reaction and not its final thermodynamic equilibrium position, precise knowledge of its activity or solution volume is not required. Indirectly, the activity of the enzyme is important since it will affect the shape of the enthalpogram which, in turn, has a small effect on the precision of the measurement of ΔT. However, small changes due to syringe delivery volume will not be a significant factor in this context. As a result, the precision of the isoperibol injection procedure for substrate determination approaches that of any direct-injection procedure for an instantaneous reaction. A typical series of isoperibol enthalpograms obtained for

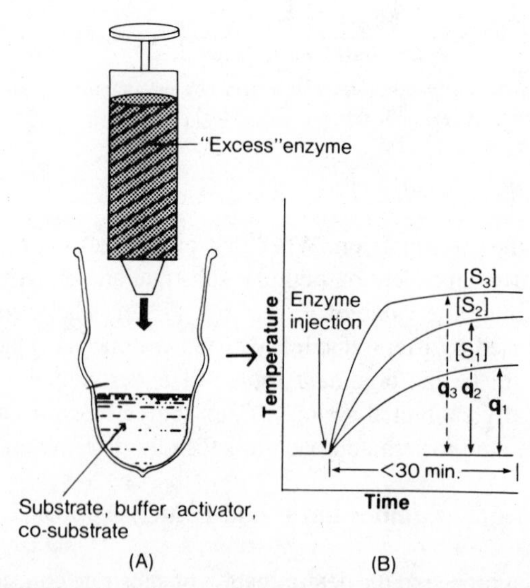

Figure 7.17. Determination of substrate concentration by the "equilibrium approach" (isoperibol calorimeter): (A) reagent mixing sequence; (B) response curves in a series of substrate determinations, $[S_3]>[S_2]>[S_1]$.

a series of substrate concentrations are shown in Fig. 7.17; q_n is the analytical signal.

The amount of enzyme needed to bring the reaction to completion can be estimated by using the stated V_{max} of the reagent enzyme as a guide. A more accurate calculation can be performed by using Eq. (7.22). Substitution of the desired percentage completion, for example, 99.9%, will determine $[S_t]$, and t is the required completion time. With prior knowledge of K_m, Eq. (7.22) can be solved for V_{max} which will have the dimensions mol L^{-1} time $^{-1}$ and will represent the minimum amount of enzyme activity required for the experiment. Calculation of $[S_0]$ from Eq. (7.44) requires that the enthalpy of reaction ΔH_R be known. If this is not tabulated, a substrate-limiting experiment must be performed using pure substrate standards so that Eq. (7.44) can be solved for ΔH_R. Calibration can be achieved by electrical heating in the usual manner or by the use of chemical reactions of known enthalpy, for example, THAM protonation. Enthalpy determinations can be avoided by the use of substrate calibration standards as shown in Fig. 7.16B.

The direct response from a heat-conduction experiment for the determination of substrate concentration would be similar to that shown in Fig. 7.18B. The total heat effect q_{tot}, can be obtained from the integrated response of the calorimeter,

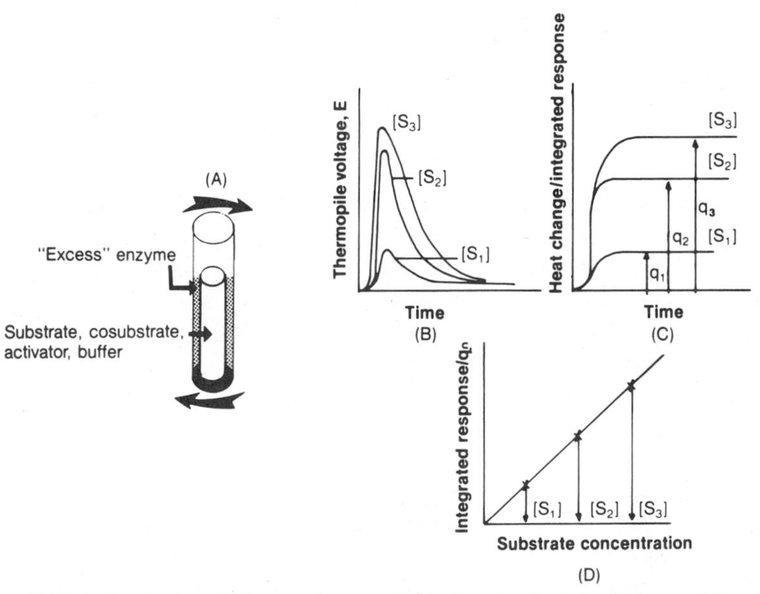

Figure 7.18. Determination of substrate concentration by heat-conduction calorimetry: (A) reagent mixing sequence; (B) direct response curves for a series of substrate concentrations, $[S_3]>[S_2]>[S_1]$; (C) corresponding "thermogenesis curves" (from Ref. 39, by permission of the American Association for Clinical Chemistry); (D) analytical calibration graph.

$$q_{tot} = \varepsilon_C A_{tot} \quad (7.45)$$

where ε_C is the calibrated energy equivalent of the calorimeter (energy/area) and A_{tot} is the area underneath the response curve. The substrate concentration can then be calculated in the usual manner, that is, $[S] = q_{tot}/\Delta H_R V$.

"Thermogenesis curves" can also be generated to give a direct reading of heat change (39) as shown in Fig. 7.18C. Both these data interpretation techniques can be used to construct an analytical calibration plot as shown in Fig. 7.18D.

7.6.3. Selected Applications

The thermochemistry of the hexokinase-catalyzed phosphorylation of glucose by adenosine triphosphate (ATP) has been studied extensively by both isoperibol and heat-conduction calorimetry, and methods have subsequently been developed for the calorimetric determination of glucose in physiological fluids (56–59). The reaction studied can be represented by

$$\text{glucose} + \text{Mg(ATP)}^{2-} \xrightarrow{\text{hexokinase}} \text{glucose-6-phosphate} + \text{Mg(ADP)}^{-} + \text{H}^{+}$$

The presence of magnesium ions is vital to the activation of this reaction and to the interpretation of thermochemical data. The latter point is illustrated by the apparent discrepancy of enthalpy assignments for the reaction represented by Eq. (7.46) in two of the pioneering studies of this reaction. Goldberg et al. had assigned a total reaction enthalpy (including protonation of THAM buffer) of -61.4 kJ mol^{-1} to this reaction in a report describing a heat-conduction calorimetric determination of glucose in human serum (56). In a later report, McGlothlin and Jordan made an enthalpy assignment of -74.89 kJ mol^{-1} ostensibly for the same reaction (57) by use of isoperibol data. Closer inspection of the experimental details of each procedure allows a rationalization of this apparent discrepancy. The phosphorylation of glucose in the absence of magnesium ions,

$$\text{glucose} + \text{ATP}^{4-} \rightleftharpoons \text{glucose}-6\text{-phosphate} + \text{ADP}^{3-} + \text{H}^{+} \quad (7.47)$$

is thermodynamically as favorable as reaction (7.46); however, it is much slower without magnesium ion "activation." Accordingly, McGlothlin and Jordan's procedure called for a substantial stoichiometric excess of Mg^{2+} with respect to ATP^{4-} to ensure that the reaction represented by Eq. (7.46) prevailed to the virtual exclusion of all other equilibria. The Goldberg et al. procedure, on the other hand, detailed a substoichiometric amount of magnesium ions. Under these conditions, other equilibria become important, namely,

$$Mg(ATP)^{2-} \rightleftharpoons Mg^{2+} + ATP^{4-} \qquad (7.48)$$

and

$$Mg^{2+} + ADP^{3-} \rightleftharpoons Mg(ADP)^{-} \qquad (7.49)$$

Evidently, the summation of Eqs. (7.48) and (7.49) leads to Eq. (7.46) with regard to $Mg^{2+}/ATP^{4-}/ADP^{3-}$ chemistry. When these additional equilibria are taken into account and the enthalpy changes associated with them included in the enthalpy assignment for Eq. (7.46), the discrepancy between the two enthalpy values can be accounted for (57). Goldberg has since discussed in detail the effect of magnesium ion concentration on the thermodynamics of hexokinase-catalyzed reactions (60). The major drawback to the acceptance of calorimetric procedures in the routine clinical analysis laboratory is the limited number of determinations which can be performed per hour, typically two or three.

For substrate determinations by isoperibol calorimetry, sample throughput can be increased by reversing the normal injection sequence, that is, by injecting substrate into enzyme. In this way successive solutions of substrate can be injected into the same buffered enzyme solution after the preceding reaction has come to chemical and thermal equilibrium. The procedure was adopted in an extension of McGlothlin and Jordan's work for the determination of glucose in serum, plasma, and, even, whole blood (59). Up to five 500-μL samples of untreated serum or plasma (three for whole blood) can be injected before the reagent solution is changed. The compromise is a decrease in precision, which as mentioned in Section 7.6.1, will be determined by the reproducibility of the syringe. However, this is not likely to be prohibitive; the coefficient of variation (rsd) for a data set over a 30-fold concentration range of glucose in sera has been reported as ±2% (59).

The determination of glucose in untreated whole blood represents a striking illustration of the versatility of calorimetric detection systems. The authors claim that glucose data could be obtained on a whole blood specimen within 5 min of sampling, suggesting that the method would be applicable for emergency use. Recovery data for this determination are shown in Table 7.11. An important feature of these data is that a protein removal procedure based on precipitation with barium hydroxide and zinc sulfate was required for all procedures except calorimetry.

TABLE 7.11. Determination and Recovery of Glucose in Human Whole Blood

Original Glucose	Glucose Added in "Spike"	Theoretical Glucose Concentration	Glucose Found		Percentage Recovery
			Glucose Oxidase[a]	DIE	
890[b]	—	890[b]	920[b]	900[b]	—
				870	—
				910	—
890	1480	2370	2380	2320	97.9
				2330	98.2
890	2870	3760	3730	3790	100.8
				3780	100.5

Source: Ref. 59, by permission of the American Association for Clinical Chemistry.
[a]According to a standard colorimetric procedure.
[b]Glucose concentration in mg L^{-1}.

The most commonly quoted chemical methods of penicillin **2**

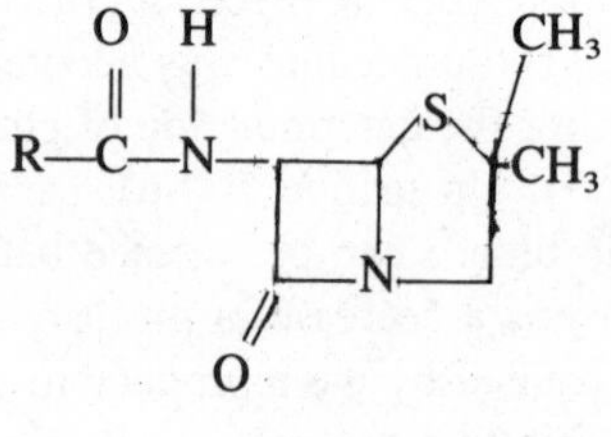

2 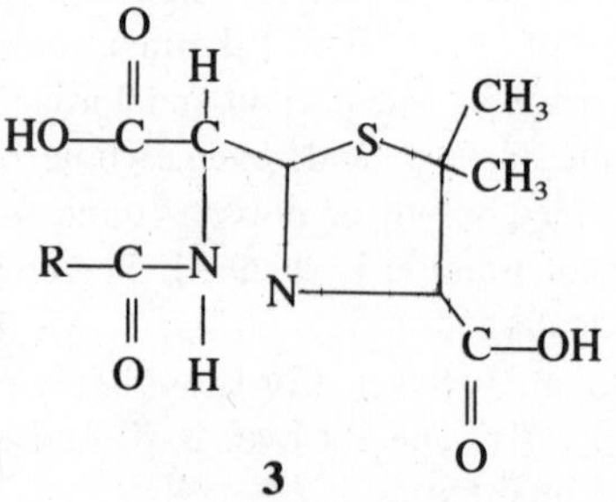 **3**

assay are based on hydrolysis of the B-lactam ring to form penicilloic acid **3** and subsequent oxidation of the sulfur functionality with iodine. Analytical data are obtained by back-titration of the iodine with thiosulfate. The fundamental problem with this determination is that it has a vague stoichiometry since the oxidation step can produce the sulfoxide or the sulfone. The hydrolysis step is however stoichiometric and can be achieved with alkali or enzymatically by penicillinase catalysis. Calorimetrically, therefore, the troublesome oxidation can be avoided, resulting in a one-step procedure (61). The enzymatic hydrolysis of penicillin G per se was determined to have a molar enthalpy of reaction of

-67.0 kJ mol^{-1}. When coupled to the protonation of THAM, the overall reaction enthalpy is -114.7 kJ mol^{-1}; micromole quantities of penicillin can therefore be determined by this procedure with acceptable precision.

Sample throughput was increased by successive injection of 500-μL aliquots of penicillin sample solutions into 200 IU of buffered penicillinase. An enthalpogram for 10 successive determinations of penicillin aliquots is shown in Fig. 7.19. The limits of enzyme reuse are defined, at least in part, by the sensitivity required; the increased heat capacity of the system after each injection means that the signal is necessarily attenuated as the number of determinations increases. At low penicillin concentrations, the number of injections into one enzyme solution will obviously be reduced. Moreover, enzyme reuse would be prohibited if significant product inhibition occurred, which would be manifested by progressive increases in enthalpogram curvature as the number of injections increased. Apparently, this is not the case with the penicillinase reaction as shown in Fig. 7.19. The thermochemistry of two other penicillin derivatives is presented in Table 7.12. Penicillin and ampicillin sodium vial preparations, dissolved in water, were determined with a precision of $<1.5\%$ rsd. Phenoxymethylpenicillin tablets containing excipients were determined with a precision of $<4\%$ rsd by the substrate injection procedure; a precision of better than 2% can be obtained by conventional enzyme injection. This reaction has been used extensively in flow reactors to monitor penicillin fermentation chemistry (see Section 7.9.7). The role of buffers in amplifying the overall enthalpy change of a reaction has been alluded to many times in this chapter. In discussion so far, it has been

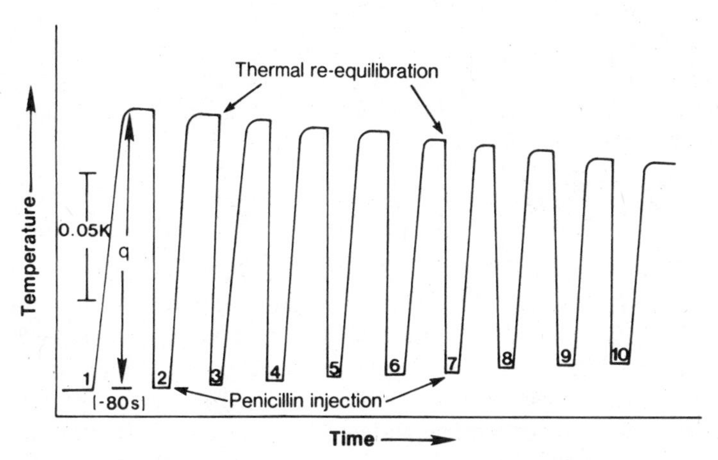

Figure 7.19. Series of enthalpograms for the successive injection of 10 penicillin samples (0.5 mL) into a buffered penicillinase solution (200 IU) (from Ref. 61, by permission of Elsevier Science Publishers).

TABLE 7.12. Reaction Enthalpy Assignments for the Penicillinase-Catalyzed Hydrolysis of Some Penicillin Derivatives[a]

Preparation	ΔH_R (kJ mol^{-1})
Penicillin G	-114.7 ± 0.6
Ampicillin sodium	-125.8 ± 1.2
Phenoxymethylpenicillin	-120.5 ± 1.3

Source: Ref. 61, by permission of Elsevier Science Publishers.

[a]Relevant Reactions:

penicillin + H_2O = penicilloic acid, ΔH_1

THAM$-NH_2$ + H_3O^+ = THAM$-NH_3^+$ + H_2O, ΔH_2

$\Delta H_R = \Delta H_1 + \Delta H_2$

assumed that the buffer's role has been simply to provide or consume protons and that the products of the enzymatic reaction per se are not affected by the choice of buffer. Jespersen has shown that, at least for the urease-catalyzed hydrolysis of urea, this is not the case (62). The hydrolysis of urea in buffered solution at neutral pH is usually represented by

$$H_2O + H_2NCONH_2 = CO_2 + 2NH_3 \tag{7.50}$$

$$\text{buffer}-H^+ + CO_2 + 2NH_3 + H_2O = 2NH_4^+ + HCO_3^- + \text{buffer} \tag{7.51}$$

irrespective of the buffer solution used. Jespersen has shown via a thermochemical study of this reaction in the presence of several different buffers, that the above reactions do indeed take place with phosphate and maleate buffer systems. In contrast, THAM and maleate buffers produce ammonium carbamate as the primary reaction product, that is,

$$H_2O + H_2NCONH_2 = H_2NCOO^-NH_4^+ \tag{7.52}$$

The delineation between these two reaction schemes can be made almost exclusively on the basis of enthalpy because they are so different. Heat of formation data allow an assignment of $+30.50$ kJ mol^{-1} for Eq. (7.50) and -24.27 kJ mol^{-1} for Eq. (7.52). Jespersen's data for the enzymatic hydrolysis of urea in different buffers are summarized in Table 7.13. The reaction products from phosphate-, citrate-, and maleate-buffered reactions can be determined directly from comparison with the theoretical enthalpy changes. Phosphate and maleate clearly have reaction enthalpies consistent with the formation of the classical products NH_4^+ and HCO_3^-. The lower (cf. theoretical) enthalpy change observed

TABLE 7.13. Reaction Enthalpy Data for the Urease-Catalyzed Hydrolysis of Urea in Phosphate, THAM, Maleate, and Citrate Buffers

	Reaction Enthalpy (kJ mol^{-1})		
	Observed	Theoretical[a]	Theoretical[b]
pH 7.5			
Phosphate	−61.29	−61.34	−24.27
THAM	−18.70	−16.86	−24.27
pH 6.7			
Phosphate	−48.24	−62.63	−24.27
Maleate	−56.65	−72.34	−24.27
Citrate	−21.92	−72.21	−24.27

Source: Ref. 62.

[a]Assuming NH_4^+ and HCO_3^- are the products.
[b]Assuming $H_2NCOONH_4$ is the product.

at pH 6.7 is due to the saturation of the reaction solution with CO_2. Similarly, the enthalpy change for citrate is unambiguously correlated with the formation of ammonium carbamate. The products formed in the THAM-buffered reaction can be deduced from a series of enthalpy measurements in mixed THAM-phosphate buffers. The principle of the experiment is best explained diagramatically by reference to Fig. 7.20. The salient feature of the plot is that there is a discontinuity in enthalpy change with the mole fraction of the buffer system. Below $f = 0.5$, where THAM dominates the system, the reaction enthalpy is essentially invariant with buffer composition. This is indicative of the formation of ammonium carbamate, a compound which is undissociated and therefore does not engage in protonation reactions with the buffer. As the phosphate buffer begins to dominate the composition of the buffer system, that is, $f > 0.5$, the enthalpy of reaction becomes dependent upon buffer type suggesting NH_4^+ and HCO_3^- formation.

The discovery of two sets of reaction products is important in the context of urease-catalyzed reaction kinetics (62). Jespersen contends that the 20–30% greater activity observed for citrate-buffered urease-catalyzed hydrolyses can be explained by the lower level of ammonium ions formed in that reaction compared to phosphate systems that produce NH_4^+ and HCO_3^-. The data are consistent with an earlier observation that ammonium ions will inhibit urease activity (63). Consistently, a similar increase in activity is observed when urease activity in THAM is compared with a phospate system. The level of uric acid in serum is small, normal levels range between 20–70 mg L^{-1} or 120–154 μmol L^{-1}.

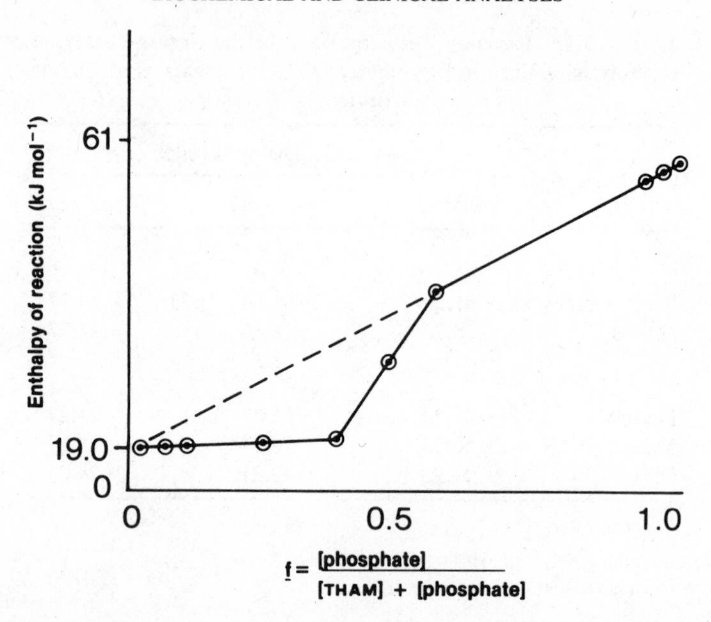

Figure 7.20. Reaction enthalpy ΔH_R for the urease-catalyzed hydrolysis of urea in a series of THAM phosphate mixed buffer systems. The dotted line represents the curve which would result if the products remained the same in both buffer systems (from Ref. 62, by permission of the American Chemical Society).

Consequently, the reaction enthalpy of the uricase-catalyzed oxidation of uric acid to allantoin is too small (ca. -20 kJ mol^{-1}) to allow the calorimetric determination of serum uric acid without the aid of a secondary reaction (64). This can be achieved conveniently by the introduction of catalase to consume the hydrogen peroxide produced in the primary reaction as shown in

$$\text{O}= \text{(N(H), N(H), N(H), NH, =O)} + O_2 + H_2O \xrightarrow{\text{uricase}} \text{HN}\ \text{(C=O)}\ \text{HN}\ \text{(C=O, NHCONH}_2\text{)} + CO_2 + H_2O_2 \qquad (7.53)$$

$$H_2O_2 \xrightarrow{\text{catalase}} H_2O + {}^1\!/_2 O_2 \qquad (7.54)$$

The total enthalpy change associated with this reaction has been determined to be -149.62 kJ mol^{-1} (64). Furthermore, this particular combination of enzymes is facile, since neither enzyme requires a secondary substrate. Additionally, catalase regenerates the oxygen required for the primary reaction. The sensitivity increase generated by this combination of reagents ($\times 7$) brings the determination within the range of a heat-conduction calorimeter. Rehak et al. (64) have determined that the calorimetric precision and accuracy of serum uric acid determinations compares favorably with a spectrophotometric Autoanalyzer technique (23.7 ± 0.9 mg L^{-1} of serum versus 23.3 mg L^{-1}).

The only interference observed with the calorimetric procedure was due to the interaction of serum protein with the surfaces in the reaction cell or with other proteins in solution. This effect was removed by washing the reaction cell with sodium hydroxide solution or a solution of a proteolytic enzyme. Another oxidoreductase—peroxidase—can be used to determine iodide ion in the presence of chloride, bromide, or fluoride (65). The reaction has been discussed in Section 7.5.5. The experimental design is simple because small variations in the concentrations of both reagents—H_2O_2 and peroxidase—will not significantly affect the precision of the data. The only restrictions are that peroxide should be in stoichiometric excess of iodide and that the level of peroxidase is sufficient to provide a well-defined enthalpogram; if these conditions are met, no reagent standardization is required. Iodide can be determined in the presence of an equal mass of either bromide or chloride with a precision better than 2% rsd. Fluoride ion does appear to inhibit the rate of the reaction, as evidenced by increased enthalpogram curvature. However, an equal mass of fluoride ion can be tolerated without significant loss in precision.

The simultaneous determination of two substrates is possible by a combination of kinetic and equilibrium procedures if the conditions can be arranged such that one substrate is consumed immediately and the rate of reaction of the second substrate is first order. This principle can be illustrated by considering the concurrent determination of free and esterified cholesterol (66). The relevant reactions are

$$\text{cholesterol ester} + H_2O \xrightarrow[\text{esterase}]{\text{cholesterol}} \text{cholesterol} + \text{fatty acid} \qquad (7.55)$$

$$\text{cholesterol} + O_2 \xrightarrow[\text{oxidase}]{\text{cholesterol}} \text{4-cholesten-3-one} + H_2O_2 \qquad (7.56)$$

$$H_2O_2 \xrightarrow{\text{catalase}} H_2O + {}^1\!/_2O_2 \qquad (7.54)$$

The amounts of enzyme can be controlled so that the redox reactions [(7.56) and (7.54)] occur quickly and the hydrolysis [(7.55)] proceeds slowly and at a rate determined by the concentration of cholesterol ester. In practice, therefore, the enthalpogram consists of a constant slope, dq/dt, superimposed on a conventional equilibrium response as shown in Fig. 7.21. Under the correct kinetic conditions, the postreaction slope is proportional to the cholesterol ester concentration. If this slope is subtracted from the entire data set either by computer or by extrapolation to zero time, the remaining heat change q_0 will be dependent on the concentration of free cholesterol. Reaction (7.56) is exothermic (-52.9 kJ mol^{-1}) and can be utilized independently for the determination of serum free cholesterol levels (3.9–7.2 mmol L^{-1}) if coupled to the peroxide decomposition reaction [Eq. (7.54)].

7.7. ENZYME-INHIBITION ANALYSIS

An inhibitor is a substance which, when it interacts with an enzyme, causes an attenuation in the rate of the enzyme-catalyzed reaction anticipated for the prevailing conditions of substrate concentration, pH, and temperature. It is a kinetic effect which will only affect the rate of the reaction and not its ultimate thermodynamic equilibrium position. The design of an experiment to study enzyme inhibition will depend on the nature of the inhibition and, related to this, the relevant enzyme kinetics. Accordingly, a brief discussion of the different types of inhibition is warranted.

7.7.1. Types of Inhibition

Enzyme inhibition can be classified as *reversible* or *irreversible*. There are characteristic empirical differences between these two types of inhibition which

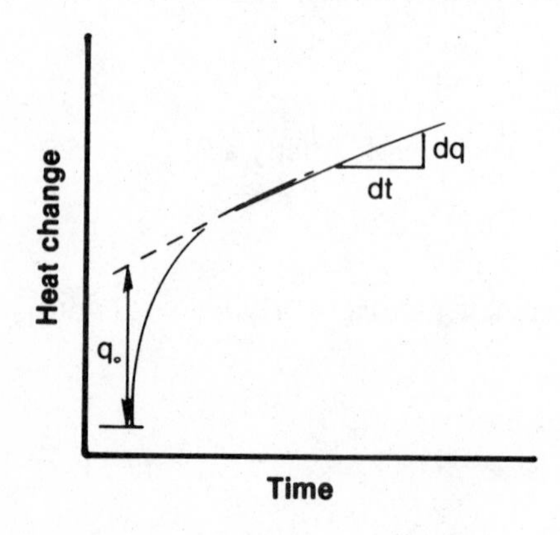

Figure 7.21. The simultaneous determination of two substrates. q_0 (extrapolated to $t = 0$) is proportional to the concentration of the substrate consumed in a fast reaction. The post reaction slope, dq/dt, is proportional to the concentration of the second substrate (adapted from Ref. 39).

demand different experimental formats. The degree of reversible inhibition can be reduced by the removal of the free inhibitor, indicating that there is a thermodynamic equilibrium between the enzyme and the inhibitor. The extent or degree of inhibition caused by a particular concentration of inhibitor can be characterized by an equilibrium constant K_I usually referred to as the inhibition constant. K_I represents the dissociation constant for the equilibrium,

$$\mathrm{E} + \mathrm{I} \underset{k_{-I}}{\overset{k_I}{\rightleftharpoons}} \mathrm{EI} \tag{7.57}$$

that is,

$$K_I = \frac{k_{-I}}{k_I} \tag{7.58}$$

Significantly, from an analytical standpoint, reversible inhibition is characterized by a definite degree of inhibition, depending on the substrate concentration, which is usually reached fairly rapidly and is thereafter independent of time, provided that the inhibitor is stable. In contrast, removal of free inhibitor does not result in a return of enzyme activity with *irreversible inhibition,* which is

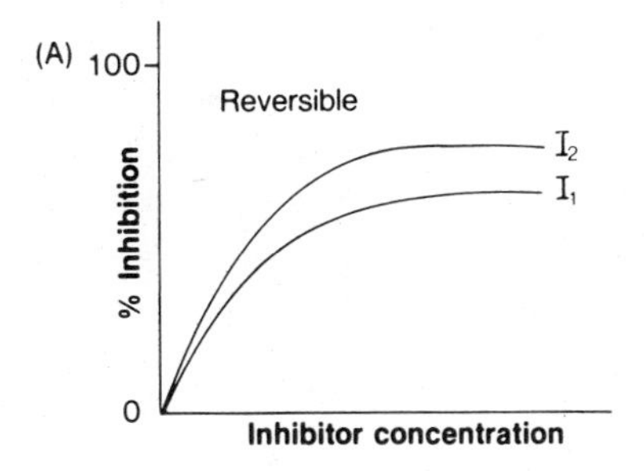

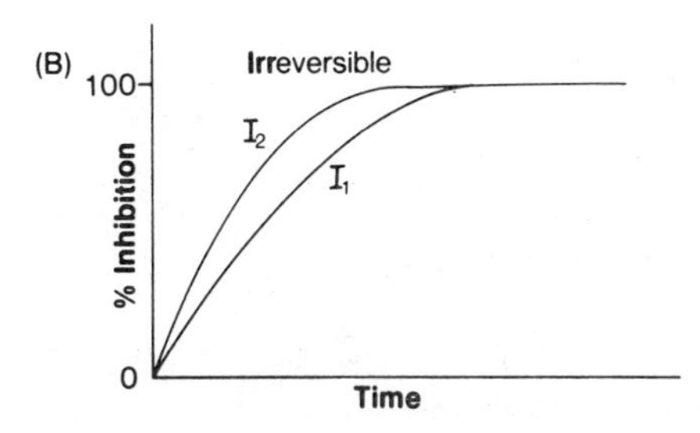

Figure 7.22. The characteristic features of reversible and irreversible enzyme inhibition. (A) Reversible inhibition; K_I inhibitor I, > K_I inhibitor I_2, constant substrate concentration. (B) Irreversible inhibition; k inhibitor $I_2 > k$ inhibitor I_1, where k is the rate constant for the reaction E + I→EI.

characterized by a progressive increase with time, ultimately reaching complete inhibition even with a very small amount of inhibitor, provided that the inhibitor is in stoichiometric excess of the amount of enzyme present. The effectiveness of the inhibitor is expressed not by an equilibrium constant but by a rate constant that determines the fraction of enzyme inhibited in a given period of time by a certain concentration of inhibitor. Irreversible inhibition experiments therefore always have a time element.

The term "irreversible" is in most cases a misnomer. Typically, such inhibitors are in fact reversible but have extremely high affinity for the enzyme. Consequently, the equilibrium shown in Eq. (7.57) still exists but lies far to the right. This equilibrium element is confirmed by the fact that large concentrations of substrate will significantly decrease the level of irreversible inhibition. The essential features of reversible and irreversible inhibition are shown in Fig. 7.22.

Reversible inhibition can be further subclassified into competitive, non-competitive, and uncompetitive inhibition. The essence of *competitive inhibition* is that the enzyme can combine either with substrate to form the productive complex, ES, or with the inhibitor to form an unproductive complex, EI, but not both (Fig. 7.23). The most common reason for this phenomenon is that the inhibitor is a substrate analogue and therefore binds precisely where the substrate should be in the active site. The outstanding characteristic of competitive inhibition, therefore, is that the inhibitor and substrate are mutually exclusive with regard to interactions with the enzyme. As a result, if the concentration of substrate is large enough, the enzyme may be entirely in the form of the active ES complex even in the presence of inhibitor, that is, maximal velocity V_{max} is always potentially obtainable.

Intuitively, the most logical alternative inhibition mechanism to competitive inhibition is one in which the enzyme can bind both the inhibitor and the substrate. As shown in Fig. 7.24, this mechanism, termed *non-competitive inhibition*,

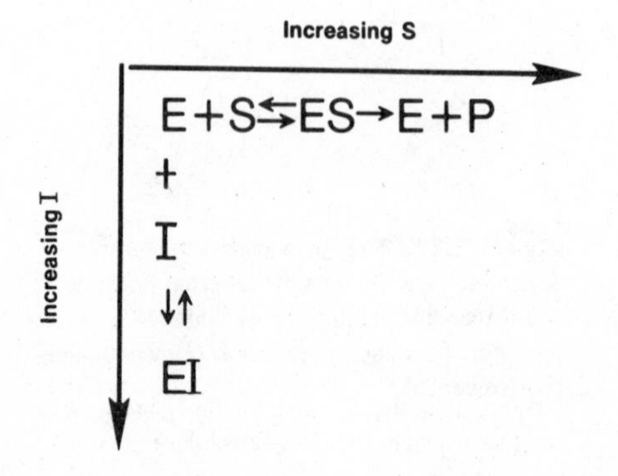

Figure 7.23. Schematic representation of competitive inhibition.

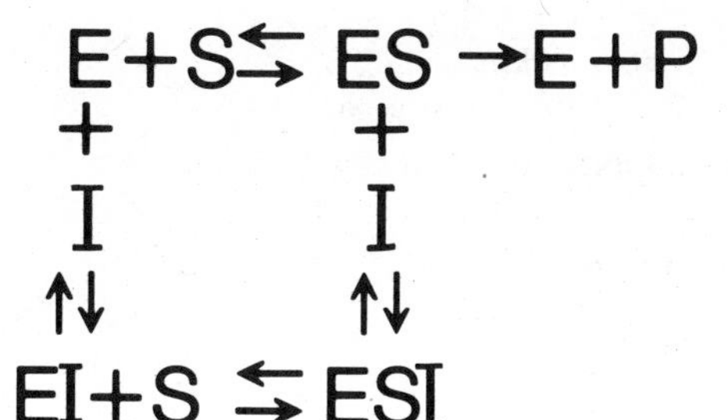

Figure 7.24. Schematic representation of non-competitive inhibition.

assumes that the inhibitor and substrate do not compete for the same site, but that the presence of the inhibitor hinders the breakdown of *ES* to products, that is, the complex ESI is unproductive. This phenomenon rarely occurs naturally, because it requires a moiety of small molecular dimensions to bind to the enzyme without sterically hindering the binding of a substrate. Indeed, the best known examples are proton and heavy-metal inhibition of enzymes containing the $-SH$ functionality. This latter aspect has been used to analytical advantage.

Other inhibition mechanisms exist including *mixed inhibition* and *uncompetitive inhibition*; however, since there are no reported analytical applications, these phenomena will not be discussed.

7.7.2. Enzyme-Inhibition Kinetics—Analytical Considerations

The inhibition of an enzyme is often a very sensitive effect. In some cases, a concentration of 1×10^{-9} mol L^{-1} of inhibitor can cause a significant attenuation of the rate of an enzyme-catalyzed reaction. The potential exists, therefore, for the development of analytical methodology to determine trace amounts of inhibitor, if inhibitor concentration can be related linearly to the degree of inhibition i, where

$$i = \frac{v_0 - v_I}{v_0} = 1 - \left(\frac{v_I}{v_0}\right) \tag{7.59}$$

In Eq. (7.59), v_0 is the rate of reaction in the absence of the inhibitor and v_I is the rate in the presence of the inhibitor, all other conditions remaining the same. The practical implementation of an inhibition analysis experiment will depend on the type of inhibition being studied. This becomes evident from consideration of the relationship between i and the other experimental parameters for each type of inhibition.

7.7.2a. Competitive Inhibition

The rate equation for a competitively inhibited reaction can be written

$$v_I = \frac{V_{max}[S]}{[S] + K_m (1 + [I]/K_I)} \tag{7.60}$$

Therefore,

$$i = 1 - \left(\frac{v_I}{v_0}\right) = \left[1 - \frac{V_{max}[S]}{K_m (1 + [I]/K_I) + [S]}\right]\left[\frac{K_m + [S]}{V_{max} + [S]}\right] \tag{7.61}$$

which simplifies to

$$i = 1 - \frac{K_m + [S]}{K_m (1 + [I]/K_I)\ [S]} \tag{7.62}$$

or

$$i = \frac{K_m (1 + [I]/K_I)}{K_m (1 + [I]/K_I) + [S]} \tag{7.63}$$

Three prerequisites limit the range of inhibitor and substrate concentration over which adequate sensitivity and precision can be maintained for the determination of a competitive inhibitor:

1. Changes in i, the degree of inhibition, should be relatively large for small changes in inhibitor concentration.
2. Ideally, the relationship between i and [I] should be linear.
3. The rate of reaction should be essentially invariant over the measurement period, a prerequisite of all kinetic determinations.

Conditions 1 and 3 clearly involve a compromise with respect to substrate concentration. The degree of inhibition, and hence analytical sensitivity, is maximized at low substrate concentration. On the other hand, maximum reaction rate linearity is obtained as [S] is increased. [S] should therefore be the *minimum* value required in order to satisfy 3. In this regard, the control rate, or the level of uninhibited enzyme activity, should be minimized. This will increase the inhibitory effect of any particular concentration of inhibitor and reduce the level of substrate required to obtain a linear reaction rate. Finally, the relationship between i and [I] only tends to linearity as [I] becomes less than K_I and $(1 + [I]/K_I)$ tends to unity.

The simultaneous fulfillment of all the above conditions leads to a severely restricted linear range (see Section 7.7.4).

7.7.2b. Non-Competitive Inhibition

Equation (7.64) details the reaction rate for a reaction exhibiting *pure* non-competitive inhibition.

$$v_I = \frac{V_{max}[S]}{K_m + [S]\left(1 + \frac{[I]}{K_I}\right)} \tag{7.64}$$

Using a similar derivation procedure as in Section 7.7.2a:

$$i = 1 - \frac{V_{max}[S]}{(K_m + [S])(1 + [I]/K_I)}\left[\frac{K_m + [S]}{V_{max}[S]}\right] \tag{7.65}$$

Therefore,

$$i = \frac{[I]/K_I}{1 + [I]/K_I} \tag{7.66}$$

or

$$i = \frac{[I]}{K_I + [I]} \tag{7.67}$$

Equation (7.67) defines the appropriate conditions necessary for the determination of a non-competitive inhibitor based on its effects on the rate of an enzymatic reaction. Clearly, the sensitivity of the inhibition effect is independent of substrate concentration. However, a minimum level of substrate is still required to maintain a linear reaction rate. A linear relationship between i and [I] only exists when $[I] < K_I$ and $K_I + [I]$ tends to K_I. A further characteristic of Eq. (7.67) is that when $[I] = K_I$, $i = 0.5$, that is, 50% inhibition is obtained. This can be used to analytical advantage as shown in Section 7.7.4.

7.7.2c. Irreversible Inhibition

If the concentration of irreversible inhibitor [I] is in stoichiometric excess of the number of moles of inhibitor binding sites on the enzyme, then the inhibition reaction $E + I \rightarrow EI$ can be treated as a pseudo-first-order reaction (67, 68) whose rate can be expressed as

$$v_I = v_0 \exp(-k_I[I]t) \tag{7.68}$$

where k_I is the rate constant for the reaction of E and I and t is the time of incubation of enzyme and inhibitor. Equation (7.68) can be arranged to

$$\log\left(\frac{v_I}{v_0} \times 100\right) = 2 - \left(\frac{k_I[I]t}{2.303}\right) \tag{7.69}$$

Accordingly a plot of $\log(v_I/v_0 \times 100)$ versus [I] will be linear with a slope of $-k_I t/2.303$ and an ordinate intercept of 2. A determination of rate constants will result from a plot of $(\log v_I/v_0 \times 100)$ versus t. The significant feature of the experiment design is that inhibition time is an experimental parameter that must remain constant in order to extrapolate the value of an unknown inhibitor concentration from a calibration graph. This is in direct contrast to reversible inhibition experiments, in which the equilibrium between E and I is reached immediately.

7.7.3. Calorimetric Methodology

The essential methodological principles of calorimetric enzyme inhibition studies are shown in Figs. 7.25A–7.25D. With one exception, namely, the titration of non-competitive inhibitors (see Section 7.7.4), the format of a calorimetric experiment to measure enzyme inhibition is the same as that described to measure enzyme activity in Section 7.5.3 for both isoperibol and and heat-conduction calorimetry. In either case, the degree of inhibition i for a particular inhibitor concentration is determined by measuring *EA* according to Eqs. (7.18) or (7.21) in the presence and absence of the inhibitor. This process is repeated for a series of standard inhibitor solutions and a calibration graph constructed (see Fig. 7.26). The incubation of the enzyme with the inhibitor is incorporated into the thermal equilibration period. For reversible systems, the length of the incubation period is not a significant factor because inhibition is essentially instantaneous; for irreversible systems, substrate is injected after a constant, predetermined time in each experiment.

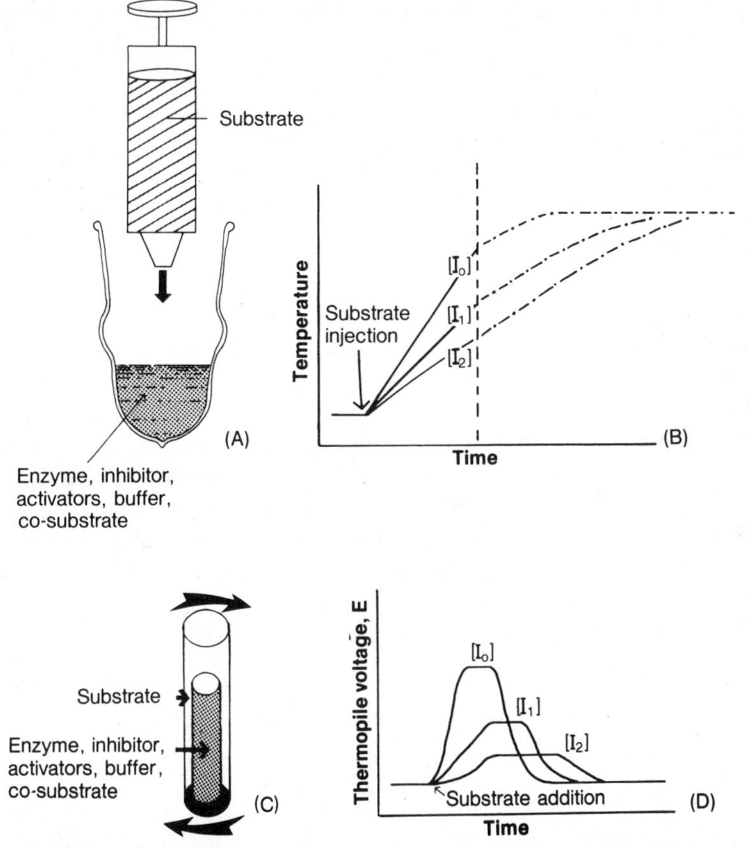

Figure 7.25. Calorimetric measurement of reversible enzyme inhibition: (A) Reagent mixing sequence (isoperibol). (B) typical recorder output for a series of inhibitor concentrations $[I_2]>[I_1]$, I_0 = no inhibitor present. The concentration of enzyme and substrate is constant in all experiments. -.- represents output if the reactions are allowed to go to completion. (C) Reagent mixing sequence (heat conduction). (D) Typical recorder output for the same series of inhibitor concentrations. Note that for heat-conduction data, the integrated response is the same for each experiment, although the maximum rate of reaction varies.

7.7.4. Selected Applications

Cholinesterase is inhibited by several classes of compounds. Reversible, competitive inhibitors are substrate analogs of acetyl or butyryl choline, that is, they are typically highly methylated and contain an ester group and a quaternary or basic nitrogen atom. Comparison of acetylcholine **4** and muscarine **5** shows that the physiologically active alkaloids could be expected to inhibit cholinesterase to a greater or lesser extent.

$$(CH_3)_3\overset{+}{N}-CH_2CH_2-O-\overset{\overset{O}{\|}}{C}-CH_3$$

4

$$H_3C-\text{(tetrahydrofuran ring, HO)}-CH_2\overset{+}{N}(CH_3)_3$$

5

An analytical method for the enthalpimetric determination of several alkaloids has been developed based on this effect (69). The relative inhibitory "potency" of a series of alkaloids is illustrated in Fig. 7.26 which represents inhibition data taken from isoperibol calorimetry data in the manner described in Section 7.7.3. These data are typical of reversible inhibition; the degree of inhibition is not always 100%, even in the presence of relatively large amounts of inhibitor, because i is determined by K_I. Analytical data were obtained by interpolation of calibration graphs prepared within the linear region of each of the curves

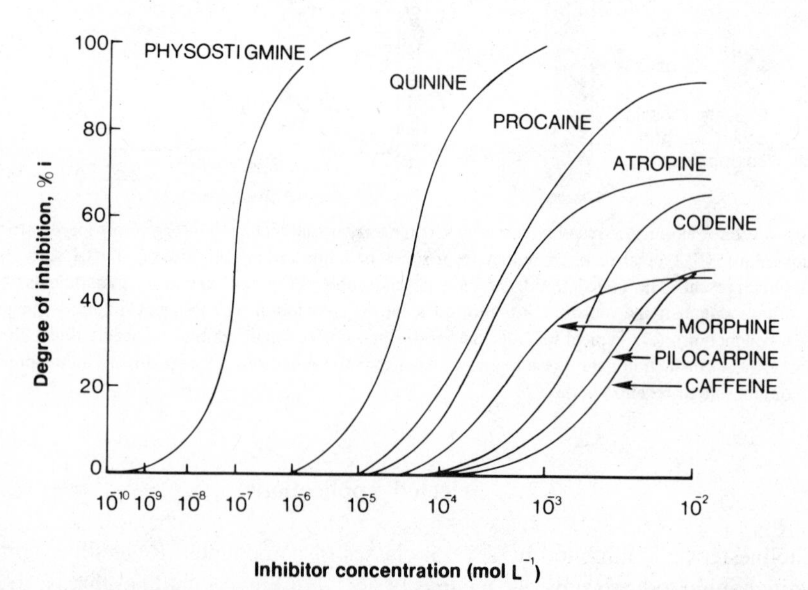

Figure 7.26. Reversible enzyme-inhibition curves, generated from isoperibol data illustrating the effect of a series of alkaloids on the ChE-catalyzed hydrolysis of butyrylcholine. Activity ChE = 5 IU, butylcholine concentration = 8.7 mmol L^{-1}. i is defined according to Eq. (7.59) (from Ref. 69, by permission of Elsevier Science Publishers).

shown. The characteristic features of an inhibition analysis, that is, potentially high sensitivity and limited linear range are apparent from the analytical data in Table 7.14. The limit of detection for the potent inhibitors is far below that of conventional enthalpimetric techniques. However, the linear range is at best one decade and often less. The biggest limitation to this (or any) inhibition procedure is the high probability of interference from many species, both organic and inorganic. The utility of calorimetric inhibition methods does not lie in the determination of inhibitor concentration in solutions of unknown composition, but rather as a universal technique to monitor inhibition mechanisms or the relative potency of a series of inhibitors on a particular enzyme-substrate system. In this context, the capability to monitor the inhibition kinetics of a reaction without perturbation by foreign species introduced as analytical reagents is a definite asset.

An example of such a study is the calorimetric determination of phosphorylated serum ChE activity and its reactivation by a pesticide poisoning antidote (70). By use of heat-conduction instrumentation it was shown that serum ChE activity, inhibited *in vitro* up to 92% by *O,O*-dimethyl-2-2-dichlorovinyl phosphate, could be reactivated up to 35% by addition of 2-pyridine aldoximemethiodide (PAM). The enzyme and inhibitor were incubated in the calorimeter cell for 15 min; this was followed by addition of PAM and a further incubation period. Enzyme activity was then monitored in the usual manner following the introduction of substrate.

A flow heat-conduction apparatus has also been used to determine the concentration of "Dimefox," a commercial organophosphorus pesticide (71) by virtue

TABLE 7.14. Linear Range and Minimum Detectable Amounts for the Enthalpimetric Determination of Some Physiologically Active Alkaloids Based on the Inhibition of Serum ChE

Alkaloid	Minimum Detectable Amount[a] (g)	Linear Range (mol L^{-1})
Physostigmine sulfate	7.4×10^{-8}	$(1.0–4.0) \times 10^{-8}$
Quinine sulfate	9.0×10^{-6}	$1.0 \times 10^{-6}–4.0 \times 10^{-5}$
Procaine hydrochloride	3.1×10^{-5}	$1.0 \times 10^{-5}–2.5 \times 10^{-4}$
Atropine sulfate	4.0×10^{-4}	$5.0 \times 10^{-5}–3.0 \times 10^{-4}$
Morphine sulfate	1.1×10^{-3}	$(1.0–8.0) \times 10^{-4}$
Codeine phosphate	1.5×10^{-3}	$3.0 \times 10^{-4}–2.4 \times 10^{-3}$
Pilocarpine nitrate	1.6×10^{-3}	$5.0 \times 10^{-4}–6.0 \times 10^{-3}$
Thiamine hydrochloride	3.9×10^{-3}	$(1.0–5.0) \times 10^{-3}$

Source: Ref. 69, by permission of Elsevier Science Publishers.

[a]Defined as the lowest mass within the linear range.

of its inhibitory effect on serum ChE. A graphical extrapolation procedure based on the pseudo-first-order reaction rate equation [Eq. (7.69)] was used as the method of calculation of inhibitor concentration. The rate constant of 7.5×10^2 mol^{-1} L min^{-1} was determined for the interaction of Dimefox with serum ChE.

A unique feature of non-competitive enzyme inhibition is that when the inhibitor concentration equals K_I, the reaction rate is precisely 50% inhibited as shown by Eq. (7.64). Therefore, if a solution of a non-competitive inhibitor is titrated into a solution in which an uninhibited enzyme-catalyzed reaction is proceeding, the 50% inhibition point can be used as an analytical "endpoint." This principle has been used for the enthalpimetric determination of Ag^+, based on the inhibition of the urease-catalyzed hydrolysis of urea (72). The measurement procedure is illustrated in Fig. 7.27. As with all kinetic determinations, the reagent concentrations must be carefully controlled to ensure the integrity of the result.

A large substrate concentration is required to maintain a constant reaction rate during the titration. Substrate dilution can be avoided by addition of substrate to the titrant. The enzyme concentration should be small enough that binding of the inhibitor is insignificant. This allows the approximation that the total inhibitor concentration in the reaction cell is equal to the free inhibitor concentration, which is necessary for the derivation of Eq. (7.64). A nanogram detection limit for Ag^+ is reported for this procedure. Interference effects severely limit the application of this procedure to the analysis of complex solutions containing Ag^+.

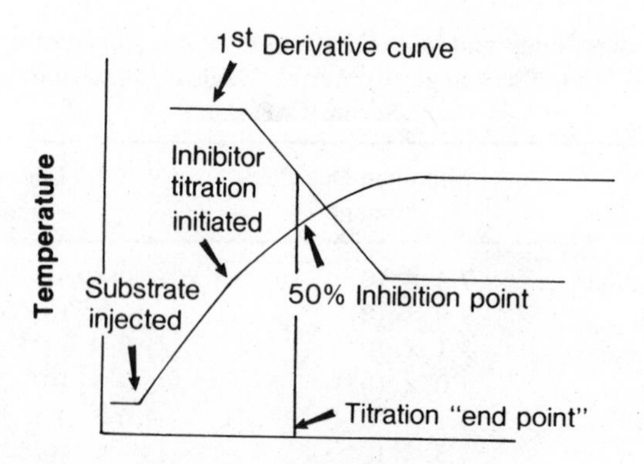

Figure 7.27. Extrapolation procedure for the determination of K_I by means of an enthalpimetric titration of a non-competitive inhibitor into an enzyme-substrate reaction mixture (from Ref. 72, by permission of Marcel Dekker, Inc.).

7.8. THE DETERMINATION OF BIOCHEMICAL CONSTANTS, K_m AND K_I

It is evident from the discussions of experimental design in Sections 7.5–7.7 that the magnitudes of K_m and K_I are important to analytical methodology. The magnitude of K_m will determine the appropriate substrate concentrations necessary to define the boundary conditions of first-order or zero-order kinetics. Similarly, the value of K_I has a direct bearing on the working linear range of an inhibitor determination. As fundamental parameters K_m and K_I are used routinely as numerical indicators of the relative affinity of substrates or inhibitors for a particular enzyme. In the area of immobilized-enzyme chemistry, both parameters can be used to compare the accessibility of the active site before and after immobilization of the enzyme to an insoluble matrix. Consequently, K_m and K_I are significant constants from both a practical and fundamental standpoint. This section will be concerned with the determination of K_m and K_I based on calorimetric data.

7.8.1. Graphical Determinations of K_m and K_I from Calorimetric Data

The working definition of K_m can be deduced from Eq. (7.8); K_m is that concentration of substrate which produces a rate of reaction equal to one-half the maximal velocity V_{max}. The smaller the value of K_m, the smaller the amount of substrate needed to achieve $\frac{1}{2}V_{max}$ and conversely. This empirical definition is the basis of the relation between K_m and substrate affinity. The asympotic nature of a plot of v versus $[S]$ does not allow an accurate assignment of K_m. Traditionally, graphical determinations of K_m have been based on the linear transformation of Eq. (7.8), that is,

$$\frac{1}{v} = \frac{K_m}{V_{max}} \frac{1}{[S]} + \frac{1}{V_{max}} \tag{7.70}$$

The so-called "double-reciprocal" or Lineweaver–Burk plot of reciprocal rate versus reciprocal substrate concentration results in the determination of K_m from the negative intercept on the abscissa. The statistical bias inherent in all the linear transformations of Eq. (7.8) have been discussed at length elsewhere (35); however, these plots still remain the most popular method for K_m determinations.

Calorimetric data for a double-reciprocal plot can be obtained from initial-rate extrapolation made at a series of substrate concentrations as described in Section 7.6.1.

Perhaps the biggest limitation of initial-slope data is that many data are ignored. With only one data point per experiment at least four experiments are required to determine K_m. The same data can be obtained from the tangential extrapolation of a calorimetric progress curve as shown in Fig. 7.28. The heat evolved at any instant, q_t is related to the substrate concentration at that instant, $[S_t]$, by the equation

$$[S_t] = \frac{q_t}{V\Delta H_R} \tag{7.71}$$

where V is the volume of the solution. If the enthalpy of reaction is not known, $[S_t]$ can be calculated as a fraction of the initial substrate concentration $[S_0]$ as represented by the heat change at that time, that is,

$$[S_t] = [S_0]\frac{q_t}{q_{\text{tot}}} \tag{7.72}$$

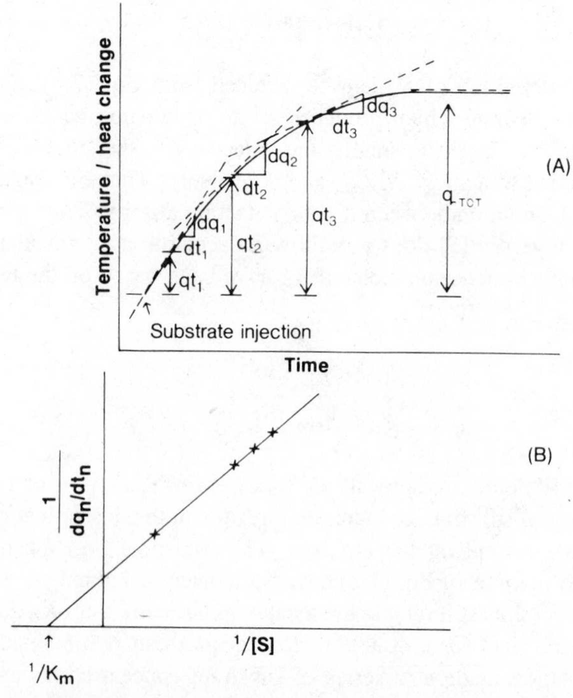

Figure 7.28. The determination of K_m by tangential extrapolation of calorimetric rate data. (A) extrapolation procedure; (B) corresponding Lineweaver–Burk plot: $[S_t] = q_{t,n}V\Delta H_R$.

As the units of q cancel in Eq. (7.72), the dimensions are unimportant. It is possible, therefore, to simply use the distance measurement on the chart and avoid a heater calibration experiment. The rate of reaction v at any substrate concentration (and therefore at any time) can be calculated by tangential extrapolation of the slope at any point t on the progress curve. Once again the dimensions of v are arbitrary, unless a simultaneous assignment of V_{max} is required. v and $[S_t]$ data can then be substituted into Eq. (7.70).

Without the aid of a computerized data collection system it is difficult to perform the tangential extrapolation procedure without subjective error. This problem can be avoided by application of the integrated rate equation, which can be derived as follows:

$$v = \frac{-d[S]}{dt} = \frac{V_{max}[S]}{K_m + [S]} \tag{7.8}$$

$$\frac{-d[S]}{dt} = \frac{V_{max}}{(K_m/[S]) + 1} \tag{7.73}$$

or

$$d[S]\left(\frac{K_m}{[S]} + 1\right) = -V_{max}dt \tag{7.74}$$

Integrating between time $t = 0$ and t and $[S] = [S_0]$ and $[S_t]$, that is,

$$K_m \int_{[S]=[S_0]}^{[S]=[S_t]} \frac{d[S]}{[S]} + \int_{[S]=[S_0]}^{[S]=[S_t]} d[S] = -V_{max} \int_{t=0}^{t=t} dt \tag{7.75}$$

gives

$$K_m \ln [S] + [S] = -V_{max}t \tag{7.76}$$

or

$$K_m \ln ([S_t] - [S_0]) + ([S_t] - [S_0]) = -V_{max}t \tag{7.77}$$

therefore

$$\frac{[S_0] - [S_t]}{t} = V_{max} - \frac{K_m \ln \frac{[S_0]}{[S_t]}}{t} \tag{7.78}$$

Therefore, a plot of $([S_0] - [S_t]/t$ versus $(1/t) \ln ([S_0]/[S_t])$ will be linear with a slope of $-K_m$, an ordinate intercept of V_{max}, and an intercept on the abscissa of V_{max}/K_m.

The generation of data for this plot is trivial since both variables, $[S_t]$ and t, are available from a calorimetric progress curve as shown in Fig. 7.28a and Eq. (7.72). In contrast to the derivative approach, however, tangential extrapolation is not required, merely the assignment of the heat change at predetermined intervals on the progress curve. In this manner, a large data set can be assembled from just one experiment.

The pitfalls in this procedure emanate from two sources. First, application of Eq. (7.78) presupposes that the analytical signal, the evolution of heat in this case, is directly related to the rate of consumption of substrate. Therefore, correction of thermal effects, in particular heat loss, is non-trivial. Mathematically, heat loss will be interpreted as an alteration in the reaction rate (assuming an exothermic reaction) and it will bias the data accordingly. The heat loss correction procedures discussed in Chapter 3 should be adhered to rigorously for accurate K_m assignments. Similarly, any inhibition of the reaction rate by products or interfering inhibitors will have the same effect.

If product inhibition is known to occur, accurate data can still be obtained by use of an ingenious plotting procedure described by Jennings and Niemann (73).

Rearranging Eq. (7.78) we have

$$\frac{t}{\ln \frac{[S_0]}{[S_t]}} = \frac{1}{V_{max}} \frac{([S_0] - [S_t]}{\ln \frac{[S_0]}{[S_t]}} + \frac{K_m}{V_{max}} \tag{7.79}$$

which generates a straight line plot of $t/\ln ([S_0]/[S_t])$ versus $([S_0] - [S_t]/\ln ([S_0]/[S_t])$ with a slope of $1/V_{max}$ and an intercept of K_m/V_{max} on the ordinate. Cornish-Bowden notes that there is a similarity between the characteristics of Eq. (7.79) and the so-called "Eadie–Hofstee" plot, which is a variant of the double reciprocal procedure and can be written

$$\frac{[S_0]}{v_0} = \frac{K_m}{V_{max}} + \frac{[S_0]}{V_{max}} \tag{7.80}$$

Therefore, the point $[S_0]$, $[S_0]/v_0$ should lie on a straight line of slope $1/V_{max}$ and intercept K_m/V_{max} on the ordinate, that is, the same straight line as that plotted from Eq. (7.79). Consequently, if the straight line obtained from the plot of Eq. (7.79) is extrapolated to the point at which $([S_0] - [S_t])/\ln([S_0]/[S]) = [S_0]$, the value on the ordinate must be $[S_0]/v_0$; this procedure is shown in Fig. 7.29. By means of this extrapolation procedure an accurate assignment of the rate of reaction at zero time can be obtained. By definition, product inhibition cannot bias data taken in this manner. K_m and V_{max} can then be determined from a secondary plot of Eq. (7.80).

The same alternatives apply to the graphical determination of K_I from the calorimetric data. The "double-reciprocal" equations incorporating appropriate factors for competitive and non-competitive calibration and the characteristics of each plot ($1/v$ versus $1/[S]$) are presented in Table 7.15. Initial rate or tangential extrapolations of calorimetric progress curves can be used to generate $1/v$ and $1/[S]$ data as discussed for K_m determinations. An additional data set obtained in the presence of a known concentration of inhibitor results in an assignment of K_I from either the intercept or the abscicca or ordinate. If K_m and V_{max} are unknown, a minimum of eight experiments is therefore required to determine K_I. K_I can be determined from two experiments if the integrated rate equations are extended to include terms related to the type of inhibition. The characteristics of the integrated rate equation plots are shown in Table 7.16. Heat-loss corrections are once again of paramount importance to the integrity of the extrapolation procedure.

The graphical extrapolations outlined in this section utilize an indirect kinetic approach to the determination of K_I, that is, the rate of enthalpy change associated with the enzyme–substrate reaction is used as an indirect indicator of the degree

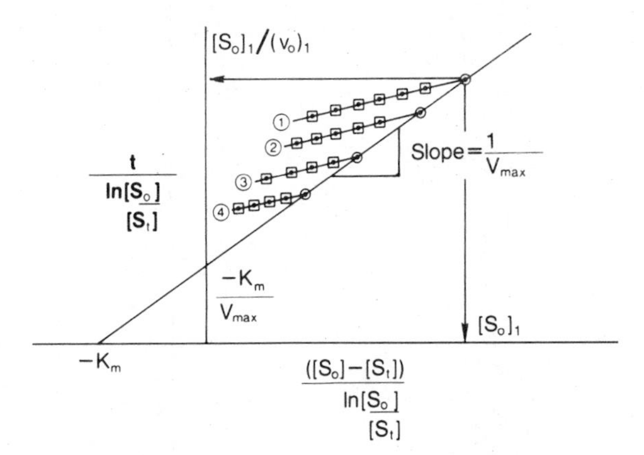

Figure 7.29. Extrapolation of integrated rate equation plot to compensate for product inhibition: ⊡, experimental data points for conventional integrated plot; ⊙, extrapolated data point, $([S_0] - [S_t]/\ln([S_0]/[S_t]) = [S_0]$. Plots 1–4 represent initial substrate concentrations, $[S_0]_1, > [S_0]_2 > [S_0]_3 > [S_0]_4$.

TABLE 7.15. Characteristics of the Double Reciprocal Equations, 1/*v* versus 1/[S/] for the Determination of K_I

Inhibition Type	Equation	Slope	Intercept
Competitive	$\frac{1}{v} = \frac{K_m}{V_{max}}\left(1 + \frac{[I]}{K_I}\right)\frac{1}{[S]} + \frac{1}{V_{max}}$	$\frac{K_m}{V_{max}}\left(1 + \frac{[I]}{K_I}\right)$	$-\frac{1}{K_m\,(1 + [I]/K_I)}$ [a]
Non-Competitive	$\frac{1}{v} = \frac{K_m}{V_{max}}\left(1 + \frac{[I]}{K_I}\right)\frac{1}{[S]} + \frac{\left(1 + \frac{[I]}{K_I}\right)}{V_{max}}$	$\frac{K_m}{V_{max}}\left(1 + \frac{[I]}{K_I}\right)$	$\frac{1 + [I]/K_I}{V_{max}}$ [b]

[a]abscissa
[b]ordinate

TABLE 7.16. Characteristics of Integrated Rate Equation Plots ($[S_o]-[S_t])/t$ versus $1/t \ln([S_o]/[S_t])$ for the Determination of K_I

Inhibition Type	Slope	Intercept (Ordinate)
Competitive	$-K_m\left(1+\frac{[I]}{K_I}\right)$	$\frac{V_{max}}{K_m}\left(1+\frac{[I]}{K_I}\right)$
Non-competitive	$-K_m$	$\frac{V_{max}}{K_m\left(1+\frac{[I]}{K_I}\right)}$

of enzyme–inhibitor interaction. A more direct method is to monitor the enthalpy change generated by the enzyme–inhibitor reaction per se and to determine the thermodynamic equilibrium constant (in most cases this will be the reciprocal of K_I) for the equilibrium $E + I \rightleftharpoons EI$. The inhibitor can be titrated either continuously or incrementally into the enzyme solution. This techique is subject to the considerations of any thermodynamic constant determination and as such has been covered in Chapter 5. The determination of K_m from flow calorimetry experiments in which a reacting solution of enzyme and substrate is pumped into a flow-through detector cell requires separate consideration.

Under zero-order conditions the instrument response ($\mathbf{E}_C$) from a flow calorimeter has been shown (74, 75) to take the form

$$\mathbf{E}_C \equiv \mathbf{E}_Z = \varepsilon_C V_{max} V \Delta H_R \tag{7.81}$$

where ε_C is an instrument calibration factor as defined in Chapter 3. In this case V is the volume of the flow-through cell and V_{max} has the dimensions mol time^{-1} L^{-1}. Under first-order kinetic conditions ($[S_0] < K_m$), $\mathbf{E}_C$ is given by

$$\mathbf{E}_C = \varepsilon_C F[S_0]\Delta H_R \left[1 - \exp\left(\frac{-V_{max}\bar{t}_R}{K_m}\right)\right] \exp\left(\frac{-V_{max}t}{K_m}\right) \tag{7.82}$$

where F is the flow rate and $\bar{t}_R$ is the average residence time in the cell. As discussed in Section 7.6.1, V_{max}/K_m is analogous to a first-order rate constant. When $V_{max}/K_m\bar{t}_R$ is small,

$$1 - \exp\left(\frac{-V_{max}\bar{t}}{K_m}\right) \approx \frac{V_{max}\bar{t}_R}{K_m} = \frac{V_{max}V}{K_m F} \tag{7.83}$$

Substituting into Eq. (7.82), we obtain

$$\mathbf{E}_C = \varepsilon_C[\mathrm{S}_0] \frac{V_{\mathrm{max}}}{K_m} V\Delta H_R \exp\left(\frac{-V_{\mathrm{max}}}{K_m} t\right) \tag{7.84}$$

Equation (7.84] predicts that the instrument response $\mathbf{E}_C$ will decay exponentially with time and that the slope of a plot of log $\mathbf{E}_C$ versus time will be V_{max}/K_m. The intercept at zero time, $\mathbf{E}_{C,\mathrm{o}}$ will vary linearly with $[\mathrm{S}_0]$ and from Eq. (7.81),

$$\frac{\mathbf{E}_{C,0}}{[\mathrm{S}_0]} = \frac{\varepsilon_C \mathrm{V}_{\mathrm{max}} V\Delta H_R}{K_m} \tag{7.85}$$

or

$$\frac{\mathbf{E}_{C,0}}{[\mathrm{S}_0]} = \frac{\mathbf{E}_Z}{K_m} \tag{7.86}$$

Beezer and Tyrrell have shown (74) that a more accurate form of Eq. (7.86) is

$$\mathbf{E}_{C,0} = \frac{\mathbf{E}_Z[\mathrm{S}_0]}{K_m + [\mathrm{S}_0]} \tag{7.87}$$

because the approximation that $[\mathrm{S}_0] << K_m$ is not always fulfilled. The linear transformation of this equation, namely,

$$1/\mathbf{E}_{C,0} = \frac{K_m}{\mathbf{E}_Z[\mathrm{S}_0]} + \frac{1}{\mathbf{E}_Z} \tag{7.88}$$

produces a familiar double reciprocal plot of $1/\mathbf{E}_{C,0}$ *versus* $1/[\mathrm{S}_0]$, which leads to an accurate assignment of K_m.

Eftink et al. (76) have argued that enzyme kinetic data from flow measurements are best analyzed by use of integrated rate equations, thereby avoiding the distortions in double-reciprocal plots due to product inhibition, reaction reversal, and the absence of initial rate conditions. This is indeed the case for tangential extrapolation of initial rates; however, the graphical extrapolation procedure based on Eq. (7.88) should provide an unbiased result since by definition it is based on the instrument response at zero time, $\mathbf{E}_{C,0}$.

7.8.2. Selected Applications

A simple direct-injection enthalpimetry apparatus has been used to determine K_m for the urease-catalyzed hydrolysis of urease by means of initial rate measurements (77). A K_m assignment of 0.027 mol L^{-1} was made from the intercept of a straight line plot (correlation coefficient 0.9918) of $1/\Delta T/\Delta t$ *versus* $1/[S_0]$. The tangential extrapolation of reaction rates at selected intervals on a kinetic injection enthalpogram has been reported for the determination of K_m for the lactate-dehydrogenase-catalyzed reduction of pyruvate to lactate (78). The resultant data were also fitted to a double-reciprocal plot. Isoperibol data fitted to the integrated rate equation have been utilized for the determination of K_m for the α-chymotrypsin-catalyzed hydrolysis of *N*-acetyl-L-tyrosine ethyl ester (79). This procedure was subsequently extended to the determination of K_I for the competitive inhibition of serum cholinesterase by quinine, morphine, and procaine (80). Flow measurements have provided K_m determinations for the urease-catalyzed hydrolysis of urease (81) and the hydrolysis of cyclic phosphates by ribonuclease A (82).

7.9. CALORIMETRIC METHODS AND DEVICES BASED ON IMMOBILIZED-ENZYME TECHNOLOGY

The instrumental and theoretical aspects of flow calorimeters in general are discussed in detail in Chapter 4. Discussion in this section will be limited to instruments and methods utilizing immobilized enzyme reactors in one form or another.

In the opening paragraph of this chapter, it was stated that attempts to introduce the calorimeter into the clinical laboratory have been the primary motivation in many of the methodological and instrumental developments in the last decade. This phenomenon is most evident in the area of flow calorimetry, where the emphasis has been on increasing sample throughput and instrument simplicity, two essential features of routine analytical instrumentation. The achievement of these objectives has been aided considerably by the introduction of immobilized or insoluble matrix-bound enzymes into the flow reactor. Two added benefits accrue from the use of an immobilized enzyme. The true catalytic properties of the enzyme can be exploited to the fullest extent in that immobilization produces a reusable reagent, significantly reducing the cost of an enzymatic analysis which can be prohibitive if carried out in "batch" fashion in homogeneous solution. Furthermore, a stabilization of the enzyme often occurs which can increase the lifetime of the reagent and enhance its resistance to changes in pH, ionic strength,etc. Calorimetric flow devices that have been developed based on immobilized-enzyme reactors can be classified into two distinct categories in terms of operational principles: (a) flow calorimeters or enthalpimeters, popularly termed "enzyme thermistors," and (b) "thermal enzyme probes."

7.9.1. Flow Calorimeters/"Enzyme Thermistors"

The nomenclature associated with this area of instrumentation is confused, primarily due to the misuse of the term "enzyme thermistor," which properly should apply to a device in which the enzyme forms an integral part of the thermistor, that is, when the enzyme is either physically or chemically bound to the thermistor itself. The term has, however, become synonymous with any flow device in which the enzyme is merely placed in close proximity to the site of an immobilized enzyme/substrate reaction. The looseness of this description has resulted in its application to many configurations, none of which are true "enzyme thermistors." There is no major difference in principle, therefore, between the single-column "enzyme thermistors" described by Danielsson et al. (83) and the flow enthalpimeter described by Schifreen et al. (84). Generally, both are classified as flow calorimeters or and their salient operational characteristics are described in detail in Chapter 4.

Nomenclature notwithstanding, the immobilized-enzyme reactor has had a considerable impact on the feasibility of clinical (substrate) analysis by calorimetric methods and it has been the most prolific area of research publication in the last 5 years.

The primary rationale behind these devices is that, by placing the thermistor in a microenvironment in which a biochemical (or chemical) reaction is occurring heat transfer is maximized resulting in increased sensitivity. Indeed, Danielsson et al. have determined that the efficiency of heat transfer is between 50–75% depending on the configuration (83). As an example, the complete conversion of 1×10^{-6} mol of substrate (1 mL of 1 mmol L^{-1} solution) having an arbitrary enthalpy of reaction of 84 kJ mol^{-1} (-20 kcal mol^{-1}) will result in a minimum temperature change of 1×10^{-2}°C. Accordingly, the temperature resolution required to determine this level of substrate with less than 1% error is 1×10^{-4}°C. The determination of substrate concentrations of 10^{-5}–10^{-4} mol L^{-1} has therefore become possible using relatively simple and inexpensive equipment. This represents a detection-limit improvement of at least an order of magnitude when compared to a conventional batch or direct-injection experiment on isoperibol equipment. The same type of sensitivity can be achieved with microcalorimetric (heat-conduction) equipment but at considerably increased expense. Moreover, between 20 and 30 samples per hour can be analyzed on the flow instrument.

It is instructive to follow the development of the original "enzyme thermistor" design, since it illustrates the advantages and limitations of each device.

The initial report citing the introduction of "enzyme thermistors" described a configuration in which the thermistor was encircled by a coil of PVC tubing containing the enzyme immobilized onto glass beads (85). The heat generated by the reaction of the sample as it flowed through the enzyme reactor was transmitted through a paraffin oil bath to the thermistor (Fig. 7.30). One-milliliter sample pulses were introduced into the flow stream by a timed three-way valve. Although able to detect a temperature change of 1×10^{-2}°C from the reaction of 1×10^{-6} mol of substrate, the design had several limitations. The small diameter of tubing used for the flow reactor (1 mm i.d.) allowed only three units of enzyme to be immobilized and restricted flow rates to 0.16 mL min^{-1}. Moreover, the time response of the paraffin oil/thermistor combination was slow. All these problems led to low sample throughput. The second development stage, shown in Fig. 7.31, is the most often reported design. The central feature of this configuration is a single-column, immobilized-enzyme reactor with a volume of 0.5–1.0 mL. The thermistor is positioned in the reactor, close to the exit. The whole unit is immersed in a thermostated water bath. The increased sensitivity obtained by placing the thermistor in the reactor bed allows the determination of $10^{-8} - 10^{-7}$ mol (1 mL of 10^{-5}–10^{-4} mol L^{-1}) of substrate depending on the size of the reaction enthalpy.

With an enzyme immobilized onto a high surface area matrix, for example, controlled-pore glass and a reactor volume of up to 1 mL, a large amount of enzyme (>100 IU) can be incorporated into the reactor. Accordingly, *in most*

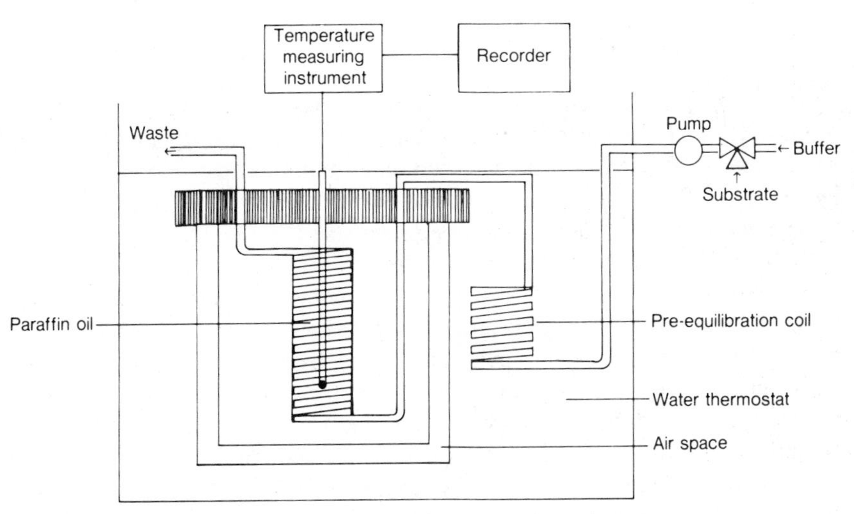

Figure 7.30. Original design of the "enzyme thermistor." (Adapted from Ref. 85, by permission of Elsevier Biomedical Press.).

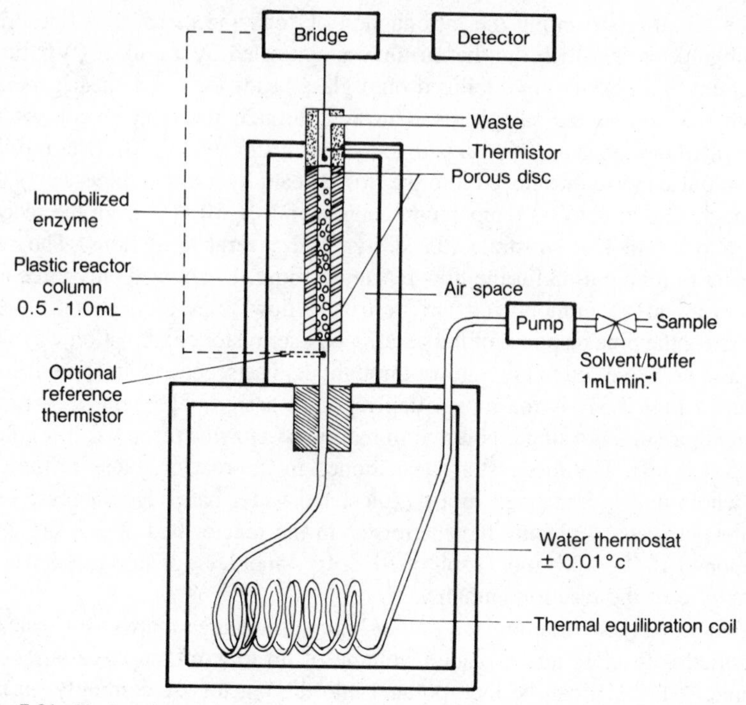

Figure 7.31. Single-column "enzyme thermistor" (from Ref. 83, by permission of Academic Press, Inc.).

instances ($[S_0] \leqslant K_m$), the sample substrate is consumed in the initial regions of the column. The method is therefore an equilibrium or "endpoint" procedure and is not rate dependent. The expected temperature change can be calculated by means of Eq. (7.44). This is an important theoretical difference between these devices and the "thermal enzyme probe" (see Section7.9.5).

Baseline drift due to fluctuations in flow stream temperature can be considerably reduced by the introduction of a second reference thermistor at the entrance to the column. Differential detection does not, however, reduce other sources of interference, namely, pump pulsation effects, hydrothermal noise caused by turbulent flow around the thermistor bead, and the contribution to the enthalpy change from the adsorption of species onto the glass in the reactor. The latter is particularly evident with clinical samples containing protein.

These effects can be minimized by making a differential measurement, not between two thermistors in one column, but between two thermistors in two separate, parallel columns. One column contains the immobilized-enzyme reactor bed as in Fig. 7.30, the other is identical in every respect except that the enzyme is omitted. This "split-flow" configuration is reported to have the highest signal-to-noise ratio (83), and it is the basis of the most recent design, which also features a constant-temperature aluminum jacket in lieu of a water-bath thermostat (Fig. 7.32). The problem of hydrothermal noise has been addressed by reverting to the placement of the thermistor external to the column at the expense of sensitivity. Thermal contact is made by mounting the thermistor on a gold capillary column through which the column effluent must flow immediately after its exit from the reactor. With some clinical samples, clogging of a controlled pore-glass reactor can occur (83). Following the example of open-tubular columns in gas chromatography, the reactor can be fabricated from a material such as nylon and the enzyme immobilized to the walls (86). In order to increase the functionality of the relatively inert nylon backbone, controlled cleavage of amide bonds is performed either by reaction with a nucleophile or by mild acid hydrolysis. The amine functionalities produced can then be derivatized with any of the standard techniques, for example, diazotization, condensation, arylation, etc. Theoretically, the final activity of a nylon column will be considerably less than the controlled-pore glass equivalent due to surface area considerations. This will inevitably result in a lower dynamic range.

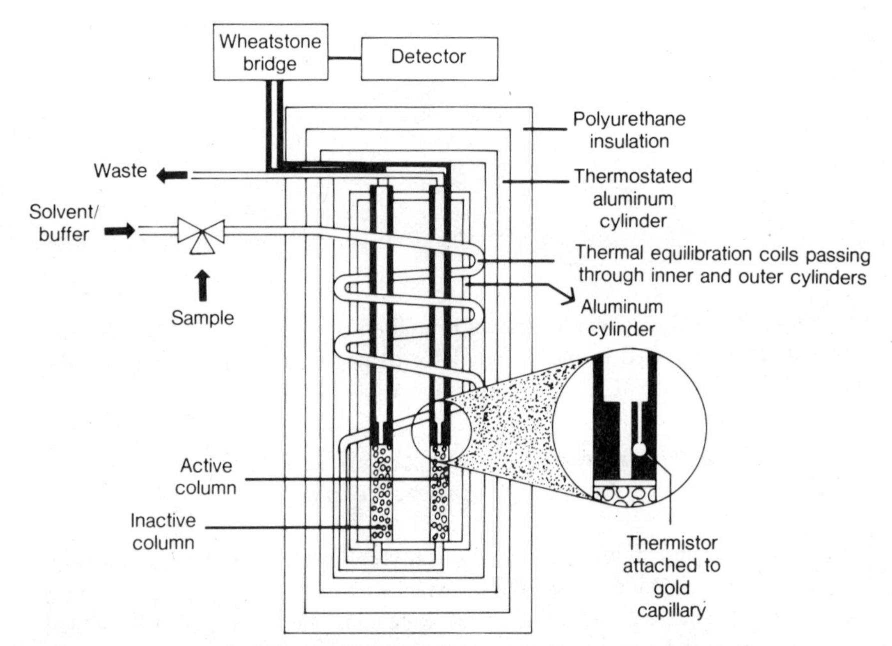

Figure 7.32. "Split-flow enzyme thermistor" with aluminum-jacket thermostat. (Adapted from Ref. 83).

A detailed discussion of the enzyme-immobilization chemistry is beyond the scope of this text. However, a large proportion of enzyme reactions are based on the use of glutaraldehyde, $OHC(CH_2)_3CHO$, as a bifunctional coupling reagent in combination with silylated controlled-pore glass.

The sequence of reactions is shown in Eqs. (7.89) and (7.90):

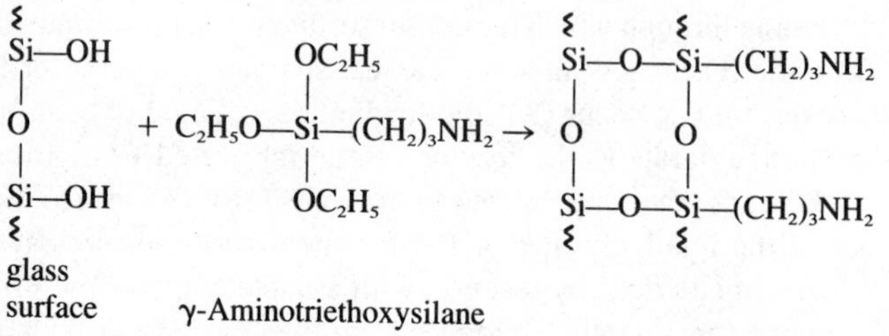

$$\begin{array}{l} \wr \\ \text{Si—OH} \\ | \\ \text{O} \\ | \\ \text{Si—OH} \\ \wr \\ \text{glass} \\ \text{surface} \end{array} + \begin{array}{c} OC_2H_5 \\ | \\ C_2H_5O\text{—Si—}(CH_2)_3NH_2 \\ | \\ OC_2H_5 \\ \gamma\text{-Aminotriethoxysilane} \end{array} \rightarrow \begin{array}{l} \wr \quad\;\; \wr \\ \text{Si—O—Si—}(CH_2)_3NH_2 \\ | \qquad\; | \\ \text{O} \qquad \text{O} \\ | \qquad\; | \\ \text{Si—O—Si—}(CH_2)_3NH_2 \\ \wr \quad\;\; \wr \end{array} \tag{7.89}$$

$$\begin{array}{l} \wr \quad\;\; \wr \\ \text{Si—O—Si—}(CH_2)_3NH_2 \\ | \\ \text{O} \\ | \\ \text{Si—O—Si—}(CH_2)_3NH_2 \\ \wr \quad\;\; \wr \end{array} + \begin{array}{l} OHC(CH_2)_3CHO + \text{ENZ—}NH_2 \rightarrow \\ \text{Glutaraldehyde} \end{array} \tag{7.90}$$

$$\begin{array}{l} \wr \quad\;\; \wr \\ \text{Si—O—Si—}(CH_2)_3\text{—N}=CH(CH_2)_3\text{CONH—ENZ} \\ | \qquad\; | \\ \text{O} \qquad \text{O} \\ | \qquad\; | \\ \text{Si—O—Si—}(CH_2)_3\text{—N}=CH(CH_2)_3\text{CONH—ENZ} \\ \wr \quad\;\; \wr \end{array}$$

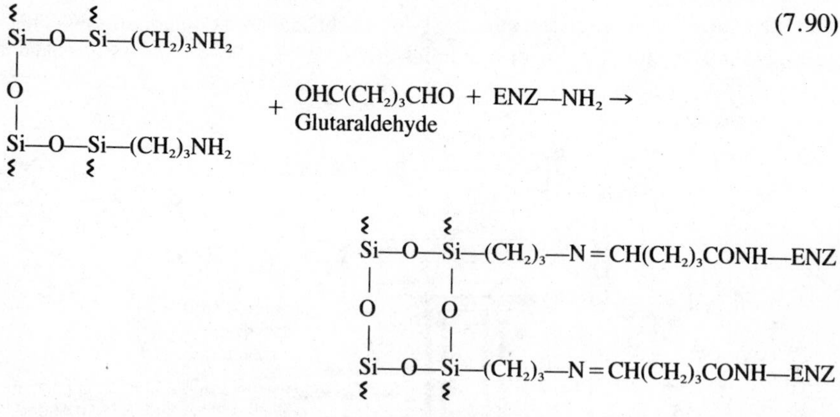

The coupling to the enzyme occurs via formation of a peptide bond between an amino function attached to the enzyme, for example, a lysine residue and the aldehyde functionality of the coupling reagent.

7.9.1a. Optimization of Enzyme Activity in a Calorimetric Immobilized-Enzyme Reactor

The sensitivity (temperature change/mole analyte) of a calorimetric enzyme reactor is obviously related to the degree or extent to which the analyte is converted to product. The smaller the fraction of substrate consumed, the smaller the signal and, therefore, the lower the sensitivity. The level of enzyme activity required to consume the substrate in a flow reactor can be calculated in much the same way as is possible for the analogous batch reactor (Section 7.5.3a) by use of the integrated rate equation (87). Equation (7.22) can be arranged to give

$$([S_0] - [S_t]) = V_{max}t - K_m \ln \frac{[S_0]}{[S_t]} \tag{7.91}$$

In the case of a flow reactor, t is the residence time of the substrate in the reactor. $[S_0]$ and $[S_t]$ are the substrate concentrations at the reactor entrance and exit, respectively.

The fraction of substrate converted to product is given by

$$f = \frac{([S_0] - [S_t])}{[S_0]} \tag{7.92}$$

Therefore, Eq. (7.91) becomes

$$f[S_0] = V_{max}t + K_m \ln (1 - f) \tag{7.93}$$

If t is expressed in terms of experimental variables, that is,

$$t = \frac{L}{F} \tag{7.94}$$

where L is the length of the reactor (cm) and F is the flow rate (cm s^{-1}), and V_{max} is expressed as the activity per unit volume of the reactor, then Eq. (7.93) becomes

$$f[S_0] = \frac{V_{max}L}{F} + K_m \ln (1 - f) \tag{7.95}$$

Given the dimensions of any reactor, Eq. (7.95) allows the calculation of the level of the enzyme activity necessary in the reactor (IU per unit volume) to achieve the desired extent of reaction, f.

As mentioned earlier, at most substrate concentrations ($[S_0] < K_m$) conversion to product occurs completely and almost instantaneously. However, at high levels of substrate, $[S_0] >> K_m$, kinetics play a role in determining response. Under these conditions, Eq. (7.95) approximates to

$$f[S_0] = \frac{V_{max}fL}{F} \tag{7.96}$$

The resultant temperature change can be calculated by substitution of $f[S_0]$ for $[S_0]$ in Eq. (7.44).

There are clearly compromises involved in the use of Eq. (7.95) to optimize a flow reactor analysis. In particular the values of F and L will have significant effects on the thermal and diffusion characteristics of the reactor. Moreover, the generation of large temperature changes as $[S_0]$ increases will also lead to nonlinear response. These aspects are discussed in more detail in Chapter 4.

7.9.2. Process Control by Means of Calorimetric Flow Detectors

The ability of the thermistor to tolerate insoluble matrices makes flow calorimetry a logical choice for process control of crude flow reactor effluents. This aspect has been the subject of several reviews (88–95). A typical experimental configuration, used for the control of a lactase (B-galactosidase) reactor, is shown in Fig. 7.33 (89). In this setup, the primary reaction is the formation of glucose and galactose from the lactase-catalyzed hydrolysis of lactose. The flow calorimetric effluent monitor contains immobilized glucose oxidase and catalase as secondary reagent enzymes. Dilution of the primary reactor effluent can be achieved by pumps P_2 and P_3. Changes in product (glucose) concentration are monitored by the output of the Wheatstone bridge, which, in turn, regulates the pumping rate of lactose into the primary reactor via a variable control oscillator (VCO). The dual-channel recorder monitors the glucose concentration (or at least a proportional signal) and the variations in flow rate, as shown in Fig. 7.34, for an operational experiment in which whey was used as the primary substrate. As can be seen, the lactose conversion can be maintained even though the activity

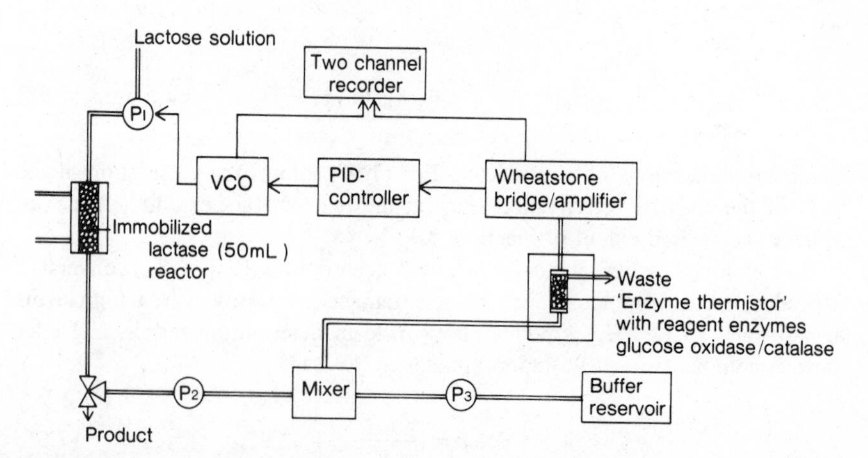

Figure 7.33. Schematic representation of instrumentation for continuous monitoring of enzymic lactose conversion by means of an "enzyme thermistor"; P_1, P_2, and P_3 are peristaltic pumps. P_1 is controlled by the frequency of a variable control oscillator (VCO). PID controller is a proportional, integrating, or derivative mode universal controller. (Adapted from Ref. 89.)

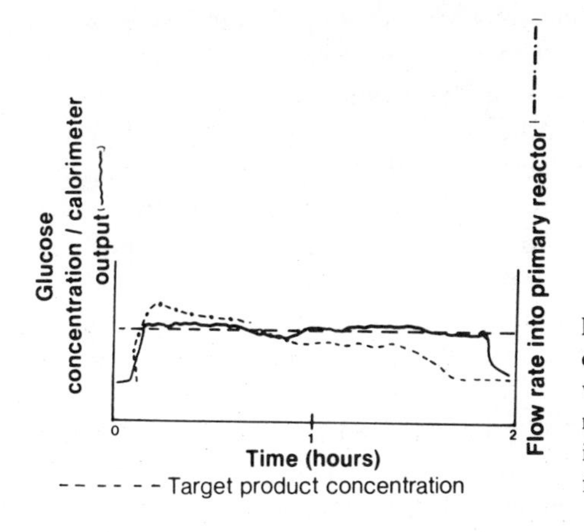

Figure 7.34. Process control data from a calorimetric monitoring device; controlled variations in flow rate of substrate to primary reactor as a function of glucose concentration in the reactor effluent and time. (Adapted from Ref. 89).

of the primary reactor inevitably decreases with time. The device was operated for 3–4 h before clogging at the analytical reactor occurred. Reactor blockage is a serious problem for the analysis of crude effluents. This problem can be approached in three ways (88). A dialysis step can be introduced to remove the high-molecular-weight fraction at the effluent, or the sample can be diluted on-line before it reaches the calorimeter. As the product concentration of a reactor is typically large, both these operations can be accommodated without exceeding the sensitivity limits of the instrument. A third alternative is to use an open tubular nylon column as discussed in the previous section.

The concept of flow calorimeters being used as large-scale process monitors is not new. However, since the pioneering work of Crompton and Cope, who produced a continuous flow analyzer in the determination of impurities in hydrocarbon streams in 1968 (96), there has been little, if any, follow-up. There is no doubt that the laboratory-scale application of flow calorimetry to the control of biochemical process streams shows considerable potential. The large-scale operation of such devices will hopefully be demonstrated soon.

7.9.3. Thermometric Enzyme-Linked Immunosorbent Assay (TELISA)

Enzyme-linked immunosorbent assay (ELISA) is a standary immunoassay technique in which an enzyme is first linked to a reference standard of an antigen or an antibody to be determined. This solution is then allowed to interact with the corresponding matrix-bound antibody or antigen. An antibody–antigen reaction occurs, and the level of enzyme activity in the supernatent liquid is then proportional to the concentration of antibody or antigen in the reference sample. Samples of unknown composition can then be determined from a calibration curve. The technique is subject to the same limitations of any assay involving an enzyme-catalyzed reaction, that is, the number of marker enzymes that can be used is restricted by the analytical methods available to monitor the enzyme–substrate reaction. A much wider choice of enzymes becomes available if a calorimetric detection system is used to monitor the enzyme activity; this can be achieved conveniently by incorporation of the immunosorbent assay into a flow reactor. This procedure has been termed "thermometric enzyme-linked immunosorbent assay" (TELISA) (97). A typical flow cycle pattern for this technique is shown in Fig.7.35.

A flow reactor containing an antibody immobilized to an insoluble support such as Sepharose is first flushed with buffer solution. A solution of the antigen to be determined mixed with a known amount of standard enzyme-labeled antigen is then introduced into the flow stream. Competition occurs for the immobilized antibody sites between the free antigen and the enzyme-marked antigen, resulting in a "mixed" matrix-bound complex. The ratio of enzyme-marked complexes to unmarked antibody–antigen pairs will be determined by the ratio of marked to unmarked antigen in the flow stream. The system is then flushed again with buffer. The analytical signal is obtained by allowing a substrate solution to interact with the complex. The resulting temperature–time profile will reflect the amount of bound enzyme, which is, in turn, inversely proportional to the amount of free antigen in the sample. After the signal has been recorded, the antibody–antigen complexes are disassociated by introduction of a glycine/hydrochloric acid solution, and the column is finally regenerated by flushing the system with buffer. The cycle can then be repeated and so on. The sensitivity of the technique can be controlled to some extent by changing the amount of antibody–enzyme conjugate added to the sample or the amount of bound antibody in the flow reactor. Antigen concentrations as low as 10^{-13} mol L^{-1} can be determined by

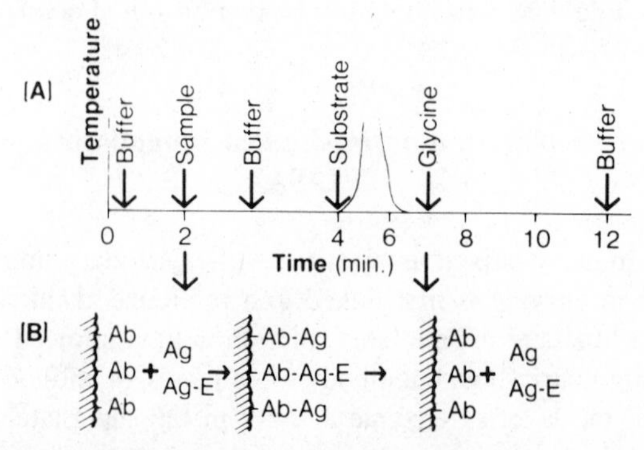

Figure 7.35. (A) Flow cycle for TELISA procedure to determine the concentration of antigen. (B) Corresponding reactor interactions (from Ref. 97, by permission of Elsevier Biomedical Press).

use of this procedure (98) with up to 100 reuses of the same immobilized antibody system.

This relatively fast and convenient method of immobilizing an enzyme–antigen conjugate to an immobilized antibody offers the potential of "reversibly immobilized" enzyme reactor columns. In theory, the conditions can be created to change the function of the reactor in a few minutes by disassociating the incumbent enzyme complex and replacing it with another (99). TELISA does not have the same sensitivity as radioimmunoassay (RIA) but it does eliminate the potential hazards of using radiolabeled compounds.

7.9.4. Enzyme-Inhibition Analysis

The most logical application of immobilized-enzyme reactors in a flow calorimeter is substrate determination. However, a change in the sequence of reagents flowing through the column allows the determination of selected enzyme inhibitors (100).

The cycle of reagents necessary to obtain a response proportional to inhibitor concentration is shown schematically in Fig. 7.36. The substrate peak is first recorded in the usual manner, by pumping a precisely timed pulse of substrate through the column. A pulse of inhibitor is then pumped for the same time period, during which time the enzyme-inhibitor interaction takes place. The analytical signal is obtained by allowing a second substrate pulse to interact with the inhibited enzyme in the reactor. The original substrate pulse will be attenuated by a factor proportional to the degree of inhibition. For a constant inhibition reaction period, the degree of inhibition will be proportional to the concentration of inhibitor (see Section 7.7).

In order to reuse the column, the enzyme-inhibitor complex must be disassociated; this is achieved by "eluting" the inhibitor off the column by means of

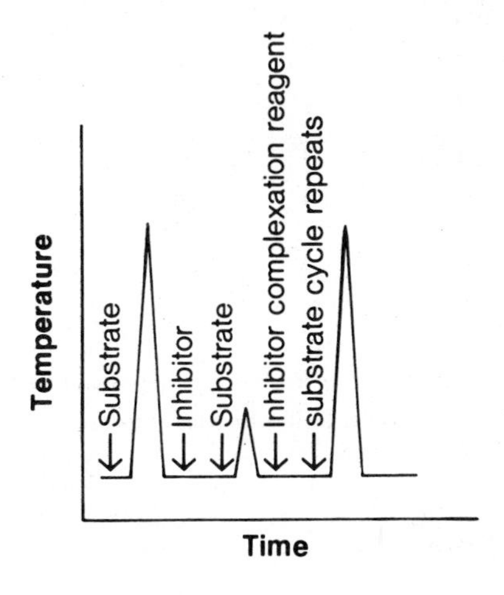

Figure 7.36. Reagent addition sequence for the determination of inhibitor concentration in an immobilized-enzyme reactor/flow calorimeter experiment. (Adapted from Ref. 100, by permission of Elsevier Science Publishers.)

a complexing agent. The cycle can now begin again with a substrate response. As discussed in Section 7.7.3, analytical calibration curves of percentage inhibition versus inhibitor concentration can be constructed. The application of this methodology is limited to enzyme-inhibitor complexes which can be easily disassociated by complexing agents. Heavy-metal inhibitors are the most logical candidates for this reason (100).

7.9.5. "Thermal Enzyme Probes"

Although often classified together the "thermal enzyme probe" and the "enzyme thermistor" are conceptually quite different.

The "thermal enzyme probe" is constructed by immobilizing an enzyme onto the surface of a glass encapsulated thermistor: One method of coating the thermistor probe is to adhere the previously insolubilized enzyme (in the form of a lyphilized gel) by means of a commercial adhesive (101).

A simple mathematical model can be derived to determine the parameters affecting the response of the probe (101). The appropriate parameters are in Fig. 7.37.

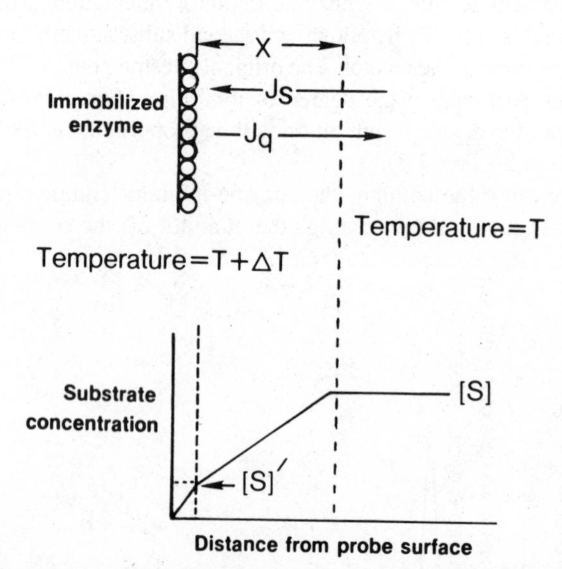

Figure 7.37. Schematic representation of the mathematical model to describe the response of a thermal enzyme probe. J_s and J_q are the substrate flux to the probe and the heat flux away from the probe, respectively; [S] and [S]′ are the concentrations of the substrate in the bulk of the solution and at the probe surface, respectively; x is the thickness of the unstirred layer adjacent to the probe (from Ref. 101, by permission of Elsevier Biomedical Press).

The steady-state substrate flux from the bulk solution to the surface of the probe, J_S, is given by

$$J_S = \frac{D_S([S] - [S]')}{x} \tag{7.97}$$

where D_S is the diffusion coefficient of the substrate and $[S] - [S]'$ is the concentration gradient of the substrate between the bulk solution and the probe surface. x is the thickness of the stationary, unstirred, boundary layer. The heat generated at the probe will be conducted away from the surface according to

$$J_q = \frac{\lambda \Delta T}{x} \tag{7.98}$$

where J_q is the heat flux density, λ is the thermal conductivity of the unstirred layer, and ΔT is the temperature gradient between the probe surface and the bulk of the solution.

The rate of the enzymatic reaction occurring at the probe is given by the familiar equation,

$$v = \frac{V_{max}[S]'}{K_m + [S]'} \tag{7.8}$$

where v and V_{max} are expressed as reaction rates per unit area. If the assumption is made that $[S]' << K_m$, that is, that the kinetics are first order, then Eq. (7.8) becomes

$$v = \frac{V_{max}}{K_m} [S]' \tag{7.99}$$

Under steady-state conditions $J_S = v$ and $V\Delta H_R = J_q$, where ΔH_R is the total enthalpy of reaction of all processes occurring at the probe. Therefore,

$$\Delta T \approx \frac{\left(\frac{D_S}{x}\right)\left(\frac{V_{max}}{K_m}\right)}{\frac{D_S}{x} + \frac{V_{max}}{K_m}} \left(\frac{x\Delta H_R}{\lambda}\right) [S] \tag{7.100}$$

If the reaction is diffusion controlled because $V_{max}/K_m >> D_S/x$, that is, all substrate reaching the probe is consumed instantaneously, then the rate of reaction is controlled by the rate of diffusion of the substrate to the surface. Under this condition, Eq. (7.100) simplifies to

$$\Delta T \approx \left(\frac{D_S \Delta H_R}{\lambda}\right) [\mathrm{S}] \tag{7.101}$$

which is the operational equation describing the response of a thermal enzyme probe. Equation (7.101) reveals some interesting characteristics of a thermal enzyme probe. Predictably, the response is directly proportional to the bulk concentration of the substrate and the overall enthalpy of reaction. The response is, however, independent of both the enzyme concentration and the thickness of the boundary layer. The above derivation assumes that $S' << K_m$; if $S' >> K_m$, then v will approach V_{max} and the amount of enzyme per unit area of the probe will affect ΔT.

Significantly, Eq. (7.101) underlines the kinetic nature of this measurement. The steady-slate response ΔT is proportional to the rate of diffusion of the substrate through the unstirred layer as represented by D_S. Since reaction occurs only in the immediate vicinity of the thermistor, only a small fraction of the sample substrate is consumed, that is, $([\mathrm{S}] - [\mathrm{S}]') \approx [\mathrm{S}]$. In effect, therefore, the probe is a non-destructive device. This is in direct contrast to the response of an "enzyme thermistor," which is an equilibrium or "endpoint" response related to the complete conversion of the substrate.

The major limitation of the thermal enzyme probe at this stage of development is sensitivity. If the total enthalpy change monitored is approximately 40 kJ mol^{-1}, the sensitivity of a probe is 0.3–3 $\times 10^{-1}$°C per mol L^{-1} (101). This compares with 10°C per mol L^{-1} for an "enzyme thermistor" configuration. To put this into perspective, a blood glucose sample diluted 10-fold would produce a steady-state temperature change of only 8.4×10^{-5}°C in the presence of triethanolamine-buffered adenosine triphosphate and immobilized hexokinase, which is close to the detection limit for a thermistor-based isoperibol flow calorimeter.

Bowers and Carr (103) have noted that an open, tubular flow system which provides quiet, laminar flow eliminates the hydrothermal noise associated with a turbulent flow system. This principle has been incorporated into a thermal enzyme probe system resulting in root-mean-square baseline noise levels of 2–3μ°C (101), a peak-to-peak ΔT resolution of 1.5×10^{-5} °C, and an improvement in detection limit for the determination of urea by a factor of 10.

The nucleation of air bubbles on or around the thermistor probe is another significant source of noise, since their presence will clearly affect the heat-transfer properties of the flow stream. Continuous nitrogen degassing of the flow stream will alleviate this problem. Fulton et al. (102) recommended a membrane interface similar to the type used in a mass spectrometer for this purpose.

Most single-column flow systems employ a differential thermistor system to minimize the effect of temperature fluctuations in the flow stream. However, response mismatch between the two thermistors can introduce additional noise. For perspective, unmatched thermistors have a common mode rejection ratio ($CMRR_T$) of about 10 (102); $CMRR_T$ is defined as the reciprocal of the temperature change measured when the temperature of both thermistors is raised by 1°C. In other words a $CCMR_T$ of 10 means that a fluctuation in the stream temperature of 0.1°C will result in an apparent recorded temperature change of 0.01°C. Matched thermistors can be obtained with a $CMMR_T$ of 10^3, which will result in registered noise of 10^{-6}°C for typical water-bath temperature fluctuations of 10^{-3}°C.

7.9.6. Miscellaneous Devices

Both the "enzyme thermistor" and "thermal enzyme probe" systems are subject to noise caused by the placement of a thermistor in a flowing stream. This effect can be eliminated by isolation of the thermistor probe from the flow stream by an osmosis membrane (104) (Fig. 7.38). The flow cell assembly consists of three parts: the enzyme reactor (A), the flow-through cell (B), and the "blank reactor" (C). Matched thermistors are incorporated into both reactors. A contains a solution of the reagent enzyme and buffer system, B, on the other hand, contains

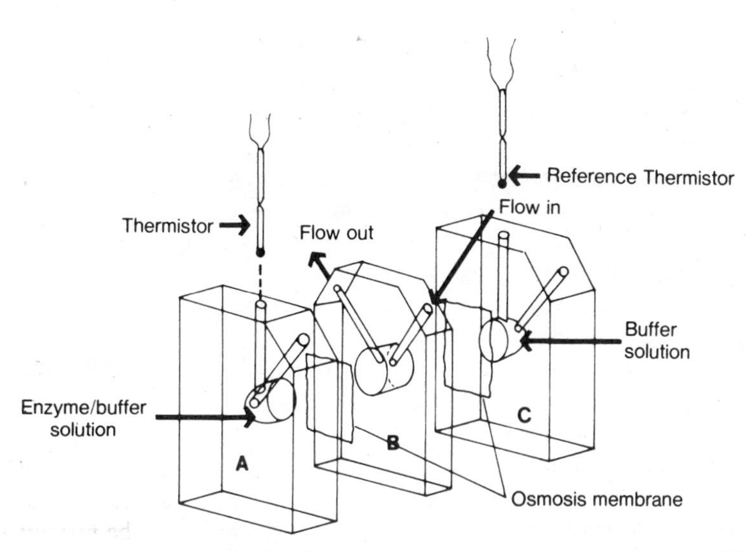

Figure 7.38. Schematic representation of a calorimetric laminar flow detector based on diffusion membranes (from Ref. 104, by permission of Marcel Dekker, Inc.).

only buffer. The flow-through cell is separated from both reactors by two $20 \times 20 \times 0.1$ mm cellophane membranes. The entire assembly is immersed in a water-bath thermostat. The analytical signal is obtained by interrupting a flow of buffer with 10 ml of sample solution. As the sample flows slowly past the membrane, diffusion of the substrate occurs into both reactors, and a differential signal is recorded as the enzyme substrate reaction occurs.

In the initial design, the application is limited by poor detection limits (1×10^{-3} mol L^{-1} glucose) and large sample volume. However, the principle of diffusion cells incorporated into "enzyme thermistor" or "thermal enzyme probe" systems has considerable potential.

An enzyme reactor based on the entrapment of enzyme within the microcavities of porous cellulose triacetate fibers has been developed (105). The fiber reactor is fabricated by adding a water–glycerol solution of the enzyme to cellulose triacetate dissolved in methylene chloride. Under vigorous stirring, the solution emulsifies and is extruded through a spinneret into a toluene coagulating bath. The resulting fibers are wrapped around a spool and introduced into the flow stream between matched thermistors. Glucose and urea determinations in the range 100–800 mg L^{-1} were made by use of fiber entrapped glucose oxidase, catalase, and urease.

The simplest method of enzyme immobilization is physical adsorption since it simply requires that the enzyme and adsorbate be brought into contact; no coupling agent per se is generally involved. Rich et al. (106) have utilized an adsorption technique as the basis of a "thermistor enzyme probe." The device is shown schematically in Fig. 7.39. A conventional thermistor is sealed in a U-shaped glass tube with epoxy cement. A small droplet of mercury is introduced into the well surrounding the thermistor tip. Urease is then physically adsorbed on to the mercury drop by immersion of the probe assembly into a urease solution for 5 min. Excess (unadsorbed) mercury is removed by washing with phosphate buffer. The mercury droplet serves as both an enzyme-immobilization site and heat-conduction medium. The temperature change generated by the enzyme–substrate reaction at the mercury–analytical-sample-solution interface is the analytical signal. In effect, the device functions as a non-destructive probe. An interesting design feature of this instrument is that the relative thermal conductivities of mercury and water (13.6:1) will bias the heat flow toward the bulk mercury volume rather than the sample solution. Accordingly, the mercury rapidly assumes the temperature of the interface. This phenomenon makes the device inherently more sensitive than a device in which the enzyme is immobilized directly on to the probe allowing unbiased dissipation of heat away from the probe with the bulk solution.

The primary disadvantages of this probe design are poor stability, resulting in very limited re-use, and insufficient precision. Without renewal of the urease,

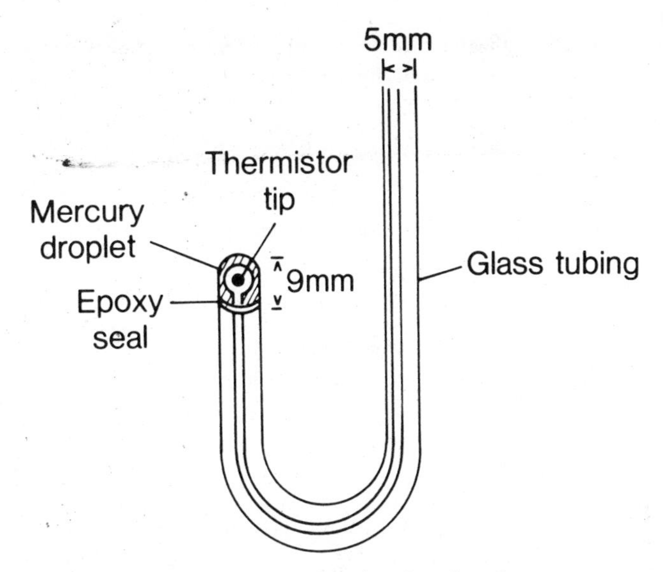

Figure 7.39. "Thermistor enzyme probe" based on the adsorption of urease on a mercury droplet (from Ref. 106, by permission of the American Chemical Society).

the probe remains active for only four experiments. It can, of course, be argued that this limitation is not prohibitive, since the activity can be restored simply by immersion in urease solution. With successive renewal after each experiment, the probe remains active for eight experiments. A new mercury surface must be introduced in order to extend the life of the probe beyond this point. Mercury surface modification and/or product inhibition are postulated as causes for this behavior. A correlation coefficient of 0.965 was obtained for a plot of probe response versus urea concentration in the range 5–30 mmol L^{-1}.

Enzyme adsorption onto a mercury surface is limited in potential since it requires the presence of disulfide linkages in the enzyme. However, the concept of a preferential heat flow toward the probe is an interesting one since it addresses the sensitivity issue associated with thermal enzyme probe measurements (see Section 7.9.5). Other variations on this theme are certainly worthy of investigation.

7.9.7. Selected Applications

Table 7.17 contains a comprehensive, but not exhaustive, list of reported applications of immobilized-enzyme devices incorporated into calorimetric detection systems of one form or another. Further information on "enzyme thermistor"

TABLE 7.17. Selected Applications of Immobilized-Enzyme Flow Reactors Incorporating Calorimetric Detectors

Analyte	Co-Substrate	Enzyme	Support	Reference
(a) "Enzyme Thermistors"/Immobilized-Enzyme Flow Calorimeters				
		(i) Clinical/Biochemical Analysis		
ATP	—	Apyrase	CPG	85
	—	hexokinase		114
Benzylamine	—	Monoamine oxidase/catalase	Sepharose and CPG	115
Benzoyl-L-arginine ethyl ester (BAEE)	—	Trypsin	CPG	82, 116
	—			
Cephalosporin	—	Cephalosporinase	CPG	109
Cholesterol	—	Cholesterol oxidase	CPG	117
	—	Cholesterol oxidase/catalase	CPG	118
Cholesterol esters	—	Cholesterol esterase/ cholesterol oxidase	CPG	118
Cyanide	Thiosulfate	Rhodanase	CPG	119
	L-cysteine	injectase	CPG	119
Glucose	—	Glucose oxidase/ catalase	Enzygel	113
	$Mg(ATP)^{2-}$	Hexokinase	CPG	120
	—	Glucose oxidase/ catalase	CPG	116–118, 121, 122
Glucose	—	Glucose oxidase/ catalase	Polyacrylamide gel	123
	—	Glucose oxidase/ catalase	Cellulose triacetate fibers	105
Hydrogen peroxide	—	Catalase	CPG	118
Lactate	—	Lactate 2-monooxygenase	CPG	111

actose	—	Lactase/glucose oxidase/catalase	CPG	117, 118
xalic acid	—	Oxalate decarboxylase	CPG	117
enicillin G	—	Penicillinase	CPG	116
riglycerides	—	Triacylglycerol lipase	CPG	124
Jrea	—	Urease	Cellulose triacetate fibers	105
Jrea	—	Urease	Enzygel	113
rea	—	Urease	CPG	116, 117, 125, 126
Jric acid	—	Uricase	CPG	118
g^{+}, Hg^{2+}, Cu^{2+}	Urea	Urease	CPG	100

rocess	Continuous or Batch Analysis	Control or Monitor	Enzyme	Support	Reference
		(ii) Process Control and Monitoring			
enicillin ermentation broth	Batch	Monitor	Penicillinase	CPG or nylon tube	128
Cellulose degrada- ion to cellubiose/ glucose	Batch	Monitor	β-Glucosidase (precolumn)/ glucose oxidase/ catalase	CPG	129
Yeast fermentation of sucrose to ethanol	Continuous	Control	Invertase	CPG coated w/ZrO_2	130, 131
actose conversion o glucose	Continuous	Control	Glucose oxidase/ catalase	CPG	88

Analyte	Antigen–Enzyme Conjugate	Immobilized Antibody	Support	Reference
	(iii) Thermometric Enzyme-Linked Immunoassay (TELISA)			
Human serum albumin (HSA)	HSA-catalase	anti-HSA	Sepharose	97, 98
Hydrogen peroxide	HSA-catalase	anti-HSA	Sepharose	99
Penicillin	HSA-penicillinase	anti-HSA	Sepharose	99
Sucrose	HSA-invertase	anti-HSA	Sepharose	99
Glucose	HSA-glucose oxidase	anti-HSA	Sepharose	99
Phenol, tyrosinase	HSA-tyrosinase	anti-HSA	Sepharose	99

Analyte	Enzyme	Flow/Batch	Support	Reference
(b) "Thermal Enzyme Probes"				
Glucose	Hexokinase	Flow and batch	Enzygel/probe	132, 101
Glucose	glucose oxidase	Batch	Enzygel/probe	101
BAEE	Trypsin	Batch	Glutaraldehyde/ probe	133
Urea	Urease	Flow	Enzygel/probe	102

Device	Analyte	Enzyme	Support	Reference
(c) Miscellaneous Devices				
"Thermistor enzyme probe"	Urea	Urease	Hg drop	106
"Enzyme-bound thermistor"	Urea, glucose	Urease glucose oxidase/ catalase	Glutaraldehyde/ probe	134
"Immobilized-enzyme modules"	Hydrogen peroxide	Catalase	Cellulose acetate tubes	135

(89, 90–111) and "thermal enzyme probe" (112) applications can be obtained elsewhere.

REFERENCES

1. J. M. Sturtevant, in C. H. W. Hirs and S. N. Timasheff (Eds.), *Methods of Enzymology*, Vol. 26, Academic Press, New York, 1972, p. 227.
2. A. E. Beezer and H. J. V. Tyrrell, *Sci. Tools* **19**, 13 (1972).
3. H. D. Brown, *Biochemical Microcalorimetry*, Academic Press, New York, 1969.
4. J. K. Grime, *Anal. Chim Acta* **118**, 191 (1980).
5. N. D. Jespersen, *Thermal Analysis, Part B, Biochemical and Clinical Applications of Thermometric and Thermal Analysis* in Wilson and Wilson's Comprehensive Analytical Chemistry, Vol. 12, Elsevier, New York, 1982.
6. A. P. Fletcher, *Am. Lab.* **5**, 42 (1973).
7. A. E. Beezer, *Thermochim Acta.* **7**, 241 (1973).
8. I. Wadso in R. Pain and B. Smith (Eds.), *New Techniques in Biophysics and Cell Biology*, Vol. 2, Wiley, New York, 1974, p. 85.

9. C. Spink and I. Wadso, in D. Glick (Ed.), *Methods of Biochemical Analysis*, Vol. 23, Wiley-Interscience, New York, 1975, p. 1.
10. R. N. Goldberg, E. J. Prosen, and B. R. Staples, *Anal. Biochem.* **64**, 68 (1975).
11. R. N. Goldberg and G. T. Armstrong, *Med. Instrum.* **8**, 30 (1974).
12. S. N. Pennington, *Enzyme Technol. Dig.* **3**, 105 (1974).
13. P. W. Carr, W. D. Bostick, L. M. Canning, Jr., and R. H. Callicott, *Am. Lab.* **8**, 45 (1976).
14. K. Levin, in H. C. Curtis and M. Roth (Eds.), *Clinical Biochemistry, Principles and Methods*, Walter Gruyter, Berlin, 1974.
15. K. Levin, *Clin. Chem.* **23**, 929 (1977).
16. N. N. Rehak and D. S. Young, *Clin. Chem.* **24**, 1414 (1978).
17. P. W. Carr, E. B. Smith, S. R. Betso, and R. H. Callicott, *Anal. Calorim.* **3**, 457 (1974).
18. E. B. Smith and P. W. Carr, *Anal. Chem.* **45**, 1688 (1973).
19. J. Jordan, "Thermometric Titrations," in *Treatise on Analytical Chemistry, Part 1*, I. M. Kolthoff and P. J. Elving, Eds. Interscience, New York, 1968, Vol. 8.
20. J. Jordan and N. D. Jespersen, *Thermochimie*, Colloques Internationaux du C.N.R.S., No. 201, Centre National de la Recherche Scientifique, (1972).
21. R. M. Izatt, L. D. Hansen, D. J. Eatough, T. E. Jensen, and J. J. Christensen in *Analytical Calorimetry*, R. S. Porter and J. F. Johnson, Eds., Plenum Press, New York, 1974, p. 237.
22. J. Steinhardt and J. A. Reynolds, *Multiple Equilibria in Proteins*, Academic Press, New York, 1969.
23. C. Bjurulf, *Eur. J. Biochem.* **30**, 33 (1972).
24. C. J. Martin and M. A. Marini, *CRC Crit. Rev. Anal. Chem.* **8**, 221 (1979); **8**, 407 (1979).
25. D. J. Eatough, L. D. Hansen, R. M. Izatt, and N. F. Mangelson, in *Methods and Standards for Environmental Measurement*, NBS Spec. Pub. 464, U.S. GPO, Washington, DC, 1977, p. 643.
26. W. W. Sukow, H. E. Sandberg, E. A. Lewis, D. J. Eatough, and L. D. Hansen, *Biochemistry* **19**, 912 (1980).
27. K. M. Kale, L. Vitello, G. C. Kresheck, and G. Vanderkooi, *Biopolymers* **18**, 1889 (1979).
28. S. J. Rehfeld, D. J. Eatough, and L. D. Hansen, *Biochem. Biophys. Res. Commun.* **66**, 2 (1975).
29. D. J. Eatough, T. E. Jensen, H. F. Loken, S. J. Rehfeld, and L. D. Hansen, *Thermochim. Acta* **25**, 289 (1978).
30. R. H. Callicott and P. W. Carr, *Clin. Chem.* **22**, 1084 (1976).
31. J. Jordan and T. G. Alleman, *Anal. Chem.* **29**, 9, (1957).
32. S. J. Rehfeld, N. Duzgunes, C. Newton, D. Papahjopoulos, and D. J. Eatough, *FEBS Lett.* **123**, 249 (1981).
33. D. J. Eatough, S. J. Rehfeld, R. M. Izatt, and J. J. Christensen, "Titration and Flow Calorimetry: Application to Proteins and Lipids," in Wilson and Wilsons *Comprehensive Analytical Chemistry*, Elsevier, New York, 1982, Vol. 12.
34. C. D. McGlothin and J. Jordan, *Anal. Lett.* **9**, 245 (1976).

35. A. Cornish-Bowden, *Principles of Enzyme Kinetics*, Butterworth, London, 1976.
36. J. C. Waselewski, P. T. S. Pei, and J. Jordan, *Anal. Chem.* **36**, 2131 (1964).
37. N. Davids and R. L. Berger, *Currents Mod. Biol.* **3**, 169 (1969).
38. N. N. Rehak and D. S. Young, *Clin. Chem.* **22**, 1177 (1976).
39. N. N. Rehak and D. S. Young, *Clin. Chem.* **24**, 1414 (1978).
40. J. K. Grime, B. Tan, and J. Jordan, *Anal. Chim. Acta* **109**, 393 (1979).
41. J. K. Grime and E. D. Sexton, *Anal. Chim. Acta* **121**, 125 (1980).
42. J. M. Sturtevant, *J. Biol. Chem.* **247**, 968 (1972).
43. J. de la Huerga, C. Yesnick, and H. Popper, *Am. J. Clin. Pathol.* **12**, 1126 (1952).
44. H. K. O'Farrell, S. K. Chattopadhyay, and H. D. Brown, *Clin. Chem.* **23**, 1853 (1977).
45. S. Rosenstein and H. D. Brown, *Biochim. Biophys. Acta* **629**, 195 (1980).
46. D. M. Yourtee, H. D. Brown, S. K. Chattopadhyay, D. Phillips, and W. J. Evans, *Anal. Lett.* **1**, 41 (1975).
47. D. L. Phillips and D. M. Yourtee, *Anal. Lett.* **9**, 235 (1976).
48. B. Chance in *Methods of Biochemical Analysis*, D. Glick, (Ed.), Interscience, New York, 1954, Vol. 1, p. 408.
49. J. K. Grime and K. R. Lockhart, *Anal. Chim. Acta* **106**, 251 (1979).
50. A. Anders, H. Schaefer, B. Schaarschmidt, and I. Lamprecht, *Arch. Dermatol. Res.* **265**, 173 (1979).
51. R. B. Kemp, *Eur. Biophys. Congr., 1st Proc.* **4**, 381 (1971).
52. M. Labadie, B. Serpaud, P. M. Laplaud, and J. C. Breton, *Analusius* **6**, 160 (1978).
53. C. D. McGlothlinn and J. Jordan, *Anal. Chem.* **47**, 1479 (1975).
54. J. Debord, M. Labadie, and J. C. Breton, *Analusius* **8**, 93 (1980).
55. J. Konickova and I. Wadso, *Protides Biol. Fluids, Proc. Colloq.* **20**, 535 (1973).
56. R. N. Goldberg, E. J. Prosen, B. R. Staples, R. N. Boyd, G. T. Armstrong, R. L. Berger, and D. S. Young, *Anal. Biochem.* **64**, 68 (1975).
57. C. D. McGlothlin and J. Jordan, *Anal. Chem.* **47**, 786 (1975).
58. R. N. Goldberg, *Clin. Chem.* **22**, 1685 (1976).
59. C. D. McGlothlin and J. Jordan, *Clin. Chem.* **21**, 741 (1975).
60. R. N. Goldberg, *Biophys. Chem.* **4**, 215 (1976).
61. J. K. Grime and B. Tan, *Anal. Chim. Acta,* **107**, 319 (1979).
62. N. D. Jespersen, *J. Am. Chem. Soc.* **97**, 1662 (1975).
63. J. P. Hoare and K. J. Laidler, *Can. J. Biochem.* **48**, 1132 (1970).
64. N. N. Rehak, G. Janes, and D. S. Young, *Clin. Chem.* **23**, 195 (1977).
65. J. K. Grime and K. R. Lockhart, *Anal. Chim. Acta* **108**, 363 (1979).
66. N. N. Rehak and D. S. Young, *Clin. Chem.* **23**, 1153 (1977).
67. W. N. Aldridge, *Biochem. J.* **46**, 451 (1950).
68. R. D. O'Brien, *Toxic Phosphorus Esters*, Academic Press, New York, 1960, p. 76.
69. J. K. Grime and B. Tan, *Anal. Chim. Acta* **106**, 39 (1979).
70. H. K. O'Farrell, H. D. Brown, S. K. Chattopadhyay, and Y. T. Das, *Anal. Lett.* **13**, 85 (1980).
71. J. Konickova and I. Wadso, *Acta Chem. Scand.* **25**, 2360 (1971).
72. J. N. Baldridge and N. D. Jespersen, *Anal. Lett.* **8**, 683 (1975).
73. R. R. Jennings and C. Niemann, *J. Am. Chem. Soc.* **77**, 5432 (1955).

74. A. E. Beezer and H. J. V. Tyrrell, *Sci. Tools* **19**, 13 (1972).
75. A. E. Beezer, T. I. Steenson, and H. J. V. Tyrrell, *Talanta* **21**, 467 (1974).
76. M. R. Eftink, R. E., Johnson, and R. L. Biltonen, *Anal. Biochem.* **111**, 291 (1981).
77. A. E. Beezer, *Thermochim. Acta* **7**, 241 (1973).
78. K. Barclay and N. D. Jespersen, *Anal. Lett.* **8**, 33 (1975).
79. J. K. Grime, K. R. Lockhart, and B. Tan, *Anal. Chim. Acta* **91**, 243 (1977).
80. B. Tan and J. K. Grime, *J. Therm. Anal.* **21**, 367 (1981).
81. A. E. Beezer, T. I. Steenson, and H. J. V. Tyrrell, *Protides Biol. Fluids, Proc. Colloq.* **20**, 563 (1973).
82. M. Tribout, S. Paredes, and J. Leonis, *Biochem. J.* **153**, 89 (1976).
83. B. Danielsson, B. Mattiasson, and K. Mosbach, *App. Biochem. Bioeng.* **3**, 97 (1981).
84. R. S. Schifreen, D. A. Hanna, L. D. Bowers, and P. W. Carr, *Anal. Chem.* **49**, 1929 (1977).
85. K. Mosbach and B. Danielsson, *Biochim. Biophys. Acta* **364**, 140 (1974).
86. P. Kirch, J. Danzer, G. Krisam, and H. L. Schmidt, *Enz. Eng.* **4**, 217 (1978).
87. P. W. Carr and L. D. Bowers, in *Immobilized Enzymes in Analytical and Clinical Chemistry*, Wiley-Interscience, New York, 1980.
88. B. Mattiasson, B. Danielsson, C. F. Mandenius, and F. Winquist, *Biotech. Bioeng., Ann. N.Y., Acad. Sci., Biochem. Eng. 2* **369**, 295 (1981).
89. B. Danielsson, B. Mattiasson, R. Karlsson, and F. Winquist, *Biotechnol. Bioeng.* **21**, 1749 (1979).
90. B. Mattiasson, B. Danielsson, and K. Mosbach, *Enz Eng.* **4**, 213 (1978).
91. B. Mattiasson, B. Danielsson, and K. Mosbach, *Food Process Eng.* **2**, 59 (1980).
92. B. Mattiasson, B. Danielson, and F. Winquist, *Enz. Eng.* **5**, 251 (1980).
93. B. Mattiasson, B. Danielsson, *Prepr.-Eur. Congr. Biotechnol., 1st* 1978, p. 27.
94. B. Danielsson, C. F. Mandenius, F. Winquist, B. Mattiasson, and K. Mosbach, *Adv. Biotechnol., Proc. Int. Ferment. Symp. 6th* **1**, 445 (1981).
95. C. F. Mandenius, B. Danielsson, F. Winquist, B. Mattiasson, and K. Mosbach, *Appl. Biochem. Biotechnol.* **7**, 141 (1982).
96. T. R. Crompton and B. Cope, *Anal. Chem.* **40**, 274 (1968).
97. B. Mattiasson, C. Borrebaeck, B. Sanfridson, and K. Mosbach, *Biochim. Biophys. Acta* **483**, 221 (1977).
98. C. Borrebaeck, J. Borjeson, and B. Mattiasson, *Clin. Chim. Acta* **86**, 267 (1978).
99. B. Mattiasson, *FEBS Lett.* **77**, 107 (1977).
100. B. Mattiasson, B. Danielsson, C. Hermansson, and K. Mosbach, *FEBS Lett.* **85**, 203 (1978).
101. J. C. Weaver, C. L. Cooney, S. P. Fulton, D. Schuler, and S. R. Tannenbaum, *Biochim. Biophys. Acta* **452**, 285 (1976).
102. S. P. Fulton, C. L. Cooney, and J. C. Weaver, *Anal. Chem.* **80**, 505 (1980).
103. L. D. Bowers and P. W. Carr, *Thermochim. Acta* **10**, 129 (1974).
104. S. P. Dufrane, F. Quertain-Defrise, M. Vanderbranden, A. Laudet, G. J. Patriarche, and J. M. Ruysschaert, *Anal. Lett.* **14**, 1269 (1981).
105. W. Marconi, F. Bartoli, F. Morisi, and F. Pittalis, *Int. J. Art. Org.* **2**, 159 (1979).
106. S. Rich, R. M. Ianniello, and N. D. Jespersen, *Anal. Chem.* **51**, 204 (1979).

107. K. Mosbach and B. Danielsson, *Anal. Chem.* **53**, 83A (1981).
108. K. Mosbach, B. Danielsson, and B. Mattiasson, *Biochem. Soc. Trans.* **7**, 11 (1979).
109. B. Danielsson, B. Mattiasson, and K. Mosbach, *Pure Appl. Chem.* **51**, 1443 (1979).
110. B. Danielsson, B. Mattiasson, and K. Mosbach, *Appl. Biochem. Bioeng.* **3**, 97 (1981).
111. B. Danielsson, *Appl. Biochem. Biotechnol.* **7**, 127 (1982).
112. J. C. Weaver, C. L. Cooney, S. R. Tannenbaum, and S. P. Fulton, *Biomed. Appl. Imm. Enz. Prot.* **2**, 191 (1977).
113. H. L. Schmidt, G. Krisam and G. Grenner, *Biochim. Biophys. Acta* **429**, 283 (1976).
114. M. Aizawa, Y. Watanabe, and S. Susuki, *J. Solid-Phase Biochem.* **4**, 131 (1979).
115. A. Svenson, P. A. Hynning, and B. Mattiasson, *J. Appl. Biochem.* **1**, 318 (1979).
116. K. Mosbach, B. Danielsson, A. Borgerud, and M. Scott, *Biochim. Biophys. Acta* **403**, 265 (1975).
117. B. Mattiasson, B. Danielsson, and K. Mosbach, *Enz. Eng.* **3**, 453 (1978).
118. B. Mattiasson, B. Danielsson, and K. Mosbach, *Anal. Lett.* **3**, 217 (1976).
119. B. Mattiasson, K. Mosbach, and A. Svenson, *Biotech. Bioeng.* **19**, 1643 (1977).
120. L. D. Bowers and P. W. Carr, *Clin. Chem.* **22**, 1427 (1976).
121. B. Danielsson, K. Gadd, B. Mattiasson, and K. Mosbach, *Clin. Chim. Acta* **81**, 163 (1977).
122. B. Mattiasson, B. Danielsson, and K. Mosbach, *Anal. Lett.* **9**, 867 (1976).
123. A. Johansson, *Prot. Biol. Fluids, Proc. Colloq.* **20**, 567 (1973).
124. I. Satoh, B. Danielsson, and K. Mosbach. *Anal. Chim. Acta* **131**, 255 (1981).
125. B. Danielsson, K. Gadd, B. Mattiasson, and K. Mosbach, *Anal. Lett.* **9**, 987 (1976).
126. L. M. Canning and P. W. Carr, *Anal. Lett.* **8**, 359 (1975).
127. L. D. Bowers, L. M. Canning, G. N. Sayers, and P. W. Carr, *Clin. Chem.* **22**, 1314 (1976).
128. B. Mattiasson, B. Danielsson, F. Winquist, H. Nilsson, and K. Mosbach, *Appl. Environ. Microbiol.* **41**, 903 (1981).
129. B. Danielsson, E. Rieke, B. Mattiasson, F. Winquist, and K. Mosbach, *Appl. Biochem. Biotechnol.* **6**, 207 (1981).
130. C. F. Mandenius, B. Danielsson, and B. Mattiasson, *Acta Chem. Scand.* **B34**, 463 (1980).
131. C. F. Mandenius, B. Danielsson, and B. Mattiasson, *Biotechnol. Lett.* **3**, 629 (1981).
132. C. L. Cooney, J. C. Weaver, S. P. Fulton, and S. R. Tannenbaum, *Enz. Eng.* **3**, 431 (1978).
133. C. L. Cooney, J. C. Weaver, S. R. Tannenbaum, D. V. Faller, A. Shields and M. Jahnke, *Enz. Eng.* (Pap. Res. Rep. Eng. Found. Conf.), 2nd, Plenum, New York, 1974.
134. C. Tran-Minh and D. Vallin, *Anal. Chem.* **50**, 1874 (1978).
135. L. J. Forrester, D. M. Yourtee, and H. D. Brown, *Anal. Lett.* **7**, 599 (1974).

INDEX

Vol. 67. **An Introduction to Photoelectron Spectroscopy.** By Pradip K. Ghosh

Vol. 68. **Room Temperature Phosphorimetry for Chemical Analysis.** By Tuan Vo-Dinh

Vol. 69. **Potentiometry and Potentiometric Titrations.** By E. P. Serjeant

Vol. 70. **Design and Application of Process Analyzer Systems.** By Paul E. Mix

Vol. 71. **Analysis of Organic and Biological Surfaces.** Edited by Patrick Echlin

Vol. 72. **Small Bore Liquid Chromatography Columns: Their Properties and Uses.** Edited by Raymond P. W. Scott

Vol. 73. **Modern Methods of Particle Size Analysis.** Edited by Howard G. Barth

Vol. 74. **Auger Electron Spectroscopy.** By Michael Thompson, M. D. Baker, Alec Christie, and J. F. Tyson.

Vol. 75. **Spot Test Analysis: Clinical, Environmental, Forensic, and Geochemical Applications.** By Ervin Jungreis.

Vol. 76. **Receptor Modeling in Environmental Chemistry.** By Philip K. Hopke

Vol. 77. **Molecular Luminescense Spectroscopy: Methods and Applications—Part 1** Edited by Stephen G. Schulman

Vol. 78. **Inorganic Chromatographic Analysis.** Edited by John C. MacDonald

Vol. 79. **Analytical Solution Calorimetry.** Edited by J. K. Grime